One Hundred Years of Agricultural Chemistry and Biochemistry at Wisconsin

One Hundred Years of Agricultural Chemistry and Biochemistry at Wisconsin

Proceedings of the 13th Annual Steenbock Symposium in Biochemistry – August, 1983, University of Wisconsin–Madison

Edited by
David L. Nelson
and
Brook Chase Soltvedt

Science Tech Publishers
Madison, WI

Science Tech Publishers, Inc. 701 Ridge St.
Madison, Wisconsin 53705, U.S.A.

Library of Congress Cataloging-in-Publication Data

Steenbock Symposium (13th : 1983 : University of Wisconsin–Madison)
One hundred years of agricultural chemistry and biochemistry at Wisconsin : proceedings of the 13th Annual Steenbock Symposium in Biochemistry, August, 1983, University of Wisconsin–Madison / edited by David L. Nelson and Brook Chase Soltvedt.
p. cm.
Includes bibliographical references.
ISBN 0-910239-24-X
1. University of Wisconsin–Madison. Dept. of Biochemistry–History–Congresses. 2. Biochemistry–Research–Wisconsin–History–Congresses. 3. Agricultural chemistry–Research–Wisconsin–History–Congresses. I. Nelson, David L., 1942- . II. Soltvedt, Brook Chase, 1959- . III. Title. IV. Title: 100 years of agricultural chemistry and biochemistry at Wisconsin.
QP511.5.W6S73 1983 89-10580
574.19'2'0711775–dc20 CIP

ISBN 0-910239-24-X Science Tech Publishers Madison, WI

Production and editorial supervision: Science Tech Publishers

Printed in the United States of America

10 9 8 7 6 5 4 3 2 1

This volume is dedicated to Evelyn Steenbock, a graduate of the Department of Biochemistry at the University of Wisconsin, whose unstinting generosity over many years has provided invaluable support for the faculty and graduates of that department.

Contents

Preface

The Thirteenth Steenbock Symposium at the University of Wisconsin-Madison Department of Biochemistry was part of that Department's celebration of its centenary. The subject of the symposium was the historical development of the Department, and the speakers were all men and women intimately familiar with the subject; each witnessed and participated in some aspect of the Department's history. Some of the talks at the symposium were personal recollections of a period or a person; others were broader descriptions of the development of a research area. This volume includes all of the papers presented at the symposium.

Although the organizing committee for the symposium tried to arrange to have speakers who would cover each of the major research groups and areas in the Department's history, for various reasons we were unable to achieve such complete coverage with the symposium talks. We have therefore included in this volume several papers previously published elsewhere: biographical essays on Stephen M. Babcock (by Aaron Ihde), E.B. Hart (by Conrad Elvehjem), Harry Steenbock (by Howard A. Schneider), and E.V. McCollum (by A.A. Rider), and an essay on the discovery of Dicumarol (by K.P. Link). We regret that we were unable to include essays on all past members of the Biochemistry Department; much history remains to be written.

Even if every past member of the Department of Biochemistry were represented in this volume, it would remain an incomplete account of biological chemistry at the University of Wisconsin. In particular, the distinguished contributions of members of the Department of Physiological Chemistry and of the Institute for Enzyme Research are not described here. As the symposium represented the celebration of one department's centennial, we decided to limit its scope to that department. Perhaps this volume will encourage someone to write a broader history of biochemistry at Wisconsin.

The Steenbock Symposium series honors the memory of Harry Steenbock, a distinguished Professor of Biochemistry at the University of Wisconsin, and is supported by funds provided by his wife, Evelyn Steenbock, to the Department of Biochemistry. We acknowledge with deep gratitude her remarkable generosity in this and other support of biochemical teaching and research at Wisconsin. Her benevolence has enhanced and extended the Steenbock contribution to Wisconsin biochemistry.

We gratefully acknowledge the University of Wisconsin Photographic Laboratory, the University Archives, the Photographic Media Center, the Photo Media Center of the Department of Biochemistry, Obma Studios (Dodgeville, Wisconsin), and many individuals for providing photographs for this volume.

Finally, Karen Davis's tireless efforts in organizing this symposium deserve special recognition and gratitude.

David L. Nelson

1

The First 100 Years

H. F. DeLuca

Department of Biochemistry
University of Wisconsin-Madison
Madison, Wisconsin

In a recent survey of graduate schools in the United States, the Department of Biochemistry at the University of Wisconsin-Madison ranked among the top five in biochemical training and research. It was actually tied with two other departments for third place. The Department of Biochemistry at the University of Wisconsin-Madison was placed in this position when the departments were first ranked by such evaluations and has continued to remain high in these rankings. There is little doubt that it can be considered as one of the top biochemistry departments in the country if not the world. The reasons for such a ranking are many, but certainly its research accomplishments must be given a large measure of credit.

Although it is difficult to select a research accomplishment with the greatest impact from the long list of achievements of the Department, quite likely the discovery by Harry Steenbock of the production of vitamin D by the irradiation process must be ranked as the most important from several points of view. First, by irradiating foods such as milk and butter, it was possible to induce vitamin D activity quickly and efficiently, thereby eliminating rickets and osteomalacia as major medical problems. Secondly, this discovery provided the basic biology that led to the isolation and identification of vitamin D. However, the impact of this discovery was not confined to the academic and medical areas. Because of the insight and capabilities of Professor Harry Steenbock, this discovery was to have a major long-term effect on scientific development at the University of Wisconsin. Professor Steenbock recognized that the irradiation process could be abused by companies interested only in financial gain. For this and financial reasons, with his own personal funds he chose to file patent applications on the irradiation process. These applications were very successful and, as history has demonstrated, it proved to be one of the most successful patents in the history of the United States. Professor Steenbock approached members of the University administration and the Board of Regents with the patents hoping to assign them to the University. The University, however, took the position that the patents belonged to the inventor and did not wish to deal with them directly. Professor Steenbock began to collect royalties but did not use them personally. He deposited them in a bank account as he continued his quest for a suitable vehicle for utilizing these funds for support of research at Wisconsin. Professor Steenbock came up with the concept of a nonprofit foundation that could invest the income from patents and thereby generate more income that could be used to support basic research at the University of Wisconsin-Madison. In his deliberations with Dean Slichter of the Graduate School, the structure of the Wisconsin Alumni Research Foundation (WARF) as we now know it began to take form. A group of distinguished alumni were approached to set up the Foundation by contributing their own funds and to become members of the board of trustees. Professor Steenbock then assigned his patents to the Foundation. The Board deemed it important to return 15 percent of the income from the patents to Professor Steenbock for his own personal use. Professor Steenbock reluctantly accepted this 15 percent. After a period of time he refused to accept any additional amounts, leaving his own personal funds to be invested by the Foundation and to be used in his own research endeavors. These funds of "personal" origin were then used to build two additions onto Biochemistry, to endow several Steenbock professorships, establish a career development fund, a symposium fund, a doctoral fellowship program, etc. The Wisconsin Alumni Research Foundation itself has grown from the early simple investments to a Foundation having in excess of $100 million in assets. It annually contributes approximately $5–6 million to the University of Wisconsin-Madison to support research in all fields, although Professor Steenbock originally designated the fields to be in the natural sciences.

It is difficult to measure the impact of this most imaginative contribution to the University of Wisconsin and to scientific development on the Madison campus. There is no doubt that WARF, in providing funds through the 1930s, '40s, and '50s, was very important in continuing to keep Wisconsin in the forefront of science. Even today, the imaginative use of WARF money gives the University of Wisconsin a major edge in supporting and attracting outstanding research and research investigators to the Madison campus. In addition, WARF is used as a model for the establishment of similar foundations by other academic institutions. Taking into account, therefore, the discovery of the production of vitamin D as a basic science discovery, its application to the elimination of rickets as a major medical problem, and the application of proceeds from this invention to the University of Wisconsin in the form of a Foundation to support research at that institution, the Steenbock discovery must be ranked as one of the most outstanding contributions made during the past 100 years in the Department of Biochemistry if not the University.

Following very closely this development at the University, the discovery of the vitamins in the Department during the first part of this century must rank as another significant development. Thus, the unusual questioning views of Professor Stephen Moulton Babcock that simple analyses of nitrogen, carbohydrate, fat, and salt could not describe the nutritive value of rations was to give rise to the famous single-grain experiment that proved there were undiscovered components in diets. This work, in collaboration with Professors E. B. Hart and E. V. McCollum, was to spawn the experiments of E. V. McCollum using the experimental white rat, in which he clearly demonstrated the existence of a substance in fats that is necessary for the growth and reproduction of animals. This factor was called "fat-soluble A." McCollum was also able to demonstrate by similar manipulations the existence of another group of substances which he called "water-soluble B" that was necessary to prevent neurological symptoms in animals. These discoveries fostered the development of vitamin isolation, identification and chemical synthesis that was to put forth the modern era of biochemistry which clearly demonstrated that chemistry could be used to describe food and life processes.

Perhaps the next most significant discovery was the isolation and identification of the vitamin niacin as the agent that cured human pellagra. This work was carried out by Professors Conrad A. Elvehjem and F. M. Strong and by D. W. Woolley. Pellagra affected some 20 million Americans in 1920, killing thousands each year. The discovery of niacin as the antipellagra agent resulted in elimination of this disease that had a major impact on the productivity of the nation. In relative importance follows the discovery by Professors Hart, Elvehjem, and Steenbock that copper plus iron is required to cure iron-deficiency anemia. This highly successful treatment was to play a major role in the elimination of nutritional anemia in humans. It also led to another successful patent for WARF.

Of major significance from a medical and financial point of view was the discovery of the anticoagulant, Dicumarol, and the synthesis of its important analog, Warfarin, by Professors Karl Paul Link and Mark A. Stahmann. The discovery of these anticoagulants was to provide the medical world important blood-thinning compounds that could be administered to patients suffering from various forms of thrombosis. This reached a particularly high note when President Eisenhower, following his heart attack, was treated with Warfarin to prevent additional clots. Of unexpected benefit was the use of these anticoagulants as rodenticides, eliminating loss of major amounts of grain and other feedstuffs to rodent pests. The rodenticide and anticoagulant uses of these compounds resulted in the second most successful series of patents held by WARF and gave to the University millions of dollars.

Another discovery of tremendous impact was that of Professors W. H. Peterson and M. J. Johnson in the development of fermentation methods that permitted massive production of antibiotics and other fermentation products. Certainly without their work, penicillin could not have been used during

World War II in the proportions that it became available. The methods developed by the University of Wisconsin fermentation group were applied universally by both industrial concerns and academic institutions, affecting the entire antibiotic industry.

Professor Hart, working with Professor Bohstedt of the Animal Sciences Department, discovered geographical areas of iodine deficiency, a condition that resulted in endemic goiter as well as hair loss of animals. This led to the development of a stable iodized salt that could be used in the prevention of these diseases.

More recently, the discovery that vitamin D must be converted to active hormones and that these active hormones and their analogs can be used to treat metabolic bone disease must also be ranked as important contributions to the first 100 years in the Department of Biochemistry. These compounds have eliminated metabolic bone diseases such as renal osteodystrophy, hypoparathyroidism, pseudohypoparathyroidism, and certain vitamin D-resistant rickets. More recently these compounds have shown great promise in the treatment of the major bone disease of old age and postmenopausal women, osteoporosis. These compounds are also yielding large sums of money to WARF and have promise of approaching the income made by the original vitamin D patents of Professor Steenbock.

There are many more discoveries of great importance that have been made and are currently being made in the Department in such important areas as protein nutrition, trace minerals, nitrogen fixation, vitamin K metabolism and function, bull sperm preservation, enzymology, membranes, sensory systems, and molecular biology, leaving no doubt of the enormous research contributions of this Department. However, one cannot examine the contributions of this Department without sitting in awe at the number and quality of its Ph.D. students who have left the University and assumed important positions elsewhere. Many of these people appear as contributors to this volume and there are many more who were unavailable to contribute. The Department of Biochemistry has graduated over 1,057 Ph.D.'s since its beginning. A large number of Masters students have also been produced to say nothing of the large number of postdoctorates and undergraduate students who have been educated and trained in the Department of Biochemistry.

The Department of Biochemistry can be very proud of its first 100 years of accomplishments in research and teaching. The current staff of the Department is ready for the next 100 years, wherein new and important biological problems are being approached with the tools of chemistry. The Department is now staffed not only by nutritional biochemists who made the initial impact for the Department, but also by prominent representatives of the newer fields of enzymology, molecular biology, plant biochemistry, endocrinology, virology, membrane biochemistry, sensory systems, natural products, carbohydrate chemistry, nuclear magnetic resonance spectrometry, and x-ray crystallography. Many important problems are currently being approached. One is a continuation of an understanding of the metabolism and function of nutrients and of biologically active substances. Another is an understanding of sensory systems. Still others are an understanding of the molecular events involved in the expression of genetic information including the broad new field of developmental biology; an approach to the understanding of how viruses carry out their replication and function; of how the basic macromolecules are able to carry out their catalytic or informational storage and exchange functions; and the isolation and identification of biological molecules both in the small and large molecular weight ranges. Finally, there is a constant awareness in the Department of how these discoveries can be applied to the problems of human disease, to the elimination of hunger, increased crop production, and to improve the life of humans. We believe that the Department of Biochemistry will continue its tradition of contributing heavily, not only to fields of historic strength, but also in these new and promising areas.

2

Rats, Fats, and History*

Howard A. Schneider

Institute of Nutrition
University of North Carolina System, and
Departments of Biochemistry and Nutrition
University of North Carolina
Chapel Hill, North Carolina

All have heard the phrase "bringing coals to Newcastle." The remark is intended to convey the sense of a useless performance, of the needless struggle to do what has already been done and, to render a final judgment, that the whole exercise is a waste of time.

My assignment has been to come to Madison and to speak to the subject "Fat-Soluble Vitamins." Before I am smudged with coal dust, and have confused the Yahara River with the Tyne, I think I owe you a few words of explanation.

Many here in the audience are residents of Madison, and it goes almost without saying that those of us who have been invited back for this centennial celebration were once Madisonians, at least for a while. In my particular instance I left Madison in August 1939, just 44 years ago. I fully expected to return to the department after a fellowship year at Cambridge, but events such as World War II took over. I think I am speaking for many when I recall that when I left, having spent both undergraduate and graduate years in Madison, I found it difficult to imagine a more desirable place to live than here. In due time, however, I found that it wasn't necessary to freeze in winter and boil in summer, that there were other places and other climes. But Madison always has a permanent and nostalgic charm for us expatriates; a charm that has more to do with our experiences here, than with, say, the eclectic architecture of a campus that never quite remembered what it had in mind.

*A revised version of this address appears in *Perspectives in Biology and Medicine* 29:393–406 (1986), University of Chicago Press.

DEPARTMENT OF BIOCHEMISTRY

College of Agricultural and Life Sciences
UNIVERSITY OF WISCONSIN-MADISON
420 Henry Mall
Madison, Wisconsin 53706-1569 USA
Telephone: (608) 262-3026/262-3040
Fax: (608) 262-3453

April 25, 1998

Biochemistry Alumni and Friends:

Enclosed is a copy of the Proceedings of the 13th Annual Steenbock Symposium in Biochemistry, **One Hundred Years of Agricultural Chemistry and Biochemistry at Wisconsin**.

Sincerely,

Hector F. DeLuca
Professor and Chair
Department of Biochemistry

hfd/cam

Enclosure

No, I prefer to think of the Madison experience much as Thoreau thought of his life on Walden Pond. He lived there only a bit over two years, and when he left and was seen again on the streets of Concord, he was accused by an acquaintance of having, by his departure, given lie to his touting of the life at Walden as the very best that a man might do. Thoreau's answer was, as is mine, "I have many lives to live."

In returning to Madison, then, we might be expected to bring some small contribution from our intervening years, even as we respond to our respective assignments. I found that my assignment to discuss the fat soluble vitamins allowed me a certain freedom of selection in what I might be reasonably expected to say. Here, then, is the framework of what I have to say.

Although the founding date of the Department of Agricultural Chemistry of the University of Wisconsin is given as 1883, I shall leave those earlier days to others and focus my remarks on the year 1913. That is not 100 years ago, but it is a respectable 70, or the proverbial three score and ten. It also happens to be my own age, and it is the year when the words "Agricultural Chemistry" were inscribed in the stone lintel over the entrance door to the first building to bear the name on this campus. If you care to look you will find it still there.

Why I chose 1913 will become clearer a bit later, but now I will choose further and say that I will not, for this audience, bring coals, and I will not set forth the details of the fat-soluble vitamins, their biochemistry, physiology, or—God save the mark—their molecular biology. What I propose is to delineate, as best I can, the historical and philosophical roots of their discovery, and for this occasion, try to understand why the fat-soluble vitamins were discovered here in Madison; why it was in 1913; and how the people who did it came to do it just then. As you can now tell, absence from Madison has made me bolder, for I wince even now from what Professor Hart and Dr. Steenbock would say to him who would talk philosophy under this roof. I tried it once, and Dr. Steenbock advised me to "wait until you are older." I think I now qualify, and can come out of the philosophical closet.

Why Madison?

In thinking about the questions raised by recognition that a certain event—say, the discovery of vitamin A—took place not far from where I stand, is to link that unique event with specifications of time (1913) and place (Madison, Wisconsin). There is a school of thought that says that such a discovery was bound to take place, sooner or later. Further, says this school, it is the total activity of the scientific community, committed to certain principles of outlook and procedure, that insures the slow but certain unfolding of the laws of nature. Progress at times may be slow, but it is certain. By these lights, then, if vitamin A was discovered in Madison, then Madison was just the locus of the last, but sufficient, effort. Science, says this view, progresses by increments, and there is nothing special about Madison in the matter of the vitamin A discovery. Madison was the scene of the final increment.

As scientists, then, by this view we are actors in scenes of frenetic activity (which may be an accurate description at times), and are like the oft-remarked monkeys that in large numbers and over long periods of time by jumping up and down on the keys of typewriters can produce all of the books in the British Museum. If, at times, the product of our own labors seems ridiculously small, then perhaps we are involved with a bad set of typewriters.

This view of the history of science, which is widely held by the general public if it thinks about it at all, is a comfort to many since it is so "democratic." Scientists, to change the metaphor, like ants, seem to be very busy in their mysterious comings and goings, but just what is on their minds is not at all clear.

Let me restore our spirits. We are not monkeys, and we are not ants. The discovery of vitamin A at Madison was not an incremental event in the history of science. It was a supremely human achievement, it was unique, and it took place at Madison for very special reasons. It was a revolution. And we are here today to celebrate our ties to that revolution.

At this point you find me still warily circling my subject, the discovery at Madison of vitamin A. But now I have introduced a word, "revolution," that reveals the plan of what follows, and reveals my debt to a historian and philosopher of science, Thomas S. Kuhn (Kuhn, 1962).

In 1962 Thomas Kuhn published his remarkable book, *The Structure of Scientific Revolutions*. It is his view of the nature, causes and consequences of revolutions in basic science concepts that, it seems to me, is the most satisfying framework for our understanding of what, precisely, we are engaged with here: namely, the mode and meaning of the discovery of the fat soluble vitamins A and D, beginning at Madison in 1913. In the years since I left Madison in 1939 I have read, from time to time, books aspiring to explain the nature of scientific discovery. I naively thought that I might find therein fruitful guides for my own exertions at the laboratory bench. But all, it seemed to me, left me standing outside the process, admonished by my betters to be struck with awe, but no better armed to understand scientific progress than before. For me, Thomas Kuhn changed all that. And since his insights are the road we will now travel, it will be expeditious to briefly sketch his scheme of things.

Most of our lives as scientists, according to Kuhn, are busy with what he calls "normal science." This means getting on with solving the problems that are reared within a conceptual framework which is silently accepted by the field of science in which we are working. Within this framework, professors and their students are all busy poking into every nook and cranny, seizing on "problems" cast up by the framework, defined in terms of the framework, and attacked by methods prescribed by the framework. "Answers" are found, and several generations of professors and doctoral candidates are busy indeed.

But, inevitably, a crisis arises, according to Kuhn. The prescribed and accepted conceptual framework does not accommodate, and provides no means of resolving, new and troubling experiences in the observations rightfully embraced by the field which is being so assiduously tilled. It is to the solution of this kind of crisis that we really should reserve the term "discovery." For, by the very definition of the situation I have just described, that former concepts would not work, resolution of the crisis inexorably must come from the new and novel. A new conceptual framework is constructed, the new framework is seen to be productive, it is adopted, and professors and students are busy once again with the many new tasks, and all are again involved in "normal" science.

This, in very brief, is the Kuhnian scheme of things. In addition to the special definition we have given to "normal" science here, there is one more term that needs introduction for our purpose. It is an old word, but Kuhn has given it a new life and utility. This is the word "paradigm." A paradigm is a model or pattern, and, as used by Kuhn, concisely conveys the sense of the framework that, in its acceptance, provides a silent compendium of how to proceed with the tasks of "normal" science. "Normal" scientists do "normal" science by following the accepted paradigm for their field. When the paradigm becomes exhausted, a crisis is at hand. The paradigm no longer works as a productive guide. Only the introduction of a new paradigm, a true discovery arrived at by an acceptance, however reluctant, of the unknown, followed by exploration of alternative and fresh frameworks, resolves the crisis and makes possible the resumption of "normal" science.

Let me now cast my story of the discovery of the fat-soluble vitamins in Kuhnian terms. By 1913 the field of nutrition was in crisis. Chemical analysis of foods (the accepted paradigm) had uncertain, and at times completely erroneous, predictive powers for the assembling of rations for farm animals, with obvious economic consequences. The crisis was resolved in terms of a new paradigm: the testing of the nutrient value of foodstuffs by feeding to an experimental animal, the rat. The new information thus gained resolved the crisis, acceptance of the new conceptual framework quickly followed, and "normal" nutritional science resumed.

Let us now return to the real world of Madison in 1913.

The crisis in nutrition, as described above, was the failure of chemical analysis of feedstuffs to provide information capable of predicting the ability of assembled rations to nourish farm animals. This failure had important and obvious economic consequence. Agricultural experiment stations smarted under this failure and the frustration at the station at Madison was as intense as any.

If the sense of crisis was slow in coming it was certainly very real by the late nineteenth century, and there now arrived in Madison, in 1888, just five years after the founding of the department here, the necessary iconoclast who would clear the way. This was the chemist, Stephen Moulton Babcock, who came to Madison from the New York State Agricultural Experiment Station at Geneva. It was there that he had become thoroughly aware of the serious inadequacy of chemical analysis to provide the information that a science of nutrition demanded. Paul de Kruif has told this story well (de Kruif, 1928) so there is no need to dwell on it. But two events occurred in the life of Babcock which, by hindsight, we can now see as forming the preconditions which led to the discovery of vitamins A and D. The first of these was, of course, the designing of the Babcock test for butterfat in milk. This put the dairy industry of Wisconsin, of the United States, of the world, on a sound economic base. And as a governor of Wisconsin, W. D. Hoard, remarked, "The Babcock Test has made more dairymen honest than the Bible has ever made." The watering of milk became a stupid thing to do when one was paid for the butterfat and not for the weight of the milk. The Babcock Test made Babcock world famous. But for our purpose here, this is not the important point. The important point is that Babcock was not only famous, he was now a heavyweight in the political considerations which always bore in on state-supported institutions. When troubles brewed in the legislature with ominous consequences for the University, smart deans and presidents asked Babcock to come with them to explain matters to the farmer-legislators of the dairy state. There Babcock was greeted with awe and listened to with respect.

Babcock's role in the story of vitamin A and D was thus two-fold: he brought the required spirit of iconoclasm to the study of nutrition and, by virtue of his fame as the designer of the Babcock Test for butterfat, he wielded considerable political power. And this power was capable of being used within the University as well as without. Its use within the University is the next step in our story.

Babcock had brought with him from the Experiment Station at Geneva an interesting idea. It was, in a sense, a retreat; but it was a retreat to the high ground. If chemical analysis was failing to provide information capable of accurately predicting the nutritional value of feedstuffs, then before hazarding all on chemical analysis one ought to look at the feeds themselves. For many good economic and empirical reasons farm animals were fed mixtures of feedstuffs, and the struggle always was to devise the most nourishing and, simultaneously, the most economic combination. Chemical analysis of the separate feedstuffs had provided a kind of common denominator for these purposes. From the chemist's view, a given level of, say, protein was achieved by calculation from the analytical results obtained from the various items. A Kjeldahl nitrogen analysis was a Kjeldahl nitrogen analysis, and if you wanted to know the protein this denoted, you multiplied by 6.25. It didn't matter to the chemist whether you were talking about wheat, corn, oats, or what. "But, ah," said Babcock, "what if it did matter?" This made agricultural chemists of the day very angry, and they appointed committees.

Babcock's answer was to design an experiment. "Try one feedstuff at a time, feed it to a cow for a long time, and see what happens." Well, people were not handing out cows, and so Babcock's idea had to wait until he got to Wisconsin. Twenty years after he had his idea, Babcock got his cow—two of them. He got them from the professor of animal husbandry at Madison, W. L. Carlyle. Babcock fed one a ration of oats, oat straw, salt and water. The other cow got a similar mixture, but only corn. In three months the cow on oats died, and the other, on corn, looked pretty peaked. Carlyle took his remaining cow back and returned her to health on the Station official ration. Dr. Babcock did not publish.

Six years later, in 1907, Babcock got his second chance, this time not with two cows but with sixteen. The results, the famous Research Bulletin #17 of the Wisconsin Agricultural Experiment Station, appeared four years later, in 1911 (Hart et al., 1911). Sixteen calves were divided into four groups of four: three groups were fed single grain rations, balanced to identical chemical analysis by various amounts of grain, gluten, stover, straw, etc., but all from the same plant: oats, corn, or wheat. The fourth group received a mixture of all three. The results, all familiar to this audience, showed that identical chemical analyses failed to predict identical nutritional results. The plants were different, unambiguously different, as sources of nourishment; and these differences were not explicable in terms of chemical analysis, for this had been carefully adjusted to equality, each group with the others. This was Babcock's iconoclastic contribution; this was the experiment that shattered the icon of the chemist as the arbiter of matters nutritional; and, in Kuhnian terms, a paradigm collapsed.

Paradigms do not collapse overnight, and scientists do not rush into the streets crying, "My God, what shall we do?" Things take a little longer. The Babcock single-grain experiment was begun in 1907 and the results were published in 1911. It is interesting that the 1911 publication did not have Stephen Moulton Babcock among its authors, although there can be no doubt that the idea was his. The 1911 authors were E. B. Hart, an agricultural chemist newly come to Wisconsin in 1906 from (where else?) the Experiment Station at Geneva, New York; E. V. McCollum, an organic chemist from Yale who had had no success in finding a job since obtaining his doctorate in 1906 and was now hired by Hart as an instructor in agricultural chemistry; Harry Steenbock, in 1907 a student in agricultural chemistry, enrolled in McCollum's first course at Madison (McCollum awarded him an "A"); and G. C. Humphrey, Professor of Animal Husbandry. Each of these men has an interesting history, but in pursuing the trail to the discovery of the fat-soluble vitamins we are led to focus our attention on E. V. McCollum. He has given us an autobiographical account (McCollum, 1964) of his career and, drawing on it, I think we can identify him as the self-aware designer of the new paradigm for nutritional research. And it is to that paradigm we now turn our attention.

E. B. Hart, successor to Babcock (who had stepped down from administrative duties), hired the young McCollum to perform the innumerable analyses incessantly demanded for the formulation of the rations descending into the stomachs of the sixteen cows. It was an analytical treadmill and McCollum soon tired of it.

McCollum started work here at Madison as Hart's subordinate on July 1, 1907. When he saw the cows on the single-plant experiment he was impressed and awed by his responsibility and opportunity. For there were amazing contrasts to be explained. The wheat-fed cows were stunted and blind. All of their calves had been born dead. The oat-fed cows were somewhat better off; although their calves were carried to term and were born alive, they soon died. Only one was a survivor. But the corn-fed cows did well, even better than the mixed-ration controls. All of the cows had been fed rations of the same chemical composition, but the results were dramatically different. What was the basis of the difference? In time McCollum found an answer to that question. But before he could do that he had to invent some new ideas and some new methods of grappling with nutritional problems. First there came a time of seasoning.

It was chemistry of a different kind which provided the seasoning of McCollum. It was the chemistry of the relationship between the young McCollum and the emeritus Babcock. Embedded in the humorous approach to life and living of the older man were the factual nuggets of a very real world. Babcock was light-hearted, but he was, at the same time, capable of communicating some serious ideas with an unforgettable impact. Here, for example, is a brief excerpt from McCollum's autobiography that shows the influence of the older on the younger man (McCollum, 1964).

> In time I shared appreciation of the humor of the advice Dr. Babcock gave Dr. Atwater, then the outstanding authority on human nutrition. He recommended that instead of feeding pigs on farm crops it would be cheaper to feed them soft coal. When such coal was analyzed by the food-analysis procedures the results indicated that it was a well-balanced food. It contained nitrogen. Most proteins are 16 percent nitrogen. Multiplying nitrogen content by 6.25 gives protein content. Soft coal contains ether-soluble substances which in the food analysis, without further identification, was called fat. Other fractions determined in the ordinary practice could be considered sources of energy. Hence by the criteria of the chemical methods of food analysis bituminous coal had high food value. Dr. Atwater did not like the analogy and was irritated by Babcock's treating a serious subject with levity.

Babcock's humor made his iconoclastic views palatable and, it seems safe to surmise, encouraged the young McCollum to an independence of thought so necessary for the invention of new ideas. But McCollum brought something of himself to the task, too. This was his indefatigable scholarship.

Of all the scientists who have spent their allotted time in this department I think it is fair to say that E. V. McCollum was the deepest read and had the greatest sense of history. There is a legend that at the end of each working day he took home a few volumes of the bound journals of publications which seemed to him might improve his understanding of the problems he faced in his work. The legend further says that he began with volume one and worked his way to the last, and then current, volume for each journal. As a graduate student here I went to the Agricultural Library to check on this. The legend was confirmed. I found that many of the older journals, in their earlier volumes, had only one entry on the card in the jacket inside the cover. That entry was "E. V. McCollum."

Now, as I understand it, McCollum did not read every paper in every journal. Rather, he browsed through the pages and when he saw a title indicating something of interest he stopped and read it, and then continued on. By means of such prospecting McCollum finally struck gold. The "gold" was the thirty-seven volumes of a German publication, Maly's *Jahresbericht über die Fortschritte der Tier-Chemie*. McCollum had seen the file of this yearbook when he was at Yale, was delighted to find all of the series at the Agricultural Library at Madison, and wound up buying the whole set to keep for study at leisure at home. And it was in these volumes that he found information that was not in textbooks or in the current journals. As McCollum remarks in his autobiography (McCollum, 1964), "From Maly's *Jahresbericht* I learned the history of constructive thought and experiment in animal and plant biochemistry between 1870 and 1907." It strikes me as significant that McCollum here links "history" and "constructive thought" with "experiment." History and philosophy ("constructive thought") were joined in McCollum with the experiments of the scientist.

And just what did this browsing in volumes running back thirty-seven years accomplish? Clearly, it led the way to the new paradigm for nutritional research that was to yield such rich returns. But let McCollum tell of his prospecting amid the gold of Maly's yearbook (McCollum, 1964):

> Leafing the pages of these volumes I came upon the abstracts published by thirteen authors between 1873 and 1906. In them were described efforts to nourish small animals, mostly mice, on diets composed of isolated and purified proteins, carbohydrates, fats, and inorganic salts. I was struck by the fact that in every instance in which small animals had been restricted to such "purified" diets they promptly failed in health, rapidly deteriorated physically, and lived only a few weeks. I made notes and reflected on all these experiments. I concluded that the most important problem in nutrition was to discover what was lacking in such diets. They contained everything that chemists, physiologists, and medical men considered essential, yet when fed to mice they proved wholly inadequate for the maintenance of life and health.

McCollum now had his great idea. He could get out from under the drudgery of those endless analyses connected with the consuming cows, and better yet, he could realistically tackle the problem of preparing adequate supplies of purified foodstuffs which would sharpen the experiments immeasurably. Small animals such as mice ate by the gram, and not by the kilo. A whole new world of experimentation beckoned. Late in 1907 McCollum broached his idea to Professor Hart, suggesting that for experiments with purified foodstuffs they use rats instead of mice, because of their omnivorous feeding habits and more convenient size. Mice were perhaps a little too small.

Hart, in a word, did not take kindly to the idea that after only a few months the new instructor was proposing work which would take his time away from the task he had been hired for, to investigate why the rations from single-plant sources differed so greatly in value. One can imagine the cold water that poured forth from the department head.

A disappointed McCollum now found his champion, the redoubtable Babcock. For Hart and the dean of the College of Agriculture, H. L. Russell, gave way to the endorsement of Babcock, and McCollum was on his way, tolerated if not whole-heartedly approved by his superiors. But when McCollum placed a requisition for two dollars' worth of wire mesh screen to make some cages for the rats, Hart refused to sign. McCollum paid for the mesh out of his own pocket. His salary at the time was $1200 a year.

Wild gray rats, trapped in the old horse barn on the station farm proved too wild, too vicious for experimental work. So McCollum bought a dozen young albino rats from a pet dealer in Chicago and started the first rat colony ever to be used for experimental purposes. I think Hart became convinced of the doggedness of his new instructor for as the colony began to outgrow the make-shift cages, he approved an allocation of fifty dollars for more and better cages to be made in the university carpenter shop.

The switch to experiments with rats as a way to investigation of problems of nutrition now seems so simple, so mundane. But let me emphasize, it was the first step, an important step, in the forming of the new paradigm.

The next step was not so easily clarified or reached. McCollum had seen the merit, indeed the strategic necessity, of designing experimental diets from purified foodstuffs. Slowly it had to be learned that chemical purification, the elimination of confounding materials brought along from the natural sources of these materials, was the crux. For as purification of such items as protein sources proceeded, or as different natural sources of fats were compared, it became evident that purification and even the choice of starting materials had profound effects on the capacity to nourish. The hay pile was now squarely in front of McCollum, but where were the needles?

As we all now know, many needles were found in due course, and we will address the discovery of the first fat-soluble vitamin, vitamin A, momentarily; but it should not be lost upon us that the seminal event was not the discovery of vitamin A, important though it was. Rather it was the working toward, and finally the clear statement of, the new paradigm for nutritional research. Like so many scientific advances, the true import became clearer with the advantage of a bit of hindsight. It is not surprising, therefore, that the clear statement of what McCollum called "the biological method of analysis" was published in 1915, two years after the watershed year of 1913, and was restated more formally in 1925 (McCollum, 1957). We can summarize it here as follows: A nutritionally inadequate diet of natural materials or assembled from chemically described, purified materials is supplemented, singly or multiply, by additions from other natural foodstuffs. The supplements identified as improving nutritional performance, usually by the criterion of improved growth, are chemically fractionated. One is led ultimately thereby to the identification of the chemical nature of the successful supplement. And for a quarter of a century that paradigm led to a series of discoveries and the era of modern nutrition.

There are several elements of the new paradigm which, I think, bear emphasis. The evidence of nourishing properties of dietary elements was sought in the favorable response of physiological parameters, such as increased growth of the weanling rat. Chemistry was used, not to provide analytical data of the foostuffs for interpretation by standards accepted *a priori*, but chemistry was used for the purification and fractionation of the foodstuffs used in the diets themselves. And with the relatively small amounts needed for small animals these requirements could be met in the laboratory.

There is another aspect embedded in the new paradigm which, I think, has simple but weighty philosophical consequences. By beginning, as one must, with diets of natural foodstuffs, the superiority of, say, diet I to diet II, by previous paradigms was sought by quantitative comparisons afforded by the results of chemical analysis for known constituents. The great power of the new paradigm was that it made no assumptions *a priori* as to the nature of substances responsible for the observed differences between diet I and diet II. They could, in the end, be substances already known, or more exciting, they could be items completely unknown in their nutritional significance hitherto. What is more, the new paradigm was completely unambiguous as to where the new substance, if such it was, lay hidden. The unambiguous operation of adding, of supplementing, made it dead certain that if favorable response occurred, then the important substance, or substances, were to be sought in the supplement. If the supplement was the hay stack, then it was up to the chemist now, by his fractionation methods, to find the needle.

Right here, allow me to comment on the role of biochemistry in these events. Just when a science of nutrition was in need of increasing sophistication and increasing powers of resolution in the methods of chemical fractionation which were now clearly needed, biochemistry, for its own ends, began just such developments for the isolation of small amounts of labile materials from natural sources. I believe that simple fact is the reason why a new science of nutrition was born and nurtured in a department of biochemistry, even if that name came later. Indeed, it was the startling nutritional discoveries made in biochemistry departments which drew favorable attention to them, and even some support. It was a matter for some preening. It has, of course, for the moment ended. Why this should be so is a matter I will comment on below. But now we must return to 1913, the first year that the new paradigm produced the memorable event known as the discovery of vitamin A.

Why 1913?

In June 1913, McCollum and Marguerite Davis sent to the *Journal of Biological Chemistry* the manuscript of their milestone paper, "The Necessity of Certain Lipins in the Diet During Growth." In this publication (McCollum and Davis, 1913) it was clearly shown that all fats were not nutritionally equal when it came to nourishing young rats on a diet assembled from purified materials. Protein, carbohydrates, and mineral salts furnished a base to which fats of various kinds were added. Butterfat and the ether extract of egg yolk supported growth, but fats such as lard or olive oil did not. This was surprising since up until then fats were regarded merely as concentrated energy sources in the diet. And on this basis all digestible fats should be equal. But McCollum and Davis had shown that they were not.

The question as to the nature of this difference was next resolved unambiguously when McCollum and Davis (McCollum and Davis, 1914) showed that the growth-promoting lipid was in the relatively small residue of the fat that was not saponifiable, "the nonsaponifiable fraction," and that this fraction could be transferred to olive oil, changing it thereby from a fat that failed to support growth into one that did. This was incontrovertible evidence that an ether-soluble growth-promoting substance, a "lipin" in the terminology of the day, was capable of being separated from its source in a natural fat, and of being transferred to another, nongrowth-supporting fat; and by this act of transferring bringing with it the growth-promoting property. History had been made, and as McCollum later exulted in his autobiography (McCollum, 1964), "We had discovered vitamin A."

The dam had been breached here in Madison in 1913, and the river of further successful investigations here in Madison to this very day swelled into a torrent from laboratories around the world. We will not detail here that flood of publications. Suffice it to say that McCollum went on, in July 1917, to the new School of Hygiene and Public Health as its first professor of chemistry. But here in Madison a new age was dawning in nutrition research. Hart and Steenbock were the leading and cutting edge of progress in the years that followed. Steenbock and his coworkers furthered the vitamin A story with an elucidation of the role of the carotenes as the provitamin A of the plant world that animals converted to vitamin A by their own metabolism. And, of course, we all know how Harry Steenbock traced the action of ultraviolet radiation in activating vitamin D precursors into antirachitic activity. The vitamin D story goes on in these laboratories even today. But it is not too early to discern the historical significance of the work here by DeLuca and his coworkers that has forged a clear link between a metabolically derived product of vitamin D (the dihydroxy derivative) and the domains of endocrinology, in which the vitamin D derivative functions as a hormone. As for vitamin A, as I am sure we will learn from other participants in this Centennial, there has been an explosion of interest and activity surrounding the notion that vitamin A (retinol) or provitamin A (beta-carotene) can function as an antineoplastic agent (DeLuca and Shapiro, 1981; Hennekens et al., 1982).

The discovery of the first fat-soluble vitamin had been announced, as described above, in the landmark paper of 1913 by McCollum and Davis. I believe that the audience here at the Wisconsin Centennial would be ill-served if I did not digress to shed what light I can on the history of Marguerite Davis, who was McCollum's assistant and collaborator for seven years, from 1910 to her unexplained resignation in 1916, just before McCollum went to his new post at Johns Hopkins. During this period Davis was the junior coauthor with McCollum in the publication of ten research articles, nine of which appeared in the *Journal of Biological Chemistry*. For these facts, as well as others in the life of Miss Davis, I am indebted to Harry G. Day (Day, 1983), professor emeritus of chemistry, University of Indiana.

Marguerite Davis remains a somewhat mysterious figure and there are many gaps in the historical record. But she is certainly entitled to a place in any history of the Wisconsin story such as we are celebrating here. For Marguerite Davis was a Wisconsin native, born in Racine on September 16, 1888. She died in that city just two days after her 79th birthday, on September 18, 1967. Originally enrolled at the University of Wisconsin, Davis transferred to the University of California at Berkeley in August of 1908 and received a B.S. degree in Natural Science on May 17, 1910. Her mother died, probably before Marguerite's graduation, and she returned to the Midwest to make a home for her father, a retired physician of Racine then living in Madison where he pursued botanical studies. This task was not completely fulfilling for Davis, and on the advice of the formidable Abby L. Marlatt, head of the Department of Home Economics here at Madison, she presented herself as a kind of freelance graduate volunteer to McCollum, who set her to learning the biochemistry of the day. But soon she learned of the new and struggling rat colony of McCollum's and offered to take on the care of it. McCollum was delighted to be relieved of the chore and with her help expanded and speeded his program of experiments. McCollum repeatedly and frequently acknowledged his debt to Davis.

Apparently the Department of Agricultural Chemistry felt no such debt, for although McCollum requested a salary for Davis each year, each year Hart denied the request. The request was denied, Hart felt, because Davis was not sufficiently trained to be placed on staff. In the sixth year Hart relented (her training was now adequate?), and Davis received $600 for her year's work. Davis resigned shortly thereafter. I do not know if the two events were connected.

It may be that this selfless dedication to her father and to McCollum's ambitious program was all of a piece with a diffident acceptance of life's burdensome struggle for an intelligent woman of her day. It is very probable that this was accentuated by the trauma, at the age of ten, of being severely burned when her clothing caught fire while she played at a bonfire. Although the record is not clear, these burns left her with a physical handicap, but apparently, from those few accounts we have, she suffered no facial disfigurement.

After stints at the University of Chicago and the Home Economics Laboratory here at Madison, where she published five articles in the period 1920–1923, Davis became an authority on vitamin A assay and was invited to set up such studies at the laboratories of E. R. Squibb and Sons in New Jersey. In 1926 Davis was at the New Jersey College of Pharmacy at Rutgers University. Thereafter the trail of Davis' life becomes shrouded, but apparently she returned to Racine and lived a life of retirement. She was not an enthusiastic correspondent and failed to answer many letters addressed to her. Racine was not unaware of her career in the founding of a science of nutrition, and took pride in her restoration, completed in 1955, of the family residence which had been built in the 1840's. The home overlooked Lake Michigan and here Marguerite Davis lived out her days.

The history of Elmer Verner McCollum, his dedicated assistant, Marguerite Davis, and the fateful year of 1913 here at the Department of Agricultural Chemistry, all bear witness to the revolution in nutritional research that was born here. Again, using the grammar of the history of science as formulated by Thomas Kuhn, we can say that the crisis of contradictions embedded in the old paradigm of chemical analysis of foodstuffs as a predictor of nutritional merit was resolved and replaced by the new paradigm of the use of a small animal, the rat, and the principles of the biological method of analysis.

The floodgates were now opened to a river of new knowledge of nutrition. We have given a small amount of time here to describing the first cargo, the discovery of vitamin A, the first fat-soluble vitamin. Indeed the whole notion of vitamins now had an operational base solid enough for the tracking down of a whole array, a process that would, as "normal science," occupy a host of professors and their laboring assistants for the next four decades. The lifetimes and the successful careers of many who are celebrants here are but the fruits of that new paradigm.

Another consequence of the new paradigm was that it led investigators into a world of much smaller dimensions than had been dreamed. The now-discredited reliance on chemical analysis had always taken satisfaction in the almost mathematical completeness of the quantification. When all of the items, the protein, carbohydrates, fats, and minerals were totted up, the total came reassuringly close to the 100 percent mark. But the new paradigm showed that the mysterious substances being brought to light lay in that tiny sliver of the unapprehended. This new world of importantly significant materials was not only qualitatively novel, it was minute. And it was ultimately recognized to be three to six orders of magnitude below what had been admitted as relevant before. This descent into smaller orders of magnitude was but another instance of the direction taken in many of the sister sciences, and probably received unconscious sanction from these trends. Thus, to name a few of these: morphology has used the microscope, light and electron, for its ends; physics has pursued the ever-smaller particles into the heart of the atom itself; and genetics did not rest until it came to a submicroscopic world of hydrogen bonds between certain special atoms in the special geometry of the double helix. The discoveries of 1913 here in Madison were of a piece with a new world of science. As Kuhn has said (1962), ". . . . paradigm changes do cause scientists to see the world of their research engagement differently. Insofar as their only recourse to that world is through what they see and do, we may want to say that after a revolution scientists are responding to a different world." After 1913, I submit, it was a different world indeed.

But now I must end. For the news of the new paradigm of 1913 is no longer news. Worse, the real news, in my view, is that the paradigm of 1913 has become exhausted. The pace of discoveries of new qualitative entities in the field of micronutrients has slowed to a snail's pace, and may have stopped altogether. The last important discovery of a new vitamin occurred in 1948 with the discovery of the chemical nature of vitamin B_{12}. That is more than three decades ago. This is not to say that nutritional scientists have not been busy. Much is to be learned about the function, the molecular roles, of the nutrients we know. And this is legitimate and should be done. What I am drawing attention to, and labeling an exhausted paradigm, is a paradigm that no longer casts up a continual array of new items, new components of the world of foodstuffs supporting the life of animals, and especially humans. As a graduate student here in the late 1930's it seemed as if this array of unknown and chemically undescribed items would stretch out in time longer than any of our professional lives could contemplate. But within a decade the list seemingly came to an end. Now what?

But this is the crisis that Kuhn warns us is the fate of all paradigms, to become exhausted as tools for discovery, sooner or later. We need a new Babcock, a new iconoclast, who will warn the onlookers at the parade that, indeed, the passing emperor has no clothes. And with the ground cleared, and with the shibboleths discarded, we need a new McCollum to point the way to a new paradigm.

Having said all this, and made bold by my personal ties as an inheritor of the Babcock-McCollum tradition, let me offer, very tentatively, one element that I think is bound to be a feature of the paradigm acoming. I can suggest it in a single word—disaggregation. What do I mean by that? I mean that, implicit in the laboratory rat model, and hidden in the simplistic assumption that a rat is a rat is a rat, is a tangle of rat genetics. Of this tangle we really know very little. If we now assume, and I think this assumption is defensible, that rat phenotypes are achieved by interaction of rat genotypes with the environment; and if nutrition is the richest and most intimate part of that environment, then disaggregating the rat genotype (or of any experimental animal, for that matter) will multiply the rat paradigm into new responsiveness, a continuum that will bring to light new items and relationships of the nutritional world.

The consequences of disaggregation are obvious. Nutritional statements will become statements linked to specific genotypes. It is no accident that, in the current obeisance to reductionism, in some universities it has become necessary to prefix some departmental names with the word "organismic," to remind the listener or reader that what is really at stake is not molecules, but live, growing, reproducing beings. Biology is in need of being reminded of that, and a science of nutrition is in similar need. The fundamental lesson of Madison in 1913 was the discernment of the proper use of experimental animals as detectors of the fine structure of the nutritional environment. That lesson I do not propose to let you forget.

The reduction of life to the spin of electrons but leaves us staring at spinning electrons, and completely at a loss as to what to do about it. The science of nutrition has goals linked to the human condition, a condition which leaves us all dependent on a thin layer of life on a very lonely planet. I am confident that a science of nutrition will, in the fullness of time, find fresh things to say about that dependency. We must strive for new beginnings, and this Centennial strikes me as a good time to remind you of this, and urge you to get on with the mission that began here a century ago.

References

Day, H. G. Letter to H. A. Schneider, Bloomington, Indiana, May 25, 1983.

de Kruif, P. (1928) *Hunger Fighters*, Harcourt Brace and Co., New York.

Hart, E. B., McCollum, E. V., Steenbock, H., and Humphrey, G. C. (1911). Res. Bull. no. 17, Wis. Agric. Exp. Sta.

Hennekens, C. H., Lipnick, R. J., Mayrent, S. L., and Willett, W. (1982) *J. Nutr. Ed* 14: 135. Also see De Luca, L. M. and Shapiro, S. S., eds. (1981) Modulation of cellular interactions by vitamin A and derivatives (retinoids). *Ann. N. Y. Acad. Sci.*: 359.

Kuhn, T. S. (1962) *The Structure of Scientific Revolutions*, University of Chicago Press, Chicago.

McCollum, E. V. (1957) *A History of Nutrition*, pp. 220–223, Houghton Mifflin, Boston.

McCollum, E. V. (1964) *From Kansas Farm Boy to Scientist*, University of Kansas Press, Lawrence, Kansas.

McCollum, E. V. and Davis, M. (1913) *J. Biol. Chem.* 15: 167.

McCollum, E. V. and Davis, M. (1914) *J. Biol. Chem.* 19: 245.

3

Stephen Moulton Babcock—Benevolent Skeptic*

Aaron J. Ihde

Department of Chemistry
University of Wisconsin
Madison, Wisconsin

Booming laughter frequently traveled down the stairwell of South Hall, startling Dean William Henry in his ground floor office and causing him to wonder what Babcock was up to now. The disturbance was a repetitious one, but the dean had finally given up warning his new chemist to consider the dignity of his professorship in the agricultural college. More commonly now, he was tempted to climb the four flights of stairs to learn what could be so humorous about a scientific experiment.

Stephen Moulton Babcock was a lighthearted man. In an obituary statement, University of Wisconsin President Glenn Frank (1943) characterized him as the "Laughing saint of science." In further remarks Frank said, "This merry man . . . was made of the stuff that gives mankind its saints and martyrs. But he was a saint without seriousness, and he could have gone to martyrdom, without a murmur of self-pity, as part of the day's work. . . . He pursued the most painstaking research as if he were playing a game. . . . He did not think it impertinent to doubt the authorities . . . he was an adventurer into the unknown to whom research was an intellectual ritual." A humble man, Babcock never took himself seriously despite the honors heaped upon him by a grateful dairy world. Nor did he take science seriously. He would sooner munch peanuts at a baseball game than listen attentively to a profound research paper at a scientific meeting. He published little. He felt relieved when others took his most profound ideas and examined them experimentally.

*Presented as the Rosetta Briegel Barton Lecture at the University of Oklahoma on April 10, 1969. Acknowledgment is due the National Science Foundation for support of this investigation. Reprinted from Roller, D. H. D., ed. (1971) *Perspectives in the History of Science and Technology*, pp. 272–282, University of Oklahoma Press, Norman, OK.

Babcock was born on a farm near Bridgewater, New York, on October 22, 1843. Following a typical upbringing on a sheep farm he became a student at Tufts College and earned the A.B. degree in 1866. At that point his interests turned to engineering and he matriculated at Rensselaer Polytechnic Institute. His studies, however, were cut short when his father's death necessitated a return to Bridgewater to undertake operation of the farm (Ihde, 1961).

The family farm provided an unstimulating life for this young man with a curious mind. Farm life posed innumerable questions, but his searching mind was out of step with the drudgery of farm operations, and he soon associated himself with Professor G. C. Caldwell at Cornell University, where he served as a chemistry assistant. In 1875 he was made an instructor in the subject, but two years later he left the post in order to undertake graduate studies in chemistry at the University of Göttingen in Germany. This chemistry department will be remembered as the one headed for four decades by the illustrious Friedrich Wöhler. Wöhler was now nearing the end of his life and no longer taught graduate students. Babcock received his Ph.D. in 1878 under Hans Hübner.

Upon returning to the United States, Babcock resumed his instructorship at Cornell, but left it in 1882 to become a chemist at the Geneva branch of the New York Agricultural Experiment Station. It was here that he worked on a gravimetric method for analysis of milk for fat, studied the size and number of fat globules in milk, and devised a viscosimeter for determination of adulteration of fats and oils. His work at the station was competent, but not particularly distinguished. It was typical of the kind of work that was being done at the time in the new agricultural experiment stations throughout the United States.

In 1888, at age 45, Babcock left his position at Geneva to become Professor of Agricultural Chemistry and Chief Chemist at the Agricultural Experiment Station at the University of Wisconsin. This position had been held during the previous five years by Dr. Henry Prentiss Armsby, an authority on animal feeding. Armsby left Wisconsin in order to become director of the agricultural experiment station in Pennsylvania, where he carried on distinguished work in energy balance in farm animals.

At this particular time, dairying was undergoing a major transition. In the past, farm families frequently owned a few cows which were milked to supply the household needs. In the season of high production, a surplus might be churned into butter or converted into cheese for local sale. However, there had just been developing a trend toward factory production of butter and cheese which was stimulating local farmers to increase their dairy herds and sell milk as a cash crop.

Since milk was sold by the hundredweight, a serious problem was developing. The honest farmer was in competition with farmers who unscrupulously added water to the milk in order to increase their income. Still others skimmed cream from the top of the milk before delivery to the cheese factory and turned the cream into butter for home use or as a source of secondary income.

Such adulteration of milk was difficult to detect with certainty. The lactometer, a specially designed hydrometer with a scale calibrated to the specific gravity of milk, could be used to detect watering because the specific gravity of milk is approximately 1.032, compared to 1.000 for water. The lactometer was also suitable for detection of skimming since milk fat, having a density of about 0.9 is the lightest portion of milk. When milk has been skimmed its density becomes greater than 1.032. The lactometer could, however, provide no certain proof against adulteration when the farmer both skimmed and watered the same batch of milk. Thus, he was able to gain in both directions while leaving the factory operator with no sure evidence of unethical manipulation.

The fat content of milk is a good indication of quality, but all of the fat tests in use in 1888 were tedious and generally required the services of a trained chemist. Scientists in various experiment stations in the United States and Europe had been seeking simplified methods for determination of the fat content of milk, but none of these modified tests was sufficiently rapid and reliable for commercial use.

When Babcock joined the Station at Wisconsin, Dean Henry soon assigned him the task of designing a suitable method for determination of fat in milk and cream. Despite his previous experience with methods for milk analysis, and despite his knowledge that others had been unsuccessful in devising a satisfactory test, Babcock set to work seeking to modify the well-known Soxhlet method for extraction of fat from foods by means of ether, with subsequent evaporation of the ether and weighing of the fat left as a residue. As a consequence of diligent developmental work, Babcock soon devised a test which proved successful for the determination of butterfat in composite samples of dairy herd milk. Being skeptical of easy success, he then sought to verify the reliability of the test by checking it on milk samples from each of the cows in the University's herd. Results compared favorably with those obtained by the slow but accurate official method for fat determination until Babcock encountered the milk of a grade Jersey named Sylvia. Sylvia's milk gave results at variance with those obtained by the official method.

Dean Henry urged that Babcock's method be published despite failure to test accurately the milk of one of the cows, arguing that cheese makers and creamery operators would be using the test on composite samples of herd milk delivered to the factory, but would seldom be testing the milk of single cows. Babcock refused, abandoned the partially successful test, and struck off in new directions.

He was aware that various investigators had utilized strong acids or alkalis to release milkfat from its normal suspension. By trying various ratios of concentrated sulfuric acid and milk he soon settled on a volume of acid which, by charring the milk sugar and protein, released the fat without damage. The fat could then be concentrated at the top of the mixture by centrifuging and brought up into the calibrated neck of a specially designed test bottle by dilution with hot water. Once in the neck the quantity of fat could be measured directly in percentage by means of forceps. The test was so simple that it could be completed in a few minutes by scientifically unsophisticated persons. The method was published in June, 1890 (Babcock, 1890). Babcock insisted that he have no remuneration for the test, but that it be released for free use by the dairy world.*

The test proved a boon to the dairy industry. Its simplicity made it possible for factory operators to run fat tests periodically on composite samples of milk delivered by their farm patrons. Payment might then be made on the basis of butterfat delivered rather than on the basis of pounds of (sometimes adulterated) milk delivered. The farmer received payment for a quality product and the cheese maker was now in a position to anticipate a standard yield from normal milk which filled his vats. Watering and skimming were abandoned since such practices were no longer profitable. The editor of *Hoard's Dairyman* stated that Babcock's test was a greater influence than the Bible in making farmers honest.

The Babcock test also played a role in improvement of dairy cattle. Testing associations in dairy states had as their purpose the evaluation of butterfat production by individual cattle in tested herds. Thus, it was possible for farmers to eliminate unprofitable animals (star boarders) and to breed for quality milk production.

Despite the many honors which were heaped upon him, Babcock tended to depreciate his role as the savior of dairying. Upon occasion he pointed out that he merely developed ideas which had been tried previously, but never carried out to the point where a quick practical test resulted. Once the test was developed he took comparatively little interest in extending its application. Such work was done by lesser associates in the experiment station, such as Fritz Woll, E. H. Farrington and J. L. Sammis. Through minor modifications of the test bottle and the methodology it was possible to apply the test to cream, skimmed milk, whey, and other dairy products (Sammis, 1924).

Babcock's next investigations were carried out in collaboration with Dr. Harry Russell, the station bacteriologist. The pasteurization process was beginning to come into widespread commercial use in order to enhance the keeping quality of commercial fluid milk and make it free of disease-producing

*The test worked satisfactorily on milk of individual cows, including Sylvia's, as well as on herd samples. It was never ascertained why Sylvia's milk gave erroneous results when using the earlier test.

Sylvia came to an ignominious end a few years later when the tuberculin test was run on all cows in the university herd. The herd was found to be badly infested with the disease, almost all animals reacting positively. It was decided to slaughter all of the cows. Autopsies confirmed the test results.

organisms. When the pasteurization process was extended to other dairy products such as cream, consumer complaints became extensive. Pasteurized cream failed to whip satisfactorily. Babcock and Russell investigated the whipping quality of pasteurized cream and showed that pasteurization caused a breakdown of fat globule clusters, thus impairing whippability. They found that whipping quality might be restored by the addition of calcium sucrate, a product obtained by treating cane sugar with quicklime.

Skeptical about the current pasteurization practices which exposed cream to very high temperatures for long periods of time, they showed that the chemical additive was quite unnecessary. Adequate destruction of disease-producing bacteria occurred when cream was pasteurized at moderate temperatures for short time periods, while whipping quality was not destroyed.

The unique collaboration between chemist and bacteriologist continued as the pair turned their attention to the curing of cheese. At the turn of the century, when factory-made cheese had become an important product of the dairy industry, the cheese was frequently of such poor quality that it was rushed to market as rapidly as possible. The consumer, therefore, was purchasing cheese with underdeveloped flavor resulting from its youngness, or cheese with harsh and unpleasant flavors developed during the fast curing process. Babcock and Russell showed that normal curing of cheddar cheese was brought about by an enzyme normally present in milk as well as enzymes added with the rennet used for coagulation of the milk. If cheese where held at a temperature below 60°F, they showed that there was a marked improvement of flavor accompanied by reduced loss of moisture as compared to the customary method of curing at environmental temperatures. The cold-curing of cheese quickly became standard in the dairy industry. By questioning customary practices in the industry Babcock and Russell brought about a superior product which benefited the industry significantly (Babcock and Russell, 1902).

Although Babcock was closely identified with the milk industry, his background, interests, and associations caused him to be aware of the broad variety of agricultural problems which might in some way be related to science.

There was much interest at this time in the scientific value of foods and feeds. If agricultural experiment stations were to prove their value they must clearly reveal to farmers how they might obtain high production with a minimum of feed input. If one looks at agricultural experiment station reports during the last decades of the nineteenth century and on into the twentieth, one finds principal attention given to such matters as the kind of feeds which will produce most rapid growth, greatest production of wool, milk, or eggs, greatest production of power in a horse, and matters of this sort.

Virtually all agricultural investigators who were interested in feed values were followers of the beliefs which had been laid down by the German chemist Justus von Liebig at the middle of the nineteenth century. Liebig was of the opinion that two ingredients were essential in foods; energy-producing materials and a source of nitrogenous food (protein) for replacement of tissues destroyed during metabolism. Liebig was also of the opinion that proteins were composed of a standard radical carrying characteristic amounts of sulfur and phosphorus.

The measurement of energy values of foods was easily undertaken. In this period when thermodynamics was undergoing rapid development by physicists, nutrition scientists eagerly jumped on the energy bandwagon and measured the energy values of foods. Respiration calorimeters were developed from the middle of the nineteenth century and respiration calorimetry was the most popular activity in nutrition studies toward the end of the century. Stanley Benedict at Connecticut Wesleyan and Henry Armsby at Penn State College were building respiration calorimeters large enough for human beings and for farm animals. Benedict even built one large enough to study the metabolism of an elephant. It was quickly recognized that carbohydrates and fats are solely sources of energy.

Proteins also serve as an energy source, but are also important for rebuilding of tissues. Since proteinaceous foods tend to be more expensive than those noted solely for their energy value, there was a marked tendency among animal feeding experts and experts in human nutrition to design diets in which emphasis was placed upon cheaper food materials such as cereal grains. Feeds high in proteins were only included to the extent necessary to provide the nitrogen necessary for replacement of that which was used in production of work, wool, eggs and milk.

Shortly before the end of the century, C. W. Langworthy of the U.S. Department of Agriculture published analytical tables of human foods not only bringing out the chemical analysis of such foods, but emphasizing the importance of spending one's food dollar for cereal grains which are high in energy while holding meats, milk, fruit, and vegetables at a minimum level. It was pointed out in the Langworthy tables that in consuming foods which are high in water, as is the case with milk, fruits and vegetables, the consumer is really wasting his food dollar because he gets much less caloric value than he does in cereal grains.

Although he was educated in Germany where food analysis was developed extensively at an early date, Babcock always remained skeptical of the importance of chemical analysis of foods. He was particularly skeptical about equivalencies of different foods and frequently raised embarrassing questions about the equivalency of proteins from different sources. He was aware that Armsby, his predecessor at Wisconsin, had been the author of a book on animal feeding where these matters were treated in a highly scientific manner. He was aware of the large amount of analytical work which was going on in experiment stations all over the world, including the one at Wisconsin where Henry was sponsoring extensive analyses of various feed materials. William Henry himself would ultimately become the author of a widely read work on feeds and feeding. Despite his friendship with Henry, and his proximity to feed analyses, Babcock remained unimpressed.

While he was still at the Geneva Station he carried out extensive analyses of dairy cattle feeds. One day he placed before his chief, Dr. Sturtevant, two analytical reports, querying Sturtevant regarding which would be the better ration for a dairy cow. Sturtevant ran his eyes over the analytical figures carefully before venturing a conclusion that there seemed to be very little difference between them. With a twinkle in his eye, Babcock ventured the opinion that this seemed very strange since one analysis represented the feed which went into the cow, the other was the excrement which came out.

At meetings of the Association of Official Agricultural Chemists Babcock frequently plagued his colleagues with questions about their faith in chemical analyses. Once he suggested that it would be possible to design a ration from purely mineral sources which would contain all necessary ingredients for a satisfactory animal ration. When challenged, Babcock suggested they look at the composition of bituminous coal. The nitrogen multiplied by 6.25 might be reported as protein. Material extractable with ether might be reported as fat. In addition, coal contained materials reportable as crude fiber and nitrogen-free extract. Presumably, coal might therefore take its place in rations recommended as animal feeds.

On another occasion Babcock inquired why leather might not be ground up and used as cattle feed, since it was rich in nitrogenous material. Despite such tendencies toward *reductio ad absurdum*, Babcock's contemporaries continued to believe that feeds might be shuffled around in animal rations as long as proper attention was given to total protein and energy.

At around the turn of the century Babcock had an opportunity to test his ideas regarding the importance of biological testing of feeds. He secured the loan of two cows, one of which was placed upon a balanced ration coming entirely from the oat plant and the other on a balanced ration coming entirely from the corn plant. Both animals soon showed obvious deterioration and when the oat-fed animal died, Professor Carlyle of the Animal Husbandry Department withdrew the other animal from further experimentation. Animal husbandry experts in those days were more concerned about the sleek appearance of their animals than about letting them be used for experiments which could potentially be hazardous to their well-being, even though such experiments might result in improved knowledge of feeding.

Growth and changes in the agricultural college led to a well-designed repetition of the single grain feeding experiment starting in 1907. E. B. Hart, a young chemist from the New York station, had now joined the agricultural chemistry department, and George Humphry had become director of the animal husbandry department when Professor Carlyle took a position elsewhere. Harry Steenbock, a hard-working farm boy who was still an undergraduate, became a member of the group. In order to have the services of a talented analyst, E. V. McCollum, a Yale Ph.D., was hired by Hart.

Babcock had no part in running the experiment, even though the basic idea of a long-term feeding experiment on a restricted but presumably chemically adequate diet was his. He had a deep interest in the results, but he was now busy seeking the answer to puzzling questions regarding metabolic water.

In setting up the single-grain experiment, Professor Hart, a superb experimenter, utilized sixteen cows over a four-year period. Starting out as heifer calves, these animals grew to maturity and went through two gestations. Four of the calves were placed upon a "scientifically" balanced ration made up entirely of parts of the wheat plant, four others received a similar ration derived from the oat plant, another four were fed on the corn plant, and the last four were fed a ration made up of equal parts of the three single-grain diets. This latter group was intended to serve as a scientific control. All of the diets contained equivalent amounts of protein, energy sources, and minerals.

After the experiment had been carried on for six months the wheat-fed animals showed marked deterioration, and such deterioration soon became evident in the oat-fed animals and those fed on the mixed ration. Only the corn-fed animals were vigorous and retained a good appearance. It was obvious at this point that chemically equivalent rations derived from different plant species were not biologically equivalent.

Upon reaching maturity the animals were bred and went through a first gestation. The wheat-fed heifers were unsuccessful in giving birth to healthy calves. One was born dead, two died during the first day, one lived twelve days. The oat-fed animals gave birth to only two calves that lived, and even these were not healthy. The animals on the mixed diet gave similar results. Only the corn-fed animals appeared healthy and gave birth to healthy calves. The animals were bred a second time and the results were similar, except that by now two of the wheat-fed cows were dead and neither of the two remaining gave birth to a calf capable of survival. During the last year of the experiment the surviving cows were shifted to another ration (wheat to corn, corn to wheat, wheat to oats, etc.). Cows shifted to the corn ration quickly showed improvement; corn-fed cows shifted to wheat or oats quickly deteriorated.

The report published in 1911 was looked upon by the experiment station with great pride, since it revealed clearly the lack of biological equivalency in feeds even though chemical equivalency was maintained (Hart et al., 1911).

In the meantime, McCollum despaired of this kind of experimentation. He quickly became convinced that chemical analysis was incapable of revealing the causes of biological deficiencies in foods. As an ardent student of the literature, he soon unearthed a dozen cases where health had been unsatisfactory when humans or animals were fed diets from a highly restricted source of materials in a high state of purity. McCollum soon concluded that the cow experiments represented a poor way to explore the problem. He said as much to Professor Hart, who chided him for so quickly losing interest in an experiment which presumably would have the greatest possible significance in the field of dairy husbandry.

One Sunday morning late in 1907, McCollum was at work when Babcock strolled into the laboratory and sat down to visit. The two men had become fast friends, the elderly Babcock finding a great stimulus in the young man from Yale who was well read in the chemical literature and seemed to be bubbling with new ideas.

McCollum revealed his discouragement with the cattle experiment, but indicated that he must keep on because of the pride which the station was taking in the studies. Babcock asked him how he would prefer to experiment. McCollum answered, "I think we should use rats instead of cows. Rats don't eat much. They have a short life span. Presumably through biological testing we might find the answers to some of these questions which chemical analyses are incapable of answering." Babcock was quickly persuaded of the sagacity of the young chemist. He told him, "I happened to see Dean Russell coming up the street as I entered the building: Let's go down and talk to him."

Babcock and McCollum turned up in the Dean's office a few minutes later. Babcock reported that McCollum had an idea worth hearing. McCollum proceeded to tell about his discouragement with the cattle experiments and his conclusion that feeding experiments should be carried out on rats. Russell rose from his chair, pounded the desk and roared, "The rat is a barnyard pest and should be exterminated. The legislature will never stand for feeding such animals. We must proceed with cattle." (McCollum, 1964.)

A few days later Babcock appeared at McCollum's workbench and, with a twinkle in his eye, suggested that McCollum proceed with his rat feeding ideas. Babcock said, "I had a talk with Professor Hart and he has no objections and the Dean needn't know about this. There is plenty of scrap feed around the experiment station so such experiments shouldn't be costly."

With this encouragement McCollum proceeded to trap wild rats in the university stables and began his rat feeding experiments. These wild rats, however, proved to be highly intractable and utterly unsuitable for animal experimentation. Shortly thereafter, McCollum purchased 12 albino rats from a pet dealer in Chicago, paying $2 for the animals. At the time of his death he had not been reimbursed for these rats, which were the beginning of the Wisconsin colony.

The rest of this history is quite well known. McCollum soon found that rats grew unsatisfactorily upon diets of highly purified proteins, carbohydrates, fats, and minerals, but that the addition to the ration of small amounts of milk resulted in healthy growth. The same observation was made almost concurrently by T. B. Osborne and L. B. Mendel at the Connecticut Agricultural Experiment Station. McCollum went on to demonstrate the presence of accessory food factors (Vitamins A and B) in butterfat and milk solids.

Babcock's name did not appear on the publication which resulted from the single-grain experiments, nor did it ever appear on any of McCollum's papers. Yet there is no question that it was Babcock's skepticism coupled with an agricultural experiment station environment which opened the door to the newer knowledge of nutrition.

Babcock retired from his position as professor of Agricultural Chemistry upon reaching seventy years of age in 1913. He had just published the results of his lengthy studies on metabolic water. These, typically, arose out of his whimsical questioning of nature. He was aware that all living creatures contain an abundance of water. Why then, can certain ones, such as clothes moths, beeswax moths, pea weevils, and confused flour beetles, survive on dry foods with no external source of water? Babcock found the answer in their use of metabolic water produced by the oxidation of dry food materials, and associated with biological mechanisms for the conservation of such moisture (Babcock, 1912).

Babcock's retirement years were spent in Madison, where he and his wife lived in their modest home a few blocks from the university. His passion for hollyhocks occupied much time during the growing seasons. He learned to drive an automobile during his eighth decade of life and enjoyed touring rural Wisconsin with Mrs. Babcock. But he was regularly seen during this period making the trek to the Agricultural Chemistry Building where he enjoyed the company of younger colleagues while pursuing studies on the nature of matter. He had developed skepticism regarding traditional views on gravity. This work was never published. After his death on July 1, 1931, his notebooks were examined by several physicists who concluded that there was nothing of significance therein.

The career of Stephen Moulton Babcock presents many enigmas. His scientific career developed slowly. At the time he came to Madison at age forty-five, he had done scientific work of only a pedestrian character. The test which brough him worldwide fame was merely patient development of approaches used unsuccessfully by others. He himself depreciated the test and looked upon the studies of metabolic water as his most significant work. His searching skepticism about food equivalencies led others to uncover the role of trace components in nutrition.

Babcock can hardly be looked upon as a scientist of the first rank. Yet, through his perseverance, his willingness to team up with members of other disciplines, and his good-natured resistance toward current paradigms, he set into motion investigations which were to have profound implications in agriculture and medicine.

References

Babcock, S. M. (1890) A new method for the estimation of fat in milk, especially adapted to creameries and cheese factories. *Bull. Univ. Wis. Expt. Sta.* 24.

Babcock, S. M. (1912) Metabolic water: its production and role in vital phenomena. *Wis. Agr. Expt. Sta. Research Bull.* 22.

Babcock, S. M. and Russell, H. L. (1902) The cold curing of cheese. *Wis. Agr. Expt. Sta. Bull.* 94. Also see Beardsley, E. H. (1969) *Harry L. Russell and Agricultural Science in Wisconsin*, pp. 49–63, University of Wisconsin Press, Madison, Wisconsin.

Frank, G. (1943) in Russell, H. L. et al., *Stephen Moulton Babcock—a Memorial to him in Observance of the Centenary of his Birth*, pp. 25–27, Wisconsin Alumni Research Foundation, Madison, Wisconsin.

Hart, E. B., McCollum, E. V., Steenbock, H., and Humphrey, G. C. (1911) Physiological effect on growth and reproduction of rations balanced from restricted sources. *Wis. Agr. Expt. Sta. Research Bull.* 17.

Ihde, A. J. (1961) in Farber, E., ed. (1961) *Famous Chemists*, pp. 808–813 (on Babcock's life) and the bibliography on pp. 828–829, Interscience, New York.

McCollum, E. V. (1964) As told to the author by E. V. McCollum on April 9, 1964. Also see McCollum, E. V. (1964) *From Kansas Farm Boy to Scientist*, pp. 118–119, University of Kansas Press, Lawrence, Kansas.

Sammis, J. L. (1924) The story of the Babcock test. *Circular 172, Extension Services, College of Agr., Univ. of Wis.*, Madison, Wisconsin.

4

Edwin Bret Hart, 1874–1953*

Conrad A. Elvehjem†

Department of Biochemistry
University of Wisconsin
Madison, Wisconsin

In a short obituary of Dr. S. M. Babcock, Professor E. B. Hart started with these words, "To have enjoyed for a quarter of a century the intimate and steadfast friendship of a professional associate and a truly great man is, indeed, a rare privilege, and yet such has been my good fortune in my association with Stephen Moulton Babcock." Similarly, I have had the good fortune of knowing Professor Hart for a period of 32 years. I first met him as a student in an elementary course in agricultural chemistry in 1921. In my senior year he called me to his office and told me to take over teaching duties in the laboratory immediately since one of the assistants was leaving for a new position. Many of his teaching and research assistants were appointed under similar circumstances. This was the beginning of my close association and intimate friendship with this great teacher, research worker and advisor—an association which continued until the time of his death. During the past twenty-five years hardly a day passed that we did not meet for a conference or just a chat.

Edwin Bret Hart was born near Sandusky, Ohio, on December 25, 1874, the son of William and Mary Hess Hart, and the youngest of their family of 14. He was raised by his older brothers and sisters because his mother died in his infancy. As a student in Sandusky High School from which he graduated in 1892, he came in contact with Dr. E. L. Moseley, a teacher of physics and biology who first interested him in science. In the words of Professor Hart, "Professor Moseley not only interested us in science, but his fine character was a glorious example of decent manhood for young men. . . ." After high

*Reprinted from *National Academy of Sciences Biographical Memoirs* 28:117–161 (1954).
†Deceased

Figure 4.1 Edwin Bret Hart

school Hart proceeded to the University of Michigan where he studied chemistry and medicine.

While at Michigan he was an assistant to a distinguished professor of analytical chemistry, Dr. E. D. Campbell, who had lost his sight due to an explosion in the laboratory. His association with Professor Campbell not only gave him early experience in teaching, but also placed him in close contact with an investigator who emphasized precision in experimental work. He often told of his relationship with Professor Campbell and frequently referred to the bicycle trips they took together. They used the then common tandem bicycle which required the same teamwork which they enjoyed in the laboratory. This ride was a daily occurrence and after a hard day in the laboratory Hart admitted that Campbell often supplied most of the power. His first scientific publication was with Professor Campbell entitled "On the Quantitative Determination of Hydrogen by Means of Palladous Chloride."

In 1897 he went to the New York State Experiment Station at Geneva, where he worked with Dr. W. H. Jordan. This gave him a new approach to experimental work. It was undoubtedly his experience at Geneva that encouraged him to go to Germany for advanced study in the field of chemistry and biochemistry. So—at the very beginning of the century he set out for Germany to study under the famous biochemist, Albrecht Kossel, who was then at Marburg. After a year Professor Kossel moved to Heidelberg and Hart accompanied him. Because of this transfer he was prevented from obtaining a higher degree. However, Professor Hart always felt that the right kind of training and experience were as valuable as degrees. He thoroughly enjoyed the student life in Heidelberg and often indulged in reminiscences of his experiences at that institution. He paid all his own expenses and purchased with his own money most of the supplies which he used in the laboratory. In later years when graduate students were dissatisfied with the small stipend which they received, he would refer to his early experiences when he had to borrow money for advanced work. He published one paper as a result of his work in Kossel's laboratory, but this experience introduced him to studies on proteins and all the intricacies of protein chemistry.

After two years in Germany, he returned to Geneva where he remained for a period of four years. In 1906 Dean W. A. Henry of the College of Agriculture invited him to come to the University of Wisconsin as Head of the Department of Agricultural Chemistry and Chemist in the Experiment Station. Undoubtedly one important factor which attracted him to Wisconsin was that Dr. Stephen M. Babcock had been at the Geneva Experiment Station from 1882 to 1888 and Hart had become acquainted with his work. In his own words "I have profited enormously by going to Wisconsin and coming in contact with such men as Henry, Babcock, Russell, King, and Van Hise."

At the New York Experiment Station, Hart became interested in two areas of research: the study of chemical changes in cheese during ripening and the importance of phosphorus compounds in plants and animals. With L. L. Van Slyke he published a number of papers dealing with enzymes in cheese, proteolytic compounds formed during the ripening of cheese, and the relation of gases and salts to chemical changes taking place in the ripening process.

This early work brought him in contact with milk and dairy products, and he continued his interest in this subject during his entire life. Shortly after coming to Wisconsin he published a simple method for the estimation of casein in cow's milk and studied the variations in the amount of casein in milks. His interest in this test was undoubtedly dependent on the fact that fifteen years earlier Dr. Babcock had developed a method for the estimation of fat in milk. In fact, the same centrifuge which Dr. Babcock had developed was used in the casein determination. Hart believed that the market price of milk should be based on the protein content as well as the percentage of fat. It is interesting to reflect that if as much attention had been given to the Hart casein test as was given to the Babcock fat test, the dairy industry would today encounter less difficulty in coping with the surplus butterfat and the high price of butterfat as compared to the other constituents of the milk. In other words, in those early days greater emphasis might have been placed on the nutritional value of milk below the cream line.

With his interest in dairy products, Professor Hart developed a course in dairy chemistry which came to be recognized as a basic training for students in dairy and food technology. When Professor Hart retired, no one in his department was capable of continuing this course, but it was fortunate that one of his early students, Dr. Hugo H. Sommer, was on the staff of the Department of Dairy Industry and continued to teach this course until his untimely death in 1953.

In 1903 Professor Hart published, with W. H. Andrews of the Geneva Experiment Station, a classical bulletin dealing with the status of phosphorus in food materials. This work clearly differentiated between the so-called bound or organic phosphorus and free or inorganic phosphorus. The following year Patten and Hart studied the principal phosphorus compounds in wheat bran and established the importance of what we know today as phytin. His interest in phosphorus continued after coming to Wisconsin and one of his first contributions from the Agricultural Chemistry Laboratory at Wisconsin was a paper by McCollum and Hart on "The Occurrence of a Phytin-Splitting Enzyme in Animal Tissues." In 1909 Hart and McCollum, in cooperation with Professor J. G. Fuller, established the role of inorganic phosphorus in the nutrition of animals. As a result of these studies McCollum concluded in 1910 that all of the phosphorus needed by an animal for skeleton, nuclein or phosphatide formation can be drawn from inorganic phosphate. This early work is especially significant in light of the tremendous developments that have taken place during the past forty years relating to phosphorylated organic compounds. In the summary of the Symposium on Phosphorus Metabolism held at Johns Hopkins University in 1951, the following statement was made: "Without question the exploration of the role of phosphorus in organisms has integrated and illuminated our understanding of the chemical nature of life processes beyond compare." Professor Hart pioneered in these early studies.

Since he had worked on proteins, especially casein, he was well aware that phosphorus was an important constituent of these molecules. He was also aware that proteins were rich in sulfur and that certain proteins contained as much as 2 percent on the dry weight basis. In 1909 Professor W. H. Peterson joined the department, and Hart and Peterson published several papers on sulfur requirements of plants and animals. It is interesting to point out that in addition to McCollum and Peterson, Professor W. E. Tottingham had also become a member of the department, and on the basis of these four men the department expanded into three areas—namely, the chemistry of animal metabolism, the chemistry of plant metabolism, and the chemistry of bacterial metabolism, with Professor Tottingham leading the work in plant chemistry and Professor Peterson developing the work in bacterial metabolism, or fermentation. A little later Dr. Steenbock joined the department and helped expand the work in the animal field.

Up until this time no specific emphasis had been placed on nutrition as such. Professor Hart had been interested in analytical chemistry, dairy chemistry, in proteins, enzymes and mineral elements; but he was now to pioneer in the new field of nutrition. A few years before Hart came to Wisconsin, Dr. Babcock had initiated an experiment with growing heifers on restricted rations. The results of these experiments had not been conclusive and when Hart arrived in 1906, Dr. Babcock presented the results of his preliminary work and suggested to Dean Henry that a more comprehensive experimental plan should be inaugurated. The results of these studies with rations from single plant sources led to the publication of Research Bulletin No. 17, which has become a classical report since it laid the foundation for most of the modern research in nutrition.

It was about this time that limited federal funds were being made available for the support of research in Agricultural Experiment Stations. To be sure, the workers in those days were plagued with requests for reports of experimental results as are our modern workers. In order to be able to show progress in these reports Hart outlined two projects, namely, Adams No. 8, which dealt with mineral metabolism and mineral requirements of animals, and Adams No. 10, which dealt with vitamins and other organic nutrients in animal nutrition. This approach was typical of Hart's philosophy—namely, to have a broad research program—one which was not confining but which allowed the investigator to modify his plan as new ideas became available. These two projects are still active in the Department of Biochemistry and modern investigators might well take cognizance of this type of philosophy. Practically all of Hart's work could be classified in these two categories.

No one individual has contributed more to research on minerals and vitamins in nutrition than Professor Hart. He was instrumental in the discovery that iodine is important in preventing goiter in animals, and as a result of this discovery the livestock industry has been saved millions of dollars. He was an enthusiastic advocate of iodized salt for humans in order to prevent simple goiter in children, and served on committees which attempted to promote legislation which would make the iodization of salt compulsory. Later he and associates developed a method for the stabilization of iodine in salt which would insure the maintenance of a standard iodine content of salt.

In 1923 he initiated a comprehensive study on the relation of iron to nutrition, and in cooperation with other members of the department clearly demonstrated, in 1928, that iron cannot be utilized for building hemoglobin unless it is accompanied by a small amount of copper. This discovery was presented by Professor Hart at the Federation Meetings in Ann Arbor, Michigan, in the spring of 1928. After his presentation he was given a standing ovation by the members. Such recognition has been given to only a few members of the American Society for Biological Chemists.

The demonstration that inorganic iron can be used for hemoglobin production only when it is accompanied by small amounts of copper was not only of importance in human nutrition but had real value in the prevention of anemia in farm animals, especially pigs. While Hart was pleased to see another discovery made in his laboratory put to immediate practical use, he was also interested in the basic aspects of this discovery and gave much attention to the mechanism by which copper functions in the living cell. Further studies on copper led him into studies on other trace elements such as manganese, zinc, cobalt and boron. He and his coworkers demonstrated that both manganese and zinc have important functions in the animal body. Extensive attempts were made to demonstrate that boron was essential in animal nutrition, but it was impossible to produce a diet low enough in boron to produce any ill effect in the animal. It was recognized, of course, that boron is an important element in plant nutrition. He also studied the importance of cobalt in preventing anemia in cows and sheep before its relationship to vitamin B_{12} was demonstrated.

He was one of the first workers to show the toxic effect of fluorine in farm animals, and these early results impressed him with the deleterious effects of this element even when used in small amounts as a therapeutic agent. Extended investigations were undertaken under his direction in 1928 to study the physiological effects of fluorine at various levels upon farm and laboratory animals. This study was brought about because rock phosphate was being introduced as an economical source of calcium and phosphorus in practical animal feeding. This rock phosphate contained appreciable amounts of fluorine and, when fed at levels high enough to supply calcium and phosphorus, also supplied toxic levels of fluorine. It is interesting that these studies were made possible by a grant from the Ruhm Phosphate and Chemical Company of Mt. Pleasant, Tennessee, which was promoting the use of rock phosphate for animal feeding. These studies showed that rock phosphate was highly unsatisfactory as a mineral supplement because of the harmful effects of the fluorine present. As a result of these studies, improvements were made in the production of rock phosphate so that a large portion of the fluorine was eliminated. Thus the original findings which appeared detrimental to the industry were used to develop procedures which were later of real benefit to the entire industry.

Professor Hart had the ability to convince industry that the facts must be obtained and their procedures modified to fit the facts.

Later, when the use of fluorine in drinking water was suggested as a means of controlling dental caries, Professor Hart was frequently called upon as an advisor and consultant in this program. As late as January, 1952, he testified before the House Select Committee to Investigate the Use of Chemicals in Foods and Cosmetics (82nd Congress) on the possible deleterious effects of fluorine. His advice at this time was typical, namely, that more extensive studies should be carried out before far-reaching conclusions could be made. Perhaps he was overly cautious in opposing the rapid application of this procedure, but his opposition was based on sound knowledge of the dangers inherent in such a program. In fact the very day before his death he was called upon for advice regarding the addition of fluorine to the drinking water in the city of Milwaukee.

He was frequently called upon for advice, but his response was always the same: secure the needed facts through research. In fact it was a question of this kind which led him to demonstrate that urea could be used as a source of nitrogen in ruminants. An industrial concern came to him with experimental data that had been reported in Germany indicating that simple forms of nitrogen might be used as a replacement for part of the protein in the diet. He searched the literature and found that every bit of evidence indicated that this was not true. However, he was not satisfied with the available facts and went to the dairy barns to get a few cows in order to set up an experiment on urea. He soon demonstrated that under proper conditions, urea can be used by ruminants, and as a result of these studies urea is used in a practical way today. After this discovery he willingly gave his time to answer problems arising as a result of adding urea to animal feeds, whether the inquiry came from an industrial company manufacturing urea or from individual farmers who might have difficulties.

Professor Hart's interest in minerals and their relation to nutrition also led him into the field of food technology. In the manufacture of condensed milk, the sterilization process involves heating the milk at a fairly high temperature for a considerable length of time. During this sterilization process the milk often coagulates, and even if the coagulum is broken up by shaking, the product is not accepted by the consuming public. Hart, together with Dr. H. H. Sommer, showed that the main factor in the heat coagulation of milk is dependent upon the composition of milk salts. They showed that this difficulty could be remedied on a commercial scale by the addition of the proper amount of sodium citrate or disodium phosphate. This observation eliminated the difficulty of coagulation, and the industry still uses this procedure.

Although Hart worked in the vitamin field most of his life, his achievements present a peculiar anomaly. It is doubtful that we can associate Hart's name with the discovery of a specific vitamin or the isolation of a new vitamin. In this respect his experiences paralleled those of Sir Jack Drummond who, though active in the vitamin field for more than a quarter of a century, has no vitamin linked to his name. Such are the unpredictable rewards of fickle fortune. However, he had a hand in studies on every one of the known vitamins. It is told that he was instrumental in helping Dr. McCollum differentiate vitamin A from vitamin B, because he felt that McCollum was spending too much money for the purchase of the sugar lactose. In the early studies McCollum used as part of the ration for his experimental animals the milk sugar lactose, but it was rather expensive and Professor Hart suggested that he use a cheaper form of carbohydrate. When this was done the ration was not only deficient in vitamin A, but also deficient in the water-soluble vitamin complex. The reason for this was the fact that lactose, as manufactured in those early days, carried with it considerable quantities of B vitamins which were present in the original milk. Hart was associated with Steenbock and others in studies on the chemical properties of the antiscorbutic vitamin or vitamin C. He was associated with Steenbock in the early work on vitamin A and especially on the antirachitic vitamin, which Dr. Steenbock later showed could be produced by exposure of foods to ultraviolet light. His interest in vitamin D was stimulated largely by the difficulty which poultry raisers were encountering in attempting to produce chicks for broilers through indoor feeding. When they were reared indoors the chicks did not grow and their legs became deformed. The condition was known as leg weakness. This deficiency was shown to be due to a lack of Vitamin D, and as a result of this work the broiler industry was saved.

He was associated with many of the studies dealing with the B vitamins. In many cases other members of his staff took the initiative in these studies, but he was genuinely interested in the progress of each study. Nothing thrilled him more than to see a new fact thoroughly established, and he was pleased when a member of his staff made these new and fundamental discoveries. Frequently he was critical of new techniques, especially if they required elaborate equipment, but as soon as he saw results from these new techniques he was as enthusiastic over the new procedure as any one else. He was always helpful to the many students who frequently came to him for advice. Often he went to the literature to find helpful hints for investigators regardless of whether they were working directly with him or with some of his colleagues. A total of 46 students received their Ph.D. degrees under Professor Hart, but he was helpful to at least three times this number of students. The following acknowledgment in a Ph.D. thesis presented shortly after Hart's death is typical of the high regard which all the students had for him. "His very presence in the laboratory and the inspiration of his personality, no less than the positive encouragement he was ever ready to supply, have been the writer's most richly rewarding experience of the years during which the work reported here was in progress." He recognized the importance of cooperation in research work, and he not only carried out this cooperation personally but encouraged other members of his staff to do likewise.

Professor Hart's pleasing voice and affable manner made him an effective public speaker both to scientific and lay audiences. He was frequently called upon to give public addresses but he turned down a very large percentage of the invitations. He realized that he could not give a good lecture without proper preparation and he never failed to take the necessary time for such preparation. Whenever he did accept a speaking engagement he would immediately proceed to work on his manuscript. Often he would have his paper ready months before the time of delivery.

He had great personal charm and a keen sense of humor which made him a delightful companion. He belonged to a dinner club in Madison for many years; he always presented his latest research findings to this group but he was equally interested in reports made by other members of the club. He had the ability to present the most difficult results in simple and understandable words. He tried to instill this ability in his students and was often critical when students handed him reports which were poorly prepared.

He was a member of the American Society of Biological Chemists and a charter member of the American Institute of Nutrition. He also held membership in the American Association for the Advancement of Science, the American Chemical Society, the Society for Animal Production, and the American Dairy Association. He was a member of Alpha Chi Sigma, Alpha Zeta, Phi Lambda Upsilon, Sigma Xi, Phi Kappa Phi and an honorary member of Phi Beta Kappa.

He received several professional awards and honors, among which were the Borden award from the American Dairy Association in 1941 and the honorary degree of Doctor of Science from the University of Wisconsin in 1949. He was elected to the National Academy of Sciences in 1944, an honor which he highly treasured; but unfortunately he never attended a meeting of the Academy. He did not enjoy meetings and when he did go to the national meetings, one or two baseball games were usually included in the itinerary. His failure to attend scientific meetings was unfortunate because he had much to contribute, and many would have liked to become better acquainted with him.

On May 15, 1931, the members of his department gave a dinner in his honor and presented him with bound volumes of his papers. This was given in commemoration of his 25 years of service as leader in research and as chairman of the Department of Agricultural Chemistry. At the time of his official retirement in 1945 his friends and colleagues gave him another dinner which was attended by several hundred people. At this time a portrait painted by Mr. C. W. Thwaites was unveiled and has been hung in the library of the Department of Biochemistry.

It took persuasion by several members of his department before he agreed to a dinner in his honor. However, he thoroughly enjoyed the experience, and his comments at the end of the program were the highlight of the evening. His final words included the following: "Science and research thus breathe the very air of freedom and offer in a day of peril and confusion, a mighty reinforcement to the free spirit of man. By bitter trial we have learned at last that this free spirit is no natural gift but must be nourished and defended or it will be lost." While Professor Hart did not like honors or special recognition, it is interesting that he went out of his way to pay special tribute to Dr. Babcock, who preceded him. A few years ago when the Babcock award was given by the food technologists in Chicago, he had agreed to talk on the life of Dr. Babcock. A few weeks before the scheduled program he was stricken with a very severe case of arthritis. In spite of this great physical handicap he was determined to proceed to Chicago to give his paper. It was only after much persuasion that he agreed to have someone else present the paper which he had written.

At this point it is interesting to relate the final outcome of this attack of arthritis. It was in May, 1950, that Hart began to suffer from sciatica. He consulted orthopedic surgeons, physical therapists and many others, but there was no improvement during a period of three months. X-rays indicated the presence of lime deposits in the lower lumbar region as the cause of the severe pain. Hart, therefore, went back to his early experiments with animals and wondered if it would be possible to remove the calcium deposits by producing an acidosis. In other words he made himself an experimental guinea pig. Instead of using hydrochloric acid, he decided to use ammonium chloride, because a larger quantity of chlorine could be ingested through the use of this compound. He took four grams of the ammonium chloride daily, two after the morning meal and two after the evening meal. He followed this treatment for three weeks and at the end of this period he found that the pain had left, he could walk normally and there was no reoccurrence of the trouble. He realized that his own case was complicated by the use of diathermy and massage during the period that he took ammonium chloride. However his improvement was much greater during the period of ammonium chloride treatment than at any other stage. This stimulated his interest in the entire field of arthritis, and during the last few years he gave much of his time to study and research on this problem. Many of his friends and total strangers wrote to him for the procedure which he used. Several individuals who tried his treatment encountered distinct improvement while others, of course, failed to find much relief.

When students returned to the Wisconsin campus, they went directly to Professor Hart's office for a visit. He was anxious to see every one of them and ask after their progress, and often gave them helpful advice that they could use when he went back to their own laboratories. After his retirement he had even more time to give to visitors, whether they were his students or distinguished scientists from other parts of the world. During his active career he was more abrupt and often would tell the student or visitor that he had no more time to give at present but that they could come back at some later time. This was especially true if you came to his office shortly before a scheduled lecture. When you sat in his lecture it would appear that he was so thoroughly familiar with his subject that he needed no notes or no frequent consultation of the literature. However, he rarely gave a lecture without reviewing his notes and developing the proper mood for a stimulating talk.

He was an outstanding teacher. As a lecturer he was pleasing, clear, and stimulating. He was a skillful and searching quiz master, immediately detecting superficial knowledge and impatient of it. In addition to the course in dairy chemistry mentioned earlier, he gave the general or elementary course in agricultural chemistry, later biochemistry. In this course he had sophomores, juniors, seniors, and often graduate students. His questions would be tempered to the level of training of the student. Nothing would please him more than to find a graduate student in his elementary course who could not answer a simple question readily answered by a sophomore. His evaluation of a student was based largely on personal contact with the student. He usually went through the formal action of giving examinations but often he would use the same questions year after year. He would diligently read the examination papers but often the grades would not necessarily depend upon the results of the examination.

Usually he had another member of the staff in charge of the laboratory part of his course but he visited the laboratory section daily. In fact, he often visited the laboratory sections in courses with which he had no connection. For many years Professor W. H. Peterson gave the other elementary course in the department, which was given largely for home economics girls. Hart would delight in visiting the laboratory in this course and asking the girls what they were doing. Usually he received intelligent answers, but one day one of the girls turned to him and said "I don't have to answer questions asked by the janitor." The reason for this reaction was undoubtedly due to the fact that he seldom wore a laboratory jacket in the laboratory but would generally take his coat off and use a small piece of toweling or cheese cloth as an apron. This apron was not changed too frequently and often he did look like a janitor. No one enjoyed this reply more than Professor Hart and he frequently told the story on himself. When a younger member of the staff took over his elementary course Professor Hart agreed, after many requests, to give an advanced course in nutrition. Students from all over the campus took this course, and discussions included many areas outside the field of nutrition. He would completely lose himself during the discussions, and often the class would not be dismissed until 30 or 45 minutes after the scheduled time for the class to close. In addition to his formal teaching he also conducted a general seminar which was equally stimulating to everyone who sat in it.

In addition to his research and teaching he was a successful administrator. He took over as chairman of the department when he arrived in 1906 and continued until 1944. In the early days his department was small, for many years not more than 4 members; there were no staff meetings and usually he made the necessary decisions with little consultation with the staff. However, as the department grew larger he reluctantly accepted the procedure of having staff meetings, although they were held as infrequently as possible and usually they were dismissed as rapidly as possible. He recognized that administrative procedure was necessary but that it took time from both teaching and research. He rebelled against unnecessary reports but agreed to do those which were absolutely necessary and prepared them in concise form and always had them ready on time. He was impatient when special problems and issues arose and he would try to settle them as rapidly as possible. In doing so he would often act without making the necessary inquiries, and he would not hesitate to make arbitrary decisions which might be distasteful to his colleagues. When he first gave the elementary course in agricultural chemistry he included material dealing with soils and bacteriology. When the professors giving related courses questioned the wisdom of this, he immediately replied that he was teaching agricultural chemistry and it was none of their business what he included in the course.

During his early years at Wisconsin and shortly after Dr. McCollum came to the department from Yale University, Dr. L. B. Mendel expressed some annoyance over the fact that the nutritional studies at Wisconsin were being conducted along somewhat similar lines to those being carried on at Yale. Hart took little time in answering this criticism in direct fashion and pointed out that there was no economic waste for two laboratories to engage in similar lines of work, since thereby the viewpoints are widened and the conclusions strengthened. While these curt answers often irritated his colleagues, those who knew him well realized that these rapid actions were made so that he would have more freedom for research work.

As an administrator he worked hard to keep his staff intact. There were few losses of personnel from his department over the 38 years that he served as its head. He expected much from his associates but he never attempted to study their work or to indicate what problems they should study. He was anxious to cooperate with workers in other departments and encouraged his staff to do the same. In fact, it was his successful cooperation with the workers in many departments that has done much to break down the departmental barriers at the University of Wisconsin.

Professor Hart served on relatively few University committees although he was an efficient committee member and worked very hard on the appointments which he accepted. One of his most enjoyable and fruitful activities was his service on the Research Committee of the Graduate School. This committee was organized in 1919 to help stimulate research in the University, and later this committee served to make grants to faculty members for research. He was an original member of this committee and served on the committee for a period of 19 years. He was greatly disappointed when President Glenn Frank came to the University of Wisconsin and failed to appoint him to this committee. However, he was reappointed in later years and during the year 1944–45 he served as its chairman. He maintained high standards in research and was not afraid to tell his colleagues that projects which they submitted did not appear to be of the quality necessary for support from the Research Committee. However, he was equally enthusiastic about good projects and would go out of his way to compliment faculty members or students who had shown real progress in their work. Some of the money dispensed by the Research Committee became available through the Wisconsin Alumni

Research Foundation. This was established to develop the discovery made by Dr. Steenbock that ultraviolet light would produce vitamin D. Professor Hart believed that income from research findings should, if possible, be used to support further research. He turned several of his discoveries over to the Wisconsin Alumni Research Foundation for development, and he was enthusiastic about using the income from these findings for further support of research.

In 1903 Professor Hart married Annie Virginia DeMille. For many years the Harts, together with their one daughter, Margaret, lived within two blocks of the laboratory. This location was very convenient since he visited the laboratory on Sundays as well as on weekdays. The family did not take part in many social activities but it was a privilege to visit in their home and to be included in their small dinner parties. For many summers the Hart family proceeded to the tip of Door County in Wisconsin where they had a cottage on Green Bay just outside the village of Ellison Bay. There were few conveniences in the cottage and during the summer they lived a primitive life. Professor Hart enjoyed sailing on the bay and swimming, as well as playing tennis on the court adjacent to his cabin. Mrs. Hart died June 28, 1936, but Margaret, together with her husband, Professor Russell Larson, lived with him in the family home. During the more recent years, he and his daughter and son-in-law enjoyed Christmas trips to Mexico, Central America, South America and Hawaii. He was fond of sports both as a spectator and as a participator. He skated and played tennis even in the later years of life, but made his sports recreation and not work.

On March 11, 1953, Professor Hart came to the laboratory as usual. Although he had been officially retired for almost 8 years, he visited the laboratory daily and continued his interest in all activities in the department. He left the laboratory about 4 o'clock in the afternoon and spent an unusually pleasant evening with his family at home. During the night he was stricken with a coronary attack and died on March 12, 1953.

If one were to write a complete story of Professor Hart's life, it would involve writing a complete history of nutrition during the first 50 years of this century. His life was exceptionally rich in outstanding achievements and great leadership and in kindly inspiration to those who had the good fortune of knowing him.

5

Harry Steenbock (1886–1967) —A Biographical Sketch*

Howard A. Schneider

Institute of Nutrition
University of North Carolina
Chapel Hill, North Carolina

Harry Steenbock, like most men of genius, was a complicated human being. These puzzling complexities of human beings are at once the biographer's delight and despair. But the delighted and despairing biographer, like the portrait painter, must finally selectively choose a suitable mode of presenting his subject. For the portrait painter this choice can range from the many, simultaneously-seen facets grasped by a Picasso, to the single-minded, strong, but far-from-simple visage of an Italian Renaissance prince or merchant set forth by an Antonello da Messina. In this memoir, set in the pages of a scholarly journal devoted to the science that was Harry Steenbock's life blood, the choice is suggested by both the man and our aims. The Harry Steenbock with whom we are concerned here is that public part of him which led, influenced and marked the science of nutrition so that, when he departed it, the field was permanently changed and the rivers of thought now flowed in the scientific landscape through channels quite new and ever-deepening. Like Antonello's 15th century merchants who look out at us as masters of their world, so, like that moment in time immobilized by the portraitist, did Harry Steenbock master his time and world for a moment of supreme significance. The time was November, 1923, and what made Harry Steenbock its master was his demonstration in the rat colony of the Department of Agricultural Chemistry of the University of Wisconsin at Madison, that the antirachitic and growth-promoting effect conferred directly through irradiation of an experimental animal by a quartz mercury vapor lamp, could be conveyed indirectly by irradiating, not the animal, but the food that the animal subsequently ate.

*Reprinted with permission from *J. Nutrition.* 103:1234–1247 (1973).

Figure 5.1 Harry Steenbock

The consequences of this simple experiment were massive and, as is now well-recognized, very far-reaching, extending not only into biological science and the medical arts, but into sociology, economics and, finally, government and the policy of democratic societies. From a narrower scientific point of view the effects were revolutionary enough. A physical entity, ultraviolet radiation, was trapped biochemically in the Steenbock experiment and thus, immobilized in the irradiated foodstuff, paradoxically escaped the constraints of time and space. For by its use, the irradiated and thus antirachitic food could exert its beneficial physiological effects by being eaten at some later time and in a setting otherwise removed from the original site of the irradiation. Such was the meaning of Steenbock's discovery: to bring into a new relationship the physical world of radiant energy and the organic world of animal life; to break thereby for animals and men and the disorders of their very bones, the bonds of time and space.

Before we move away, both forward and backward in time in our consideration of Harry Steenbock, it seems worthwhile to examine a bit longer the ingredients of this high moment in the history of nutrition.

To demonstrate the beneficial effects of radiation that we have been discussing, it was first necessary that the experimental rats be made rachitic and subnormal in their growth. Although much then remained to be learned and has since been learned in this particular experimental artifice, a very simple ration sufficed in this initial experiment: hog millet, 84 parts; casein, 12 parts; and a mixture of inorganic salts, 4 parts. Hog millet has almost an arcane sound a half-century later, but in 1923, Wisconsin farmers fed it to their swine; now it is no longer grown in that state. That hog millet should have been the major ingredient in the ration for the experimental rats of this important experiment gives us a clue to the influences that were at work so that a disease of growing children, rickets, was brought under control not in a hospital laboratory, nor in a medical school, nor in a medical research institute—but in an agricultural experiment station. It was no accident that the brain and hand that conceived and guided the experiment were engaged in an agricultural experiment station at the University of Wisconsin. Harry Steenbock had his very beginnings in Wisconsin's soil, and his roots ran back to his forefathers' Germany and to Sweden.

Harry Steenbock was born on August 16, 1886, on a farm in Charlestown township, Calumet County, near New Holstein, Wisconsin. A sister, Helen, had preceded him by two years. The parents, Henry and Christine Steenbock, following the German peasant custom, had received the farm from Henry's parents, Johannes and Abel, upon the younger Steenbock's marriage in 1883. The elder Steenbocks then retired to New Holstein. Harry's grandfather, Johannes Steenbock, had come a long way. At his marriage, in 1852, he already had served seven years in the Prussian Army and, disillusioned by the revolution of 1848 and its aftermath, had emigrated to the United States from the small village of Itzeho, near Kiel, the capital of Schleswig-Holstein. Along with other German immigrants, he purchased land in Wisconsin for a reputed 50 cents an acre. An earlier ancestor, a General Marcus Steenbock of Sweden, an honored defender of Sweden from invading Danes, had emigrated to North Germany and Schleswig-Holstein in 1776.

When Harry Steenbock was three years old his father sold the 120 acre farm in Charlestown township and on the advice of friends in Iowa set out for Davenport in that state. He planned to raise corn and pigs, but the arrival of the Steenbocks in Iowa unhappily coincided with a bad outbreak of hog cholera. The elder Steenbock made his decision quickly, reversed his course, and settled in Chilton, Wisconsin, not far from the land he had just left. He invested his capital in a building housing what was then known as a saloon, and what now would be called a tavern. The Steenbock children, Helen and

Harry, began their schooling in Chilton, a German Catholic community with some English and Irish minorities. The Steenbocks were Lutherans. English was the language of the public schools, but the Steenbock children began their schooling, Harry in kindergarten, only able to speak the German of their household, and were scoffed at by the other children for their lack of sophistication.

A fire burned down the butcher shop that adjoined the Chilton building owned by Henry Steenbock, and both buildings were lost in the blaze. A farm was sought, and a more familiar life-style entered into again. The new farm of 150 acres was found six miles to the southeast of Chilton about a mile from the edge of the village of New Holstein. At the crossroads adjacent to the driveway to the farmhouse was the local public school, District No. 3, New Holstein. It was indeed the one-room, eight-grades, one-teacher school of the legend of rural America. Here, amidst 80 pupils, with 10-minute classes, and long study intervals while the teacher made her rounds of the others, Harry Steenbock began his schooling.

It was a typical boyhood in rural America at the turn of the century, with some boyish mischief, simple rural pleasures and the farm tasks that a small boy could shoulder. Six of the eight grade school years were under the tutelage of a Martha Sell who overlooked some of the mischief and stimulated young Harry to read all of the books the school could offer, encompassed in a single bookcase. Grade school years ended on a pragmatic note: no ceremonies, but the graduates dutifully took the county examinations. Harry's sister, Helen, had preceded him by a year to the Chilton High School, seven miles away and the brother now joined the sister in driving back and forth in the fall and spring by horse and buggy. When the Wisconsin winter set in and a seven-mile buggy trip twice a day became impractical, Harry and Helen Steenbock and two of Helen's girl friends rented a small set of rooms in Chilton and thus set up a winter base.

Brother and sister completed high school in three years and Helen, a year ahead, stayed on the extra year to manage the tiny household for her brother through the harsh Wisconsin winter. In the spring of 1904, as Harry Steenbock's graduation from high school approached, the unexpected happened. A German immigrant who had saved his earnings as a driver of a Milwaukee brewery wagon appeared unannounced in the Steenbock driveway and offered to buy the farm. Henry Steenbock put the question to his seventeen-year-old son. "Do you want to continue with farming, Harry? Do you want to be on this farm, or do you want me to sell it?" On Harry's answer, the farm was sold and as the summer ended, the Steenbock's moved to Madison, to a small house on Park Street, and the young Steenbocks, Helen and Harry, entered the University in the fall. Harry enrolled in the College of Agriculture, a candidate for the degree of Bachelor of Science in Agriculture, which was attained in 1908. The record of those undergraduate years shows that Harry Steenbock was far above average in his studies, and that chemistry had become a dominant interest. In his senior year, an issue of *The Student Farmer* had a paragraph which reported that, "Harry Steenbock, '08, Agric, has isolated a bacterial organism which has in rare instances produced as high as 3% acid in milk. The common lactic acid organism will not produce over 1%." It was a modest, but propitious beginning.

The young graduate in agriculture, with no real financial resources, sought some means of support as he entered upon graduate study. Harry Steenbock had come under the appraising eye of Professor Edwin B. Hart, the chairman of the Department of Agricultural Chemistry, and he offered the new graduate his first employment, assistant in agricultural chemistry. The offer was accepted and made official by a letter to Professor Hart on May 29, 1908, from the new Dean of Agriculture, the formidable Harry L. Russell. The remuneration for the new assistant was fixed, "at a salary of $720 annually, payable in tenths, to begin July 1, 1908."

Joining the Department of Agricultural Chemistry as an assistant was an important step for the young farm boy, soon to be 22. He would now, by work and daily association, come under the influence of a newly assembling group of inspired researchers who, in the emerging field of experimental nutrition, would set a pace that would draw the best from him and, by competition and frank rivalry, strike sparks of excellence in all. Professor Hart had come to Wisconsin only a few years earlier, in 1906, at the invitation of the then dean of the College of Agriculture, W. A. Henry. Hart came from the New York State Experiment Station at Geneva, and so followed the path set before him by his predecessor, Dr. Stephen M. Babcock, who had been at Geneva from 1882 to 1888, and who had become internationally famous at Wisconsin as the designer of the Babcock test for estimating the butterfat content of milk, a test that provided a practical basis for the economics of milk production.

If any man can be identified as the single most important influence in the scientific career of Harry Steenbock, that man would have to be Professor Edwin B. Hart. In 1908, when Harry began his graduate work, Professor Hart was his senior by eleven years. Hart had already begun the selection of men for his growing department and, as events proved, showed uncanny skill in identifying and attracting a panel of men who, in their respective fields of biochemistry and nutrition, "wrote large for the world to see."

Considering Hart's influence on Harry Steenbock, it seems useful to describe, if only briefly, the character and personality of the man. Hart was a very practical man. In the terminology of today, one would say that he understood the meaning of relevancy. Characteristically, as head of a department of agricultural chemistry in what was fast becoming the leading dairy state in the nation, Hart taught the course in dairy chemistry, contributed research in dairy chemistry, had an interest in the field for all of his life, and when he retired in 1945 left a void in the teaching of dairy chemistry that was impossible to fill. Hart's influence on Steenbock proceeded from a broad base.

It is one of fate's ironies that although Hart was an early and important worker in the field of the vitamins for more than a quarter of a century, his name is not associated with the discovery or isolation of any one of them. And this in the face of a list of almost 400 publications over a span of more than half a century. It may be, of course, that with his predilection for the practical a chance was lost for the construction of scientific monuments, or at least monuments of archival visibility. Be that as it may, Hart left other monuments, and Henry Steenbock was one.

Although Harry Steenbock was to develop into and become his own man, so to speak, Hart's influence probably sensitized Steenbock in terms of a style in addressing problems in experimental biology and in a continuing awareness of problems "from the field." Farmers encountering difficulties on the soils of Wisconsin always found Hart an attentive listener. Steenbock, a shy man, never sought such encounters and turned more and more to theoretical problems in research; but there is little doubt that Hart, in the early years, made Steenbock aware of the practical applications of science, an awareness that led to some very interesting consequences in Steenbock's middle years.

It is important to record another man who, as Harry Steenbock began his graduate work, was a newcomer to the department, but as events showed, was already destined to become one of the important figures in the newly emerging science of experimental nutrition. Elmer Vernon McCollum had come to Madison just a year earlier, in 1907, from Yale University and now at 29 years of age, was the subordinate to Hart. As McCollum himself has related in his autobiography, what he knew then about nutrition he had learned from Chittenden and Mendel at Yale and a few seminars on topics then under discussion by certain German investigators. Indeed, when he came to Wisconsin, McCollum had never analyzed a food by chemical methods or conducted an animal experiment. In the Yale laboratories he had done organic synthesis under Dr. Treat B. Johnson and worked on protein analysis under Dr. Thomas Burr Osborne. It was, however, Professor Lafayette B. Mendel who had recommended young McCollum to E. B. Hart as a one-year "post doc" in his laboratory.

E. V. McCollum was an excellent foil to the practical Hart. From a Kansas boyhood, the Yale graduate had become a sound scholar with a flair for effectively sieving the world's scientific literature in the fields of chemistry and biology. He was an indefatigable reader and it is quite probable that McCollum's scholarship and sense of scientific history were important ingredients of the success of the Wisconsin group that Hart was assembling. There was another important contribution that McCollum made at this time with the enthusiastic encouragement of Professor Emeritus Babcock and the reluctant consent of the always-practical Hart. In the early months of 1908, as Harry Steenbock was finishing his senior year as a student, McCollum established the first rat colony in history for nutritional experiment by purchasing at a cost of six dollars from his own pocket a dozen albino rats from a pet-stock dealer in Chicago. The consequences of this innovation and the attendant birth of the biological method for food analysis by the use of experimental animals are of obvious importance and need not be labored here. It was significant, however, for Harry Steenbock that his lot was now being cast as a scientist with men capable of innovation and of wide scholarship. Indeed, it was in Harry's senior year that he was enrolled with Hart's approval as one of two students in McCollum's first lectures in the biochemical aspects of nutrition. The University records show that McCollum had a very high estimate of the young student's performance.

The path to the doctorate took Steenbock eight years. In 1910 he received the master's degree in agriculture at Madison and in 1916, the Ph.D. In the interval there was a "Wanderjahr" spent in 1912–13, first at Yale with Professor F. B. Mendel and then at Berlin under Professor Carl Neuberg. The work with Neuberg on the formation of the higher alcohols from aldehydes by yeast was published, in German, in *Biochemische Zeitschrift* in 1913 and 1914. It was a far cry from the colloquial German of the farm at New Holstein to the prim scientific German prose composed in the German capital. Harry Steenbock had come a long way.

On May 26, 1916, at 1:30 PM, Harry Steenbock's examining committee assembled and questioned him on his thesis, *Studies in Animal Nutrition.* Professor Hart was the major professor, of course. The first minor, physiology, was represented by Professor J. A. E. Eyster, and a second minor, organic chemistry in forest products, by S. F. Acree. The other members of the examining committee were Professor H. C. Bradley, a physiological chemist, and Professor of Chemistry, Victor Lenher. From the perspective of his later work, Steenbock's doctoral thesis was unexciting. Part of it was a meticulous examination of the occurrence in alfalfa hay of a pyrol derivative, stachydrin, which caused errors in the estimation of some amino acids by the cumbersome methods of the day. The published paper was a competent, craftsman's job and appeared in the *Journal of Biological Chemistry* in 1918. It is safe to say that only a few specialists noted its appearance.

But seven years earlier, in 1911, only a year after he had won his master's degree, Harry Steenbock's name had been listed, along with E. B. Hart, E. V. McCollum and G. C. Humphrey, as an author of the famous University Bulletin No. 17, *Physiological Effect on Growth and Reproduction of Rations Balanced from Restricted Sources.* This publication was the first published scholarship to bear Harry Steenbock's name, and it is the first paper in his collected papers that he, in later years, had bound and kept on his office shelf. This paper, it also turns out, was not only the cornerstone of Harry Steenbock's scientific career but was, some would say, the beginning of the modern era of the science of nutrition. On that estimate, the paper itself merits a closer look.

The idea for the "single grain ration experiment," as it came to be called, came from Hart's predecessor at Wisconsin, Dr. Stephen M. Babcock, who in his retirement continued to provide encouragement and counsel to his younger successors. Babcock had become disenchanted with chemical analysis as a means of predicting the nutritional value of a ration for farm animals and, in brief, had come to emphasize the then heretical notion that there might be nutritionally important substances that chemical analysis of the day failed to reveal. Babcock, well aware of the historical tradition of feeding mixtures of foods in practical rations for farm animals, conceived the idea of examining the possible effect of this practice of mixing food sources by formulating rations, which passed the tests of chemical analysis, from the separate parts of a single plant. Indeed, Babcock had performed a preliminary trial of this approach with two cows with encouraging but, for statistically obvious reasons, unconvincing results. In May, 1907, the larger experiment was begun with sixteen heifer calves divided into four lots of four. Lot I received a corn ration made from corn stover, corn meal and gluten feed: Lot II received a wheat ration of wheat straw, wheat meal and wheat gluten; Lot III received an oats ration of oat straw and rolled oats; and Lot IV received a mixture of equal parts of the three other rations. By properly adjusting the proportions of the ingredients within each ration, each separate ration was rendered analytically identical to the others. But chemical identity so arranged was now shown, over the full life cycles of the cows, to result in vastly different nutritional performance. The experiment lasted four years. As the report summarized, "Animals fed rations from different plant sources and comparably balanced in regard to the supply of digestible organic nutrients and production therms were not alike in respect to general vigor, size and strength of offspring and capacity for milk secretion."

The declining order of excellence in the physiological value of these rations, as the experiment demonstrated, was corn, oats, the threefold mixture, and wheat, with wheat a very poor fourth indeed. This is not the place to go further into the details of this pioneering experiment. Suffice it to say that the range of "physiological excellence" that the experimenters observed extended from the sleek vigor of the corn ration cows and their calves, to the gaunt, rough-coated wheat-fed cows and their prematurely born and dead-at-birth calves. An elaborate attempt was made by the young Wisconsin investigators to find the cause of the marked difference in the response of the animals to these restricted diets by autopsy, analysis of the tissues of the calves that were born, and by analysis of the feeds and excreta of the animals in the four groups. But all such attempts to find an explanation came to naught. None of the known methods of biological chemistry led to a satisfactory understanding. The experimenters were left confronting the unknown.

The reason for discussing here, at some length, the first publication to bear Harry Steenbock's name among the authors, is the sheer importance of this initial participation of the young investigator. It was a true beginning. Not only for Steenbock, but it was a beginning too, for his seniors. McCollum, tiring of preparing feed for cows in such huge quantities for such long periods of time, hit upon the idea of using a smaller animal, the rat, instead of cows. And the rest, as is said, became history. But the overriding significance of the single plant feeding experiment was that it destroyed the established paradigmatic method of conducting the erstwhile science of livestock feeding by means of chemical analysis and the consultation of tables of "digestibility" and "fuel value." A new paradigm came to be—had to be—fashioned from the lessons of Bulletin 17. It was the abandonment of chemical analysis as the arbiter of the prediction of nutritional value. Chemical analysis was to be used still as an initial and preliminary approach. But the final decision now would come from the feeding trial in an experimental animal. This was indeed, in terms of Thomas Kuhn's analysis of scientific revolutions, a bona fide revolution. It was above all a methodological revolution and toppled the contemporary hero-figure of the meticulous, German-trained chemist from his authoritative base. In nutrition, it was now clear, there were matters beyond the chemist's ken. And, by hindsight, it was as well a successful revolution because in its simple new paradigm of the appeal to animal experimentation, the following course of half a century of enthusiastic work resulted in an astonishing thrust into new and undreamed of territory, and in the conquest of certain human diseases, too familiar not to itemize here. Bulletin 17 was Harry Steenbock's beginning. It was also the beginning of much else.

Thus begun in 1911, Harry Steenbock's productive years as a scientist stretched over more than half a century. His list of publications includes more than 250 scientific papers in journals of the front rank, and the notebooks of the laboratory record more than 135 names of graduate students who worked for their advanced degrees under his guidance. Forty-eight Ph.D. and fourteen terminal M.S. degrees were awarded under his name. But, in one way or another, it was the work published in Bulletin 17 that set the tone, and the years of productivity fulfilled the promise.

Harry Steenbock's scientific productivity is to be found in scientific journals, and not, from his own hand at least, in books. In this aversion to book writing, Steenbock followed Hart who also, remarkably, never wrote a book. Perhaps the situation was too fluid and they were both so completely involved in the onward rush of their new science that they never paused for the perspective that a book would have demanded. Harry Steenbock did make one try at popularization in a piece, "Vitamines and Nutrition," which appeared in *The Scientific Monthly* in August, 1918. The events of World War I had drawn attention to the world plight in the matter of food, and the new knowledge of the rather mysterious "vitamines" had caused some anxieties among the public. Steenbock strove to dispel these anxieties, apparently uncomfortably aware that as a popularizer for science he would have to write in metaphors and perhaps be rather cavalier with scientific detail. It was not his forte, and he never tried this avenue again. One may speculate also that, caught up in the patriotic necessity of reassuring his fellow citizens, Steenbock disregarded his cautious nature and made some generalizations on the basis of slender, and as it turned out, misleading evidence. It comes as a puzzling incident to the modern reader of that 1918 paper to read, "At one time, there was a tendency to associate etiologically other conditions of malnutrition, such as scurvy and rickets, with a deficiency of specific vitamines. Evidence so far presented does not support this contention." If one acknowledges Harry Steenbock's stout conservatism, perhaps the remark is not so puzzling after all.

The productive years followed three main themes—the ramifications of vitamins A and D and the investigation of nutritional anemia. In 1918 Steenbock began a distinguished series of papers in the *Journal of Biological Chemistry* titled, "Fat Soluble Vitamine." The series continued for 16 years and totalled 42 separate contributions. Paper XIII in this series, published in June, 1923, still spelled it "vitamine," but paper XIV, published in November of that same year, adopted the modern spelling and the title thenceforward became "Fat Soluble Vitamins." The series had barely begun when Steenbock reported at the April 1919 meeting of the American Society of Biological Chemists in Baltimore that he believed he saw a tendency for vitamin A to be correlated in foods with yellow pigmentation. It was a relationship to which Steenbock returned time and again, but although there was much evidence he could marshal to support his idea there were serious stumbling blocks that defeated him. After ten years of the debate, illumination was finally gained through the work of Moore in England, that carotene, a yellow pigment of the plant world, was converted by the liver of the ingesting animal to the almost colorless vitamin A. Carotene was indeed provitamin A. In 1932 Steenbock wrote a detailed review of the vitamin A story for the *Yale Journal of Biology and Medicine*. The history of this phase of nutrition research reveals that Steenbock's persistence in his yellow pigmentation theory, sometimes in the face of negative results of his own and of others, resulted in the final understanding of a truly complex but important relationship between plants and animals with great consequences for the latter. In Steenbock, the years of work on vitamin A had two consequences: it led him to his work on vitamin D; and it led him to his first, and abortive, struggle with the social consequences of biological discovery.

In order to understand Harry Steenbock's sensitivity and prescience in the interface between science and society, certain social and cultural factors need to be examined. Wisconsin was a leading dairy state, and the University, where Steenbock was now forging his lifework, had a heavy commitment to the dairy farmers. We have already alluded to the deep involvement of Hart and Babcock with dairying and dairy chemistry. Steenbock's first publication, the famous Bulletin 17, was based on nutritional investigations with dairy cows, and Dr. Babcock, already a legendary figure, had become internationally famous for his devising of a practicable test for estimating the butterfat content of fluid milk and thereby had put the dairy industry on a firm economic base. Milk was now bought and sold on the basis of the butterfat it contained. But butter itself, the monarch of Wisconsin agriculture, was uneasy on its throne. During the shortages of World War I, margarine, a cheap substitute for butter made from animal and vegetable fats invented by a French chemist in 1869, had risen in the American marketplace as a serious challenge to the dairy industry. Almost fortuitously the new knowledge of the vitamins, fat-soluble vitamin A in particular, had restored butter to a commanding position. Butter contained vitamin A, margarine (oleomargarine, by law, in the U.S.) did not. Harry Steenbock's publications demonstrated this fact backwards and forwards, and one must not underestimate the satisfaction in Wisconsin that butter rightfully could lay claim to having a superior dietary value over oleomargarine. In the early 1920s, however, as Harry Steenbock's work led to an improved understanding of the chemistry of vitamin A and his increased ability to manipulate and concentrate it in the laboratory, the realization grew that vitamin A might be added to oleomargarine and thereby make it the nutritional equal of butter. These thoughts struck hard at much that Harry Steenbock held dear. He resolved to act, and characteristically, within the framework of the American entrepreneurial system. After consulting Dean of Agriculture H. L. Russell, the Board of Regents was petitioned to take out patents to safeguard the welfare of the dairy industry. But matters dragged on, much to Steenbock's disgust. An initial draft of February, 1923, languished in patent attorney offices until August, 1924. The Regents really were not interested. Harry Steenbock, though, had learned a lesson. Never again was he to relinquish to others, without preparation and safeguards, the control of moral imperatives which he felt in himself.

But other events, of even greater importance, were now bearing down. In his great concern for improving the quantitative measurement of vitamin A by means of biological assay in experimental animals, Steenbock, like others, had become aware that the fat-soluble vitamin was really composed of two entities: the growth-promoting vitamin A and the antirachitic entity that was to be named vitamin D. But vitamin D had growth-promoting qualities, too, when supplied to a rachitic animal. Not to control vitamin D in assays of vitamin A was to allow significant errors in estimates of the latter inasmuch as the quantification depended on measurements of body weight gain of the experimental animals. If increments of vitamin A were to be measured, then all else had to be supplied which would be needed for the growth of the young animal. Steenbock took advantage of Huldschinsky's demonstration in 1919, that ultraviolet light could be used to cure a rachitic animal. Previously, only cod-liver oil was a recognized agent capable of preventing or curing the faulty calcification of bone as observed in rickets, but cod-liver oil also contained vitamin A, which confounded its use in devising vitamin A-free diets. In using ultraviolet light as an antirachitic agent while he pursued his goal of measuring vitamin A with ever greater precision, Steenbock stumbled into a world of phenomenology from which he emerged with international fame. Fortune had favored the prepared mind.

The fundamental, puzzling phenomenon that Harry Steenbock unraveled, which confronted him repeatedly in different ways and was exhibited in a variety of circumstances, was that the beneficial effects of ultraviolet irradiation of an experimental animal could be transferred by that irradiated recipient to another unirradiated animal by the mere act of their occupying a common space, either simultaneously or, even more puzzling, in series, provided the irradiated animal preceded the unirradiated. Much of this became explicable ultimately by the recognition that excrement and perhaps other effluvium, such as salivation, could convey the infinitesimal amounts of physiologically active material from the irradiated to the unirradiated recipient, not unlike fomites conveying infection. But it was in the struggle toward this understanding that Steenbock instructed his assistant, Archie Black, to irradiate the hog's millet in the experiment described at the beginning of this memoir, and food, the very substance of nutrition, became the storehouse of the antirachitic effects of ultraviolet light. A food previously lacking in the antirachitic vitamin, by its irradiation with ultraviolet light now had the vitamin activity. Harry Steenbock, as a popular writer of the day put it, had trapped the sun.

The momentous discovery was published as a short note in *Science*, September 5, 1924, and in more detail in the September, 1924, issue of the *Journal of Biological Chemistry*. But in publishing these two papers Harry Steenbock also included a sentence which was to dominate his life thenceforward, and which was to be both a source of pride and a trial. For at the end of the paper in *Science*, "The Induction of Growth Promoting and Calcifying Properties in a Ration by Exposure to Light" and as an unnumbered footnote to the paper in the *Journal of Biological Chemistry*, "Fat Soluble Vitamins. XVIII. The Induction of Growth-Promoting and Calcifying Properties in a Ration by Exposure to Ultra-Violet Light," Harry Steenbock wrote this sentence: "To protect the interest of the public in the possible commercial use of these and other findings soon to be published, applications for Letters Patent, both as to processes and products, have been filed with the United States Patent Office and will be handled through the University of Wisconsin." That sentence brought him personal abuse and praise, and public vilification and applause. It brought his beloved University, in spite of its hesitant leadership, the financial resources which were to sustain its claims to scientific excellence in the Great Depression which lay ahead. The last phrase of that important sentence that was to dominate the rest of Harry Steenbock's life was, "through the University of Wisconsin." That was overly optimistic as events were to show.

Once again Harry Steenbock had a discovery on his hands that alarmed him with its potentialities, and once again he felt it was the dairy industry and the economy of the state that were threatened. In an unpublished manuscript now in the archives of the University of Wisconsin, Harry Steenbock set down, in January 1929, his remembered thoughts on the matter. After a brief description of the frustrating and unsuccessful attempt to have the University gain some control over the prior vitamin A issue, his own description of events was as follows:

> In the meantime another matter came up. The writer discovered that ultra violet light acting upon all naturally occurring foods, but especially upon fats and certain compounds found in fats, could endow them with properties which acted as stimulants for bone calcification. In other words, he discovered a process of treating foods which prevented the disease known as rickets. Inasmuch as rickets is a disease widely affecting both man and beast at the growing age, this discovery promised to be of as great, if not greater, importance than the former. It presented numerous opportunities for practical application and again the oleomargarine manufacturer was concerned. By the use of this process the oleomargarine manufacturer can make his product superior to butter in its bone calcifying property. He can justly claim that this dietary property is far more important than that due to its vitamin A content because rickets is far more prevalent than the symptoms resulting from Vitamin A deficiency.
>
> Foreseeing the possibility of commercial application, the writer decided upon immediate action resolving that if the University did not decide to protect his interest and that of the public, he would take the matter into his own hands and patent it himself. A conference with Dean Russell and President Birge brought out the fact that the University was no more prepared to handle the matter than before, and all they could assure the writer was that the University, no doubt, would reimburse him for the expense incurred in securing a patent, provided that such expense would not be greater than a few hundred dollars. As President Birge expressed it: "The Board of Regents cannot be expected to allot money for a patent application when it is not certain that it will receive something for such an expenditure."

Harry Steenbock was now obliged to move on his own. The first task was to obtain the patents. The process was begun by his hiring a Chicago patent attorney and an omnibus application was filed on June 28, 1924. The Patent Office raised objections of various kinds, a not unusual procedure, and ultimately four patents were granted. The first, the so-called "Basic" patent, no. 1,680,818, on the antirachitic product and process, was issued on August 14, 1928, and running for the allotted 17 years, expired on August 13, 1945. Three other patents, derived from the original application, were granted on August 9, 1932 (no. 1,871,135 and no. 1,871,136); and on October 13, 1936 (no. 2,057,399). These last three patents were ended by their dedication to the public on January 11, 1946. Canadian patents issued on July 9, 1929 and September 9, 1930, were similarly dedicated to public use on November 30, 1945. Harry Steenbock sought and obtained these patents, and when they were issued he assigned them to the Wisconsin Alumni Research Foundation. But when Harry Steenbock dug down into his pocket for $660 in June, 1924, to hire the patent attorney, a Mr. John Lee of Dryenforth, Lee, Chitten and Wiles of Chicago, there was no such foundation. Harry Steenbock had started on a path that, although unmarked, would lead into new territory. The new terrain called for new choices.

The formulation of what ultimately emerged as the Wisconsin Alumni

Research Foundation came as a novel response to a challenging situation. Steenbock was well aware that obtaining a patent, even if financed without the Regents' help by money out of his own pocket, was in itself no adequate solution. The problem was broader than that. Patents have to be defended against infringement, the rights of the public have to be protected from fraudulent or ignorant use of the discovery, licensed users of the patented information have to be regulated and monitored, and the funds, hopefully earned by the wise implementation of all this, have to be managed and husbanded. Who would do all this? It certainly was beyond the capabilities and, many argued, not even within the appropriate scope of a university professor. Perhaps it was even more inappropriate for a university. The very act of seeking a patent was not without criticism in itself, and some of that criticism came from within the University and the scientific community in general. There were those who felt that no gain properly should accrue to a person working in a university laboratory in the scholarly pursuits of his profession. Somehow, such gain was not in the genteel traditions of the professorial ranks. Harry Steenbock had his answer for that; indeed, several answers. He pointed out that other University scholars authored books as a result of their scholarly pursuits in University buildings, aided by University secretaries, and no one questioned their rights to copyrights and royalties. A second part of his answer was conclusive: Harry Steenbock would, for himself, not accept any financial benefit from the discoveries he was trying to patent. But the dilemma as to how to administer the patent, once it was obtained, remained.

The fact that Steenbock was trying to patent his discovery of the activation of foods to antirachitic activity by means of ultraviolet light was becoming well known, and as he had anticipated, commercial interests soon were beating a path to the discoverer's door. Probably the most important of these approaches came from the Quaker Oats Company which, for exclusive rights during the seventeen-year life of the patent, offered a contract which would have brought to Harry Steenbock nearly a million dollars. It was probably in the discussions with the Quaker Oats Company that Mr. Carl S. Miner, of the Miner Laboratories of Chicago, entered the stream of events. Miner's laboratory, from time to time, performed some scientific work for the Quaker Oats Company. He was well-aware of the importance of the Steenbock discovery, and, in a way as unselfish as Steenbock's, he, in conversations with Steenbock, pointed out that if some agency could be developed along the lines of a trust, then the monies that hopefully would accrue could be used for the financing of basic research in the University. The results of research would thus nourish further research.

Steenbock next wrote to other universities to see what could be learned from their experiences. Columbia University, a private university, had established an Administrative Board of University Patents "to enable the University itself to share in the benefits of the patents, to the end that the funds at its disposal for the promotion of research may be increased."

At the University of Minnesota, a state university, the Board of Regents had accepted the assignment by Dr. E. C. Kendall of his patents on the active principle of the thyroid gland. It was specified by Dr. Kendall that his object in assigning the patent was to enable the compound to be manufactured so as to do the greatest amount of good for the greatest numbers and that the net proceeds should be used to further research as well as to recompense the assignor to an extent not to exceed 10 percent of the net proceeds resulting from the sale of the preparation.

At the University of Toronto, Banting, Best and co-workers, assigned their patent rights on the preparation of insulin to the Board of Governors of the University. Their rights in England were assigned to the Medical Research Council. These organizations in turn issued licenses for the manufacture of insulin under direct control of a laboratory managed by the Insulin Committee. In this way the public was protected against the manufacture of poor preparations and was also protected against extortionate charges. The proceeds from the licenses were used to pay the expenses of the control laboratory, and the remainder was used to fund research fellowships in the University.

Fortified by these precedents, Harry Steenbock proposed to Dean Russell the formation of a trust company composed of friends of the University to which he could assign his rights if the Regents continued to delay.

The Steenbock discovery of the use of ultraviolet light in the irradiation of foods was not only a triumph for the College of Agriculture and a source of pride to its Dean, H. L. Russell, but it was also a matter of deep interest to the Graduate School and its Dean and Chairman of the University Research Committee, Charles S. Slichter. Although Slichter was not involved in the struggles with the Regents, which had been conducted by Dean Russell and President Birge on Steenbock's petition, he was aware of the situation and proffered his help. He invited Steenbock to lunch at the Madison Club and heard the whole story including the idea of a separate Trust Company. It was the latter idea which particularly captivated him. In Slichter, with his wide acquaintanceship among the University alumni in Chicago and New York, Steenbock now had an able advocate who saw the practicality of Steenbock's idea—an idea capable of breaking the deadlock with the Regents by the stratagem of providing a free-standing device, a foundation, which, with the blessings of the Regents, would relieve them of their politicized inability to move forward on the issue of accepting responsibility for a patent. Steenbock wanted a "Trust Company," as he put it, but had led the cloistered life of a researcher and lacked the contacts with the world of moneyed affairs that the situation now seemed to require. Slichter, through his wartime years of service to the National Research Council, had many acquaintances in the huge industrial complex that had been brought into contact with the academic world in the war effort, and he understood the politics of fund raising. Slichter lost little time and traveled to Chicago and New York, returning with pledges from alumni he had interested in the undertaking. He himself advanced money for the application for foreign patents.

In April, 1925, the Regents began to move and referred the matter to their Executive Committee. As expected, the Committee invited the two deans, Russell and Slichter, to give them the benefit of their judgment. On May 8, the Executive Committee met in a special meeting with the following action:

Special Meeting of the Executive Committee of the Board of Regents—University of Wisconsin, May 8, 1925:

Deans Slichter and Russell presented a plan with a reference to the administration of the results of University Research, particularly as applied to those researches that are of a character and of such importance as to warrant patents being taken out on the same. Such patents may be taken:

1. In the interest of the public at large to prevent exploitation or monopoly, or
2. Assigned by members of the University staff for the direct benefit of the University.

The University recognizes that its organization is not well suited to attend to the details of patent procedure; to defend patents in litigation and conduct the necessary business of completing the commercial utilization of patents. Immediate need for the consideration of this matter lies in the fact that the results of experimental work of Doctor Harry Steenbock of the Department of Agricultural Chemistry of the College of Agriculture are such that patents should be secured at once to protect these discoveries. Doctor Steenbock has indicated his willingness to assign his interest in large measure to the University, on the understanding that the returns, if any, will be mainly made available for the prosecution of scientific research.

In providing a working plan by which this and similar matters can be handled, it is proposed to create an organization on a broad enough basis so as to embrace any other propositions of a similar nature that may arise in the future.

It is proposed to organize a non-profit-sharing corporation or trust, the necessary capital of which will be contributed by alumni and friends of the University, the management of such corporation to be placed in the hands of Trustees. This corporation will be empowered to receive assignments of patents and all rights in patent applications; to secure patents in this and foreign countries; to carry on further work with reference to the commercialization of the patent processes; to take all the necessary steps to defend and utilize these patents in a way that will best subserve the interests of this foundation. The subscribers to this fund will be repaid their advances from the avails arising from the sales of rights, royalties, and the like, after expenses incurred in the transaction of necessary business operations of the corporation are defrayed.

It is understood that the primary purpose of the corporation is to administer the patents intrusted to its jurisdiction in such a way that research in the University may be supported from the avails in excess of sums deemed necessary by the Trustees for the most beneficial development of the patents. It is understood that members of the University Staff who may assign patents secured in their name to this corporation for the ultimate benefit of the University research will be permitted to nominate the fields of research to which such funds shall be allotted.

The Executive Committee adopted the fundamental principles of the plan presented as affording a practicable means of the possible endowment of research through funds arising from the scientific efforts of its staff members who are willing to assign patents for such purposes.

Adopted by the full Board of Regents, the sticky problem had been resolved by what was in reality a social invention in itself—an independent organization, tied to the best interests of the University by the bonds of loyalty of dedicated alumni from the world of affairs, and providing a clear, University-approved track for the handling of similar matters as they should arise. What formally emerged in a Certification of Incorporation on November 16, 1925, was the Wisconsin Alumni Research Foundation. On February 18, 1927, before the 1928 patent, the first one, actually was granted, Harry Steenbock formally assigned the patent to the Foundation for the sum of 10 dollars. He was 41 years old and half of his life still lay ahead.

This is not the place to deal further with the history of the Wisconsin Alumni Research Foundation. Suffice it to say that it was brought into being by the imperative of the discoveries of Harry Steenbock and his straight-backed insistence that scientific discoveries, so important in the modern world, should nourish the growth of science through more research. Other faculty members found, through the years, opportunities for management of their discoveries in the Wisconsin Alumni Research Foundation. The Steenbock patents resulted in capital accumulation in the Foundation over the years, before their expiration, of a total of approximately fourteen million dollars. By 1973, the University of Wisconsin had received, with no strings or directives attached for their spending, a total of over 65 million dollars. It is obvious that the Foundation has been a success and, just as obvious, that it has had the benefit of the astute management of a dedicated board of alumni-trustees.

In the early days of the patent and the Wisconsin Alumni Research Foundation, Harry Steenbock accepted no income from his patents. Modest in his personal needs, he often stated that his university professorial salary was perfectly adequate for his bachelor life. Photography claimed a few leisure hours, but even then the hours in the darkroom were spent just down the hall from his office in the Agricultural Chemistry Building. Later on, under the urging of the Trustees, he accepted fifteen percent of his net patent royalties in order to set a precedent so that other faculty patent holders, with lesser salaries but with families to support, could accept some royalties from their discoveries. But when the earnings of the vitamin D patents became, for those days, astonishingly large, Steenbock accepted only a fraction and arranged that the bulk would be used for the furtherance of research in the University.

Many important events and research discoveries lay ahead but, by hindsight, there can be little doubt that the ultraviolet light discovery and the social invention of the Wisconsin Alumni Research Foundation were the high-water marks of Harry Steenbock's scientific career. Other discoveries should not be lost sight of. There was, for example, the important collaboration with J. Waddell, C. A. Elvehjem and E. B. Hart on the investigation of nutritional anemia. This work led to the important discovery of the indispensable role of copper in the curing of nutritional anemia by supplements of iron. In the telescoping habits of history, it is Professor E. B. Hart who is traditionally credited with this discovery, but it is evident that others, including Steenbock, were involved. Credit and priority of discovery were always highly regarded by Steenbock and, in this instance, he felt ill-used by the outcome. There was some tension for a while and although the copper discovery was announced in 1928, much remained to be studied and Harry Steenbock returned again and again to the processes of hemoglobin formation. The nutritional value and biochemistry of cereals, the chemistry and mode of action of vitamin D, the nutritional effects of lipids and vitamin E—and again and again—the never-tiring theme of vitamin A, provided scope for many active years. Always interested in the processes of the calcification of bone, he was the first to produce osteolathyrism experimentally.

The years of World War II, as in all parts of American life, reduced the number of students in the Steenbock laboratory and the output suffered. After the war, although he was approaching the age when most men retire, Harry Steenbock began a fresh expansion of his laboratory, increased the number of graduate students, learned the newer techniques of chromatography and the use of isotopes. It was an astonishing demonstration of the intellectual vigor that characterized Harry Steenbock for all of his life.

During the War, in 1942, Harry Steenbock's father had died, and in 1946, Christine, Harry's mother, had followed her husband. The dutiful son was now left alone in the family home for the first time in over 40 years.

In the years of his maturity, on March 6, 1948, Harry Steenbock married Evelyn Van Donk who, too, had Wisconsin origins and who had assisted him in the laboratory and received her master's degree in the department at Madison. She had gone to work during the war years at the Lederle Laboratories at Pearl River, New York, but now returned to Madison where she entered with Harry Steenbock into a happy married life, first in the Steenbock home on West Lawn Avenue, and then in their comfortable home on Ottawa Trail in Nakoma, a Madison suburb.

Harry Steenbock, in the golden years of his life, now mellowed and relaxed from the tense years of striving and accomplishment. With his wife Evelyn he enjoyed travel, collected art, was a founder of the Madison Art Foundation, a patron of music in Madison and a major benefactor of the Wisconsin Academy of Sciences, Arts and Letters. It was the latter's sponsorship of the Junior Academy, which fostered the interest of young students in science, that attracted his interest. After his death the Academy had astonished reason to gratefully recall Harry Steenbock's supporting interest, for in his will he bequeathed it, after numerous philanthropies, the residue of his estate, which, when calculations had been completed, came to about a sixth of what was reported in the press as a total of around six million dollars.

Harry Steenbock's career as a professor of biochemistry on the faculty of the University of Wisconsin did not end at the usual mandatory age of 70. In 1956, when that event was ostensibly dictated by the calendar, the Regents of a grateful University, in an unprecedented move, refused to permit his stepping down and so he remained as an officially active member of the faculty to the end. Along the way he had received a cherished honorary D.Sc. from his Alma Mater in 1938 and the same degree from Lawrence College in 1948. He was a founding member of the American Institute of Nutrition (1928), and a recipient of its Borden Award in Nutrition in 1959. He was elected a Fellow of the American Institute of Nutrition in 1958. He belonged to numerous national and international scientific societies but, again characteristically, never sought office in any of them, and deflected it when it was proffered to him. He was devoted to scholarship and learning. From monies he had declined as an individual, and left in the care of the Wisconsin Alumni Research Foundation, on his nomination, appropriate sums were made available for the support of the Steenbock Library of Biochemistry and for the endowment of the Steenbock Research Professorship in Biochemistry. The Steenbock Memorial Library for Agricultural and Life Sciences on the Madison campus of the University of Wisconsin was the result of a gift from the Trustees of the Wisconsin Alumni Research Foundation from the general funds of the Foundation.

Philosophically, Harry Steenbock was an empiricist and, in science, an intellectual descendent of Justus von Liebig, whom he much admired. The rigorous, no-nonsense approach of the German scientific giant was much to his own taste. Long working hours, day and night, in the laboratory were considered as a norm. Gaiety and the light touch did not easily become him, and his austere dignity was seldom relieved. In rare moments, however, he revealed an inner wistfulness and a silent appreciation of the more lighthearted. His own high standards permitted no less for others, but—and this was the paradoxical irony of the man—with his insistence on personal freedom he contrarily tolerated neither a lapse in himself nor in others from those almost impossibly high criteria. Nor would he accept, in the slightest degree, instruction from others as to what those standards should be. He demanded complete freedom of choice, but at the same time his character goaded him to but one choice—that of the highest imaginable standard of discipline and what was, culturally speaking, the Protestant work ethic. In religion he was brought up as a Lutheran, but Martin Luther, the lusty peasant's son, would have appalled him. Upon Harry Steenbock's emergence from a Wisconsin farm onto a world stage of science, in his code, in his life, in his relationships with his fellows, he was more the spiritual kin of John Calvin. It was in the theology of the latter that expression had been found in northern Europe for the stimulation of an economy emerging from a medieval, agrarian world into the newborn capitalism, by the Protestant—Calvinist virtues of thrift, industry, sobriety and responsibility. Martin Luther disliked the world as he found it, and strove for a return to simplicity and the elemental. John Calvin coped with the world as he found it, in all its flawed complexity, but sought to improve the edifice by holding tirelessly before it the virtues that he believed would result in betterment. In his self-contained way, Harry Steenbock knew what those virtues were, and whatever the struggle, he reached for them.

On Christmas Day, in 1967, at the age of 81, Harry Steenbock died in Madison, Wisconsin, and there he was buried.

6

Elmer Verner McCollum (1879–1967)—A Biographical Sketch*

Agatha Ann Rider with Ernestine Becker McCollum

Department of Biochemistry
Johns Hopkins University
School of Hygiene and Public Health
Baltimore, Maryland

"My first nutrition experiment," Dr. McCollum was wont to recount in later years, "was performed collaboratively with Mother." It was her discovery that feeding young Elmer scraped apples improved his health and disposition, both of which suffered on the boiled milk and potato diet which she had deemed it necessary to give him when, seven months after his birth, she once again found herself to be pregnant. This treatment for scurvy occurred just in time to save a life whose mission was to be that of improving the eating habits of many of the inhabitants of the world.

The life of Elmer Verner McCollum began on March 3, 1879, in the farm home of Cornelius and Martha Kidwell McCollum near Ft. Scott, Kansas. Like many of the "greats" in the American life of that period when the population of the United States was largely rural, Dr. McCollum spent his early and formative years on a farm. But, unlike many of his contemporaries, he used those years to develop his imagination, his ingenuity, and his life-long habit of constant reflection. His mother's love of learning and the books which his three older sisters brought home with them from college inspired a longing for study and knowledge which was almost insatiable. His brother Burton, 16 months his junior, was his constant companion in games, in the raising of a pet pig, in the catching of rats for which their father paid a bounty. The boys' efficiency with a trap of their own invention taught them a basic lesson in economics when the supply of dead rats exceeded their father's ability to pay.

The successful running of a farm requires certain practices which are also

*Reprinted with permission from *J. Nutrition* 100:1–10 (1970).

Figure 6.1 Elmer Verner McCollum

essential in a competent investigator, such as attention to detail, adherence to a schedule, astute observation, patience and resilience. These attitudes, developed in the young boy, stood him in good stead in his later years as a nutritional investigator par excellence. A top-flight scientist also needs a certain serenity of spirit, an ability to put aside unpleasant thoughts in order to consider creatively the problem at hand. Serenity of spirit was one of Dr. McCollum's most outstanding attributes and probably received its start in his early days when he memorized poetry while plowing. This store of poetry was to prove invaluable in later times when he recited it in order to while away endless hours in the hospital following major surgery, first for intestinal obstruction, then for a detached retina (which, as he said, they were unable to "glue back again.")

Dr. McCollum's early schooling was spotty since the length of the school year depended on the ability of the farmers in the area to pay the salary of a teacher. Nevertheless, in due course, young Elmer completed high school in Lawrence, Kansas. The most memorable feature of these years was his purchase of a set of the Encyclopedia Britannica which he eventually read in its entirety. He confessed, though, that he just skimmed the parts dealing with higher mathematics.

He was duly admitted to the University of Kansas with advanced credit in the fall of 1900. Here his determination and ambition again showed themselves. To support himself he delivered newspapers, and nightly lit and at midnight extinguished the gas lamps for the city of Lawrence. He entered college with the idea of becoming a physician, but the chemistry courses taught by Dr. Edward Bartow, Dr. Edward C. Franklin, and Dr. Hamilton P. Cady filled him with ambitions to investigate the chemistry of living things, and he turned his energies to becoming an organic chemist. His ever-fertile mind led him to "cook up" eight new quinolines, purify them, and describe their most obvious physical characteristics. His unusual talents were recognized in his being elected to membership in Sigma Xi while still a junior in college. He received the B.A. degree in 1903 and an M.S. from the same University in 1904.

By the end of his stay at the University of Kansas, Dr. McCollum had decided that an academic career would offer him the most satisfaction and that a Ph.D. degree would be a necessity. Perusal of the chemical journals revealed that Professors Wheeler and Johnson at Yale University were synthesizing various purines and pyrimidines. He had learned from Hammarsten the biological importance of these compounds and so decided to venture to New Haven. When he arrived there in 1904, he had a scholarship to pay his tuition and $82 in his pocket. He promptly acquired a job teaching elementary chemistry at the Y.M.C.A. evening school and "inherited" students to tutor from a friend, Bill Cramer. These two sources of income coupled with an appointment as laboratory assistant and winning the coveted Loomis Award soon enabled him to save money. His research for the Ph.D. dissertation was done under the direction of Treat B. Johnson and centered on purines and pyrimidines. It is of interest that more than 60 years ago, as today, the components of nucleic acids were receiving much attention from investigators.

On completion of the work for the Ph.D. in June, 1906, Dr. McCollum accepted a position as a postdoctoral fellow in the laboratory of Dr. Lafayette B. Mendel, then the foremost physiological chemist in the United States. During this period he attended Dr. Russel Chittenden's course in physiological chemistry, the first in that subject to be given in this country. Through laboratory work, lectures, and extensive reading he became familiar with the knowledge then available in this field.

In 1907 on the advice of Prof. Mendel, E. V. McCollum accepted a position with Dr. E. B. Hart at the Wisconsin Agricultural Experiment Station. In this way was launched a career in nutrition and biochemistry which was to prove productive indeed.

Early in any scientific career there are factors at work which determine its outcome. Among these may be listed the personal characteristics of the investigator, the influences of his elders, his superiors, and his peers upon him, the resources available, and, of course, the ripeness of the times for the new ideas which these call forth. Certainly Dr. McCollum possessed the imagination, the inventiveness, the perseverance, the scholarliness, and the motivation which are integral parts of the successful scientist. His contacts with

Dr. Stephen Babcock, the inventor of the Babcock determination for butter fat and then one of the foremost workers in the field of foods and nutrition, proved of inestimable value. Dr. McCollum always considered it his great good fortune that Dr. Babcock in his status of Professor Emeritus had the time and interest to spend many hours in the laboratory of the young investigator, giving him the benefit of his wholesome philosophy, considerable knowledge, and vast experience in the field, encouraging him in his original thinking, and championing his cause in establishing the first rat colony for nutritional research in this country. But let it not be forgotten that the young scientist was receptive to advice and ideas from his mentor. Without this interplay, Dr. Babcock's role would have availed little.

It is interesting to look back on the state of nutritional knowledge in the early 1900's, when Dr. McCollum was starting out. Much work had been done on the role of carbohydrates in energy production by Atwater and others; diets had been defined in terms of their protein, fat, and carbohydrate content; Osborne, by characterizing the amino acid content of various proteins, had pointed out their differences. Eijkman and Grijns had induced beri-beri by feeding polished rice to chickens. Comprehensive experiments in nutrition under laboratory conditions using purified diets and the rat as an experimental animal were, however, few and far between.

Pavlov's experiments were the talk of the day. In 1909 in a paper entitled "Nuclein Synthesis in the Animal Body" (*Amer. J. Physiol.* 25: 120) Dr. McCollum wrote: "The psychic influence of palatability is one of the most important factors in nutrition, either human or animal." He then went on to report his experiences feeding rats diets composed of "12% protein (edestin or zein), 75% carbohydrate (starch and cane sugar), 5% ash of milk, 5% butter fat, 2% $Ca_3(PO_4)_2$, and 1% NaCl. This was mixed with finely divided filter paper and made into a dough with a little water. The mix was dried in an oven at 100°, cut into pieces, and preserved in a Mason jar." His report goes on:

> The rats ate this with apparent relish for about a week, after which there was evidence of a waning appetite. The sugar content of the food was changed when they again ate more readily. At this time the food was baked thoroughly and a portion fed in this form. At one time slightly caramelized sugar was used to give new flavor to the food. At another the food was moistened with water distilled from a strong cheese which was finely ground. This water possessed in some degree the cheese flavor and caused the rats to eat with more relish. Good results were frequently obtained by leaving fat out of the food entirely for a few days, changing it as much as possible by the methods mentioned above, then relieving the rats of these flavors by feeding the simple food mixed with fresh butter fat. This invariably induced good consumption for a day or two. On some days the ration was presented flavored with a trace of banana, celery, cinnamon, lemon, or vanilla flavors obtained from the commercial articles. The rats generally ate the ration on such occasions, but it cannot be determined to what extent the consumption was induced by these substances.
>
> As time went on it was found that when the mixed foods were not eaten readily

> pure edestin would be consumed with avidity, but only for one feeding. Glucose was frequently given separately, and considerable quantities were eaten. In one instance toward the end of the experiment bacon fat, freshly rendered and filtered through paper, induced a hearty consumption when every other means failed. Cellulose, ground charcoal, and bone ash were given at different times to regulate the condition of the feces. Care was also taken to change the content of the ration in sodium chloride at intervals in order to secure the change in taste which it afforded.

As we look at this work from the vantage point illuminated by some 60 years of additional experience and research, the deficiencies in the diets used are apparent but few, if any, among us could surpass that early investigator in inventiveness, ingenuity, unquenchable enthusiasm, perseverance and indomitability.

Much careful and painstaking research followed. It was not all successful, of course. Neither did all of it escape criticism from the more senior workers in the field. One of the qualities frequently exhibited by Dr. McCollum was his tenacity in defending an idea which he was convinced was sound. In scientific matters, frequently, one experiment well-conceived and well-executed can lay to rest years of wrangling. E. V. McCollum was the one to carry out this task. Criticism served to spur his efforts to unravel the mysteries of the relationships of dietary components and to resolve the apparent discrepancies between his results and those of others. And so, by the end of his ten-year sojourn at Wisconsin, he had published some of the most fundamental discoveries to be made in the field of nutrition. Foremost among them is the 1913 publication of the discovery of "fat-soluble A." This was followed in 1915 by the announcement of "water-soluble B." Of equal importance was his development of the Biological Method of Food Analysis which quickly replaced the chemical methods then in vogue. Through its subsequent use the major discoveries in the field of nutrition were made possible. During that decade of his life he also established the fact that purines could be synthesized by the rat from some complexes contained in the protein molecule and that inorganic phosphorus could be utilized in nucleic acid formation. He studied the advantages of feeding mixed diets to livestock, began to investigate the importance of various mineral elements in the diet, and looked into the supplementary relationships between foodstuffs. Impressed by the evidence obtained from his experiments on rats, he began to be concerned about the human diet and so he urged the increased consumption of milk and leafy, green vegetables. He felt the need to instruct people in the importance of proper diet for health and turned to home economists, dietitians, and to the general public for support in demonstrating the practicality and importance to health of his new concepts of an adequate diet.

The discovery of the essentiality of a fat-soluble and a water-soluble nutrient attracted widespread attention and interest. He advanced rapidly from the post of Instructor to that of Professor and his "visibility" (as he often phrased it) so increased that in 1917 he was invited to deliver the Harvey

Lectures. This was a signal honor to be conferred on one so young and so, in high spirits, Dr. McCollum set off for New York. He was invited to stay at the home of Dr. and Mrs. Graham Lusk. It was during this visit that Dr. Lusk handed him a letter from Dr. William H. Howell, the noted physiologist of The Johns Hopkins Medical School. In presenting it to him Dr. Lusk said, "I know what's in that letter. Don't fail to comply with the request." The letter was an invitation for Dr. McCollum to visit Baltimore on his way back to Madison, to discuss with Dr. Howell a matter of "considerable importance." Dr. McCollum complied and learned, much to his great surprise and satisfaction, that he was being offered the position of Professor and Head of the Department of Chemical Hygiene at the newly created School of Hygiene and Public Health at The Johns Hopkins University. Here he would be associated with illustrious men, indeed. Dr. Howell was to give up his post as Dean of the Medical School to become Assistant Director and Professor of Physiology at the new school and Dr. William H. Welch, the noted pathologist, was leaving the Medical School to become its Director. It was most certainly a tempting challenge. It would furnish the young investigator an opportunity to work in close connection with some of the best medical brains of the day and to publicize even further the cause of good nutrition as a part of the picture of Public Health. He accepted the new institution and its faculty with enthusiasm and they accepted him, although not without some qualification. As Dr. McCollum was about to board the train for Madison, Dr. Howell in his quiet way said: "We had just one misgiving about appointing you to a professorship, McCollum. You look so "frail." Indeed, at 6 feet and 127 pounds he was not the picture of health and, yet, when he died at the age of 88, he had spent more time on this "mundane sphere" (as he called it) than had any other of the original faculty. He was its last survivor.

Setting up a new department, helping to mold the policies of a new school as well as being master of the direction in which his research would go, proved exciting tasks for Dr. McCollum. Again he made the fullest use of his opportunities. The proximity to the School of Medicine made possible collaboration with pediatricians and pathologists and led to the discovery of vitamin D and the unraveling of the cause of rickets. From 1917 until his retirement in 1946 more than 150 papers issued from his laboratory. In these investigations were described the symptoms of magnesium and manganese deficiency. They helped to elucidate the roles of strontium, fluorine, aluminum, zinc, sodium, cobalt, potassium, calcium, and phosphorus in the diet. They encompassed studies on dental caries, thiamin, riboflavin, and vitamin E. They reflect the zeal and the wide-ranging curiosity of an ever-active and enquiring mind.

As his publications increased, as new discoveries poured from his laboratory, so his stature as a world leader in nutrition grew. With this advance and ever-widening recognition, came added responsibilities. Dr. McCollum always exhibited a sense of moral obligation for improving the health of mankind. This led, in 1918, to the publication of the first edition of his famous textbook, *The Newer Knowledge of Nutrition.* In this he used the phrase "Protective Foods," a catchword which was to do much to popularize the principles of wise eating practices. The fifth and last edition of this text was published in 1939. With J. Ernestine Becker (later Mrs. E. V. McCollum) he wrote the small, popular book, *Food, Nutrition and Health,* which has occupied a prominent spot in the education of dietitians, home economists and the laity.

There were many demands by outside organizations for a share of Dr. McCollum's time and efforts. He played an integral part in the founding of the Merrill-Palmer School of Motherhood and Homemaking. He became the leader and consultant during its stay in Baltimore of the National Dairy Products Corporation Research Laboratory and as such devoted an hour or more each day and one evening a week to its affairs. He also served as Nutrition Editor of *McCall's Magazine* and of *Parents' Magazine.* This was aside from his activities on the editorial boards of the *Journal of Nutrition, Journal of Biological Chemistry,* and *Nutrition Reviews.*

Beginning with his work on the Hoover Food Administration in 1917, Dr. McCollum constantly served on committees devoted to the dissemination of knowledge and the practice of sound nutrition. Among these were the Permanent Commission of Nutrition of the League of Nations, the Food and Nutrition Board of the National Research Council, The U.S. Pharmacopeia Advisory Board, the National Dairy Council, the Cereal Institute, the Nutrition Foundation, etc. To each he gave unstintingly of his efforts, never being content with a second-rate performance in anything he undertook.

In spite of the heavy demands of his research, he took time to teach courses in the universities of California, Utah, Missouri, Colorado, and Ohio.

Time passed and it was 1944, the year in which under the rules of the School of Hygiene Dr. McCollum should retire. However, due to the wartime conditions, he agreed to stay until 1946 when a successor could take over.

Retirement is a time which may bring out the best in a man. In a person's later years one sees the culmination of the experiences and outlooks of the past. By this time character has been defined and a philosophy tried, accepted, and matured. A few personal reminiscences (A. A. R.) of this period will serve to portray the flavor and the unique qualities of the man, E. V. McCollum.

One morning in late 1945, as was his wont, Dr. McCollum came into my laboratory, pulled up a chair, asked how things were going and then posed the most flattering proposal which I could have received. He had some funds which had been given to him through the years for his own personal investigations and which, in the true spirit of a Kansas farmer, he had saved for a "dry spell." He believed that the time had come to use this money. He wondered whether I would take a chance on working with him in the laboratory which the "big boys" (as he called them) were providing for him on the Homewood Campus after his retirement from the Johns Hopkins School of Hygiene in July 1946. This date coincided with the completion of the study course for my Sc.M. degree. He always felt obligated to find jobs for his students, and he said if I would consider this it would suit him nicely.

We had already begun working on "new and novel" methods for the separation of amino acids from non-aqueous solutions, a subject which had interested him for many years. The publication in 1944 by Martin and Synge of the method of paper chromatography, whereby amino acids in mixtures could be readily identified, had provided the tool needed for further exploration of the problem. Since he was now retiring and his reputation had been made, Dr. McCollum could afford to spend time on problems which might take a long time to solve.

No one would have hesitated to say "Yes" to such an offer. To be able to work closely with such an active, imaginative mind was undoubtedly the opportunity of a lifetime—no, of several lifetimes. And so began an association and collaboration, an exposure and stimulation, which has been of unparalleled value and satisfaction. Very seldom in these days of team research, where the pace is often a frantic one, does a young person have the chance to work quietly but productively with one of the most original thinkers of our time.

And how did he work? He never lost the simple uncluttered approach to problems he had exhibited as a youngster. Having read of Martin and Synge's use of air-tight chromatography chambers in which to carry out their separations, he proceeded to build his own from window sashes purchased at a nearby lumber yard, an aluminum tray for the bottom and a piece of window glass as the cover. The whole thing was coated with paraffin to seal the joints. On another occasion, realizing the need for a reading stand which could be fastened to the side of a desk or table, he invented and made one. It had a wooden screw with which the angle of the book could be adjusted as well as provision for raising and lowering the level of the reading matter.

His approach to the administrative problems of our little laboratory was also novel. For the first two years he paid me a monthly salary in one installment on the fifteenth of the month. He explained that if I would trust him for the first half of the month, he would trust me for the second half. With the appearance of outside grants our laboratory personnel expanded to the grand total of three, where it remained for about eight years before diminishing to two, and then to one as Dr. McCollum became older and less willing to ask young people to risk their careers in an endeavor which might end at any time. Our endeavor did end, finally, in the fall of 1964, when my husband's job required us to leave Baltimore for a while.

Dr. McCollum's filing system remained his own. Two large eleven-by-fifteen-inch cheese boxes held the laboratory business correspondence and laboratory reports in chronological order, the oldest being on the bottom. A careful rummaging through these boxes would turn up a wanted document in fairly short order.

Dr. McCollum was an inveterate reader of wide interests—and, what is more, a thinker. His procedure in reading a scientific paper was to read it through quickly to get the "gist" of it. He then asked himself four questions:

1. What did the experimenter ask?
2. How did he go about finding the answer?
3. What was the answer?
4. What is the next experiment to be done in this line?

It was the asking and answering of this fourth question in particular that separated him from the "run of the mill" investigators. The ability to think and plan creatively is one which must be constantly used if success is to come.

This habit of constant reflection was well shown in the topics which he would discuss in informal conversation. I can remember hearing him talk about the durability of the earth's magnetic field, the state of the United Nations, the fate of mankind if we continue to send the inorganic elements in the sewage out to sea, the poetry of Emily Dickinson, Plutarch's Lives, and reminiscences of his early scientific days—all in one sitting.

The chemical research problems he handled in his inimitable way. Once his interest in a question was aroused, ideas tumbled from him in great profusion. He was not over-awed by what others might have thought or done. "It's worth a try" was his response to a new idea and more often than not his approach was a novel one. Not even for a genius of the McCollum caliber, though, did all ideas work. If my response to his "How goes the battle?" was not as cheering as we had hoped for, he would remind me that "Life is not all beer and skittles" and, optimistically, plan the next assault. He would now view the question from a new angle and soon would come another new approach and still another and another until the problem yielded to the onslaught of a wealth of ideas.

Six papers and three patents emanated from these endeavors, but laboratory research, intriguing though it was, was not all that occupied Dr. McCollum during these two decades of retirement. He spent 10 years in the research and writing of his monumental *History of Nutrition,* which is an exhaustive review of the important work in the field up to 1940. Much time and effort were also expended in the writing of his autobiography, *From Kansas Farmboy to Scientist.* Never one to do a slip-shod piece of writing, he devoted countless hours to his Biographical Memoir of Stanley Benedict for the National Academy of Sciences, to his sketch of Alfred F. Hess for the national Dictionary of Biography, to a paper about his researches with Edwards A. Park, and to various other reminiscences of the early days of nutrition research.

From his earliest days Dr. McCollum was not one to shy away from an unpopular stand on an issue, once convinced that he was right. So we see him in 1916 in a paper in the *Journal of Biological Chemistry* (24: 491) objecting to the use of the term "vitamine" in these words:

> The prefix *vita* connotes an importance in biological processes paramount to that of certain other absolutely indispensable organic complexes, among which are a number of amino acids. We feel that this term is not in harmony with a conservative tendency in the nomenclature of biological chemistry which should avoid the employment of a term which carries the idea that one indispensable complex is of greater importance biologically than another one equally indispensable. Furthermore, the evidence of the presence of an amino group in the substances under consideration is too slight to warrant the use of the ending *amine,* which carries with it a definite meaning in organic chemical nomenclature.

Surely a sound argument in a battle still being waged in the second edition of *The Newer Knowledge of Nutrition* in 1922, but, alas, the cause was a losing one.

His stand against the enrichment of white flour with niacin, thiamin, riboflavin, and iron, begun in the 1940's and carried on with more or less vigor to his dying day, is another example of the stubbornness of his convictions in the face of overwhelming odds. He believed that white flour should be enriched—yes—but with nonfat milk solids and defatted corn and wheat germs which would then replace *all* the minerals and vitamins except the fat-soluble ones which had been removed in the milling. He felt that it was his strong stand on this issue which was responsible for his being gradually eased off the Food and Nutrition Board of the National Research Council.

He was adamant in his objections to and criticism of the manufacturers of vitamin pills, whom he equated with the old-time hawkers of patent medicines in their activities in foisting their products on the American public. It is ironic, of course, that it was his early researches which made their existence possible.

In his later years he became much concerned with the loss of inorganic elements to the sea as a result of our modern methods of sewage disposal. He was especially worried about the waste of potassium and he wrote 100 letters to people in high places, such as public health officials, conservationists, scientists, civic leaders, and the like. His cause elicited little sympathy. The replies which came were few in number and, in general, failed to see the problem as a serious one. Some suggested that perhaps crops would be developed which did not need potassium. The writers of these letters, he was quick to point out, failed to grasp the fact that it might require some time to evolve a form of *homo sapiens* which did not require this element.

Dr. McCollum exhibited many other worthy attributes aside from those which directly concerned his scientific career. Foremost among these were his gentle humor, ready wit, dignity, kindliness, and interest and concern for others. Many a student has benefited immeasurably from attentions which went far beyond those expected of him; acquaintances and friends of all ages and from all walks of life have profited from his thoughtful and considered advice and help. In 1944 when he received the first Borden Award, he sent the $1000 prize money to the University of Kansas to establish the E. V. McCollum Student Aid Fund. In the intervening years this fund has grown to more than $40,000, a large portion of it from his subsequent awards.

On November 15, 1967, Elmer Verner McCollum "joined the angels," as he would have put it, but his memory lingers on. Not only in the tangible evidences of his having passed this way such as the McCollum-Pratt Institute of The Johns Hopkins University, or in his portrait hanging in the lobby of the Welch Medical Library, or in his sculptured portrait in the Rochester Academy of Medicine's Hall of Fame, or in the McCollum Hall at the University of Kansas, but also in the hearts and lives of those whom he has touched. He lives still in the play of children saved from the crippling effects of rickets, in the researches of his students, and in the smile that lights the face of an associate or acquaintance as he remembers some well-turned phrase, some bit of subtle humor. Above all Dr. McCollum will be remembered as one who was not afraid to blaze a trail and expose new horizons in the field of research. In this he was akin to Robert Frost for whom he had great admiration. Nowhere is this better expressed than in these lines from Frost's poem "The Road Not Taken":

> I shall be telling this with a sigh
> Somewhere ages and ages hence:
> Two roads diverged in a wood,
> and I—
> I took the one less travelled by,
> And that has made all the difference.

So it did for Elmer Verner McCollum, innovator and scientist extraordinaire.

7

Microorganisms and Nutritional Research at Wisconsin*

Esmond E. Snell

Departments of Microbiology and Chemistry
University of Texas
Austin, Texas

It is a common but valid observation that unusually rapid progress in a field often results from introduction of a new experimental methodology or point of view. I hope to illustrate this observation by recounting some of the advances in nutritional knowledge made possible by investigations of bacterial growth requirements at Wisconsin. These advances lie in three areas: (1) contributions to the discovery and identification of several vitamins and related growth factors; (2) the development of convenient microbiological methods for their assay; and (3) contributions to knowledge of their metabolic roles.

Interest in bacterial nutrition at Madison appears to have arisen from the extensive investigations by Professors W. H. Peterson and E. B. Fred and their colleagues of the fermentative metabolism of lactic acid bacteria and clostridia. The first such nutritional study of which I am aware was published by **Tatum, Peterson, and Fred (1935)**,** who isolated asparagine from potatoes as an unusually effective supplement to the peptone then used as a nitrogen source for growth of *Clostridium butylicum*. It was later noted **(Langlykke et al., 1937)** that peptone also supplied something other than amino acids that was necessary for this organism. It was my good fortune to arrive in Madison in 1935, just as these efforts were getting underway, and to select W. H. Peterson as

*A revised version of this address will appear in *Annual Review of Nutrition* 9: in press, 1989.

**To assist the reader in distinguishing Wisconsin contributions to the subject matter of this article from contributions originating elsewhere, citations of work from Wisconsin are in bold face; no gradations in importance are thereby implied. It should also be clear that a short article that deals with a large subject can cite only a cross-section of the relevant literature and that this article emphasizes the "Wisconsin cross-section."

my major professor. He assigned me the problem of determining the nutritional requirements of certain lactic acid bacteria, a problem which required many years more than the four I spent under his guidance at Wisconsin.

It is difficult to today's biochemists to appreciate the many hurdles associated with the initial definition of the trace nutrients required by either microorganisms or animals. Imagine a time when the basis for nutritional requirements was unknown and their complexity unsuspected; when differences in the nutritional requirements of different species of a single genus, or even different strains of one species, also was unsuspected; when nutritionally important substances such as amino acids were mostly unavailable from supply houses; when present criteria of purity were lacking, and even supposedly pure compounds frequently contained impurities sufficient to obscure a given nutritional response; and when purification procedures were primitive by today's standards. Small wonder that 27 years elapsed between the discovery of "bios," a trace nutrient required for yeast growth (Wildiers, 1901), and identification by Eastcott (1928) of inositol as one of its components; and an additional two years before thiamine (first isolated by Jansen and Donath in 1926) also was found to be required for yeast growth (Williams and Roehm, 1930). Some of these early studies with yeast have been reviewed by Williams (1941).

Where were we to begin investigation of the lactic acid bacteria? These organisms were known to require complex media for growth. Of the presently known B vitamins only thiamine was available in pure form. Riboflavin had just been isolated in Europe, and Orla-Jensen et al. (1936) showed that it and additional factors of unknown nature were required for growth of certain lactic acid bacteria in a charcoal-treated whey medium.

The experimental approach we chose was to use the crude medium shown in Table 7.1, which did not permit good growth of our test organism, *Lactobacillus delbrueckii,* and determine what supplements to it were necessary for growth. A potato extract was an excellent source of unidentified growth-promoting factors (Figure 7.1), one or more of which was not precipitated by mercuric acetate in sodium carbonate (Neuberg's reagent), could be extracted in part from the dried, acidified extract with ether, and was partially replaced by sodium acetate. In retrospect, this ether extract could have contained lipoic

Table 7.1 Composition of Medium I

	Percent
Bacto peptone	0.5
Glucose	1.0
Mineral salts (K^+, Na^+, Mg^{2+}, Fe^{2+}, Mn^{2+}, Cl^-, PO_4^{3-}, SO_4^{2-})	

Snell, Tatum and Peterson (1937).
Limiting factor: acetate (lipoate?).

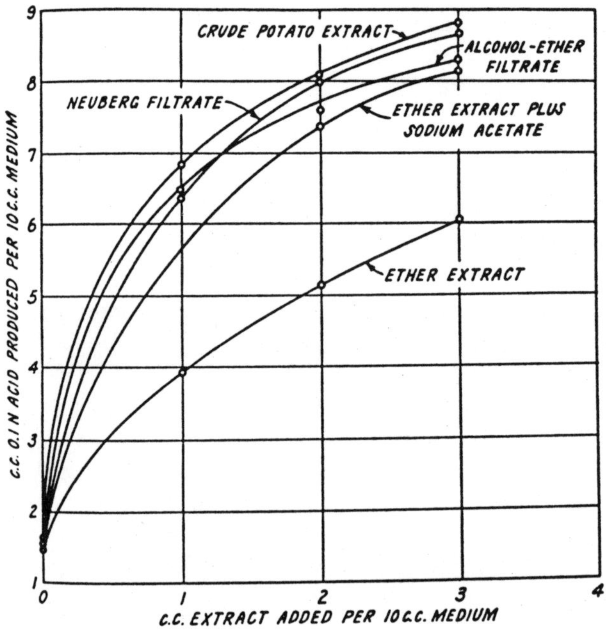

Figure 7.1 Effect of various fractions of a potato extract on growth of *Lactobacillus delbrueckii* in Medium I (from **Snell, Tatum, and Peterson (1937),** with slight modifications).

acid, biotin, and pantothenate, all of which are required under certain conditions by *L. delbrueckii;* at the time, however, the importance of the result lay in the demonstration of the peculiar efficacy of acetate (in the presence of the ether extract) in promoting growth, which led to its inclusion in all later media for these organisms. Much later (Figure 7.2), this observation was followed up to yield an assay procedure for an acetate-replacing factor (Guirard et al., 1946) by use of which lipoic acid was eventually isolated (Reed et al., 1951a, b) and shown to be catalytically involved in acetate production from pyruvate (for reviews see Reed, 1957; Gunsalus, 1954). As shown in Figure 7.2, the same substance was isolated independently by a group at Lederle Laboratories as a growth factor for *Tetrahymena geleii.* These studies eventually led to its synthesis (Bullock et al., 1952, 1954).

When sodium acetate was added to Medium I (see Table 7.1), relatively good growth of *Lactobacillus casii* was obtained. However, when the peptone was treated with 1 N NaOH for 23 h at room temperature, addition of two substances was necessary for growth. One of these was riboflavin, which was already known from the work of Orla-Jensen (1936) to be required by certain lactic acid bacteria. The partial "genealogy" of this vitamin is shown in Figure 7.3. It illustrates the major contributions made by nutritional biochemists,

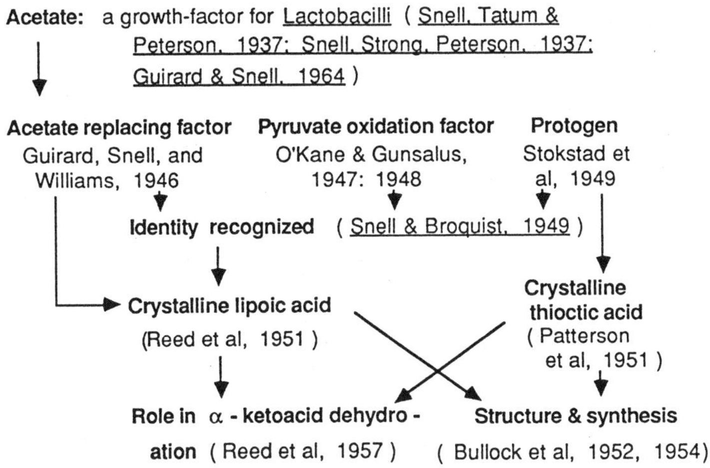

Figure 7.2 A partial chronology of findings related to the discovery, isolation, structure and function of lipoic acid.

enzymologists and organic chemists in elucidating the nature and function of a vitamin and the silliness of oft-heard priority claims to "discovery" of a given vitamin or other substance. Such claims may be valid in a limited area, and useful for impressing a University administration or granting agency. But

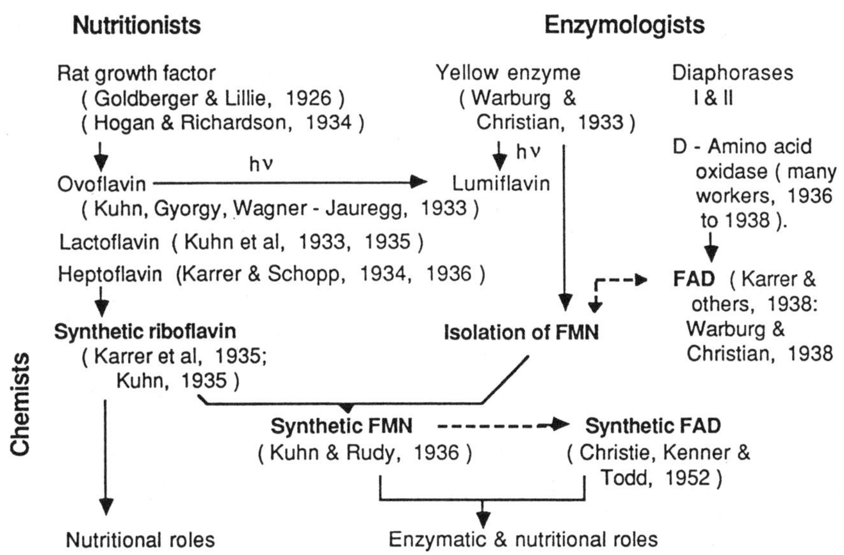

Figure 7.3 A partial "genealogy" of riboflavin and its derived coenzymes. For references cited in the figure, see Robinson (1951) and Wagner and Folkers (1964).

Table 7.2 Composition of Medium II

	Am/100 ml
Sodium hydroxide-treated peptone	0.5 g
Glucose	1.0 g
Sodium acetate	0.6 g
Cystine	0.01 g
Riboflavin	0.01 mg
Mineral salts	

Snell, Strong and Peterson, 1937, 1938, 1939.
Limiting factor: pantothenate.

in an actively developing field several laboratories using independent approaches usually are involved, and another group is almost always there to complete the job in case the "leader" slips.

Addition of riboflavin and sodium acetate to the sodium hydroxide-treated peptone medium (Medium I) gave Medium II (Table 7.2). This medium permitted no growth in the absence of an unidentified factor supplied by yeast or liver extracts, which I spent almost two years as a graduate student in purifying. Toward the end of this period, we exchanged concentrates with Professor R. J. Williams, and found his concentrates of a yeast growth factor (named "pantothenic acid" although they, too, were not pure) were highly active for our test organisms, and vice versa. The chronology of microbiological studies in this area is shown in Figure 7.4. Williams' group had started earlier and was substantially ahead of ours in activity of their concentrates. They also discovered that pantothenic acid was an amide of β-alanine. It was some consolation, however, that the Merck group, which isolated pure pantoyllactone for structural work and the initial synthesis of pure pantothenic acid (Stiller et al., 1940a, b), did so from concentrates of the vitamin prepared by a similar procedure to ours.

Some nine years later, my group at Wisconsin discovered another unidentified factor necessary for growth of *L. bulgaricus* in a medium that contained pantothenate (Figure 7.4). This time we were able to see the problem through to completion: large amounts of pantothenate replaced the factor for *L. bulgaricus,* indicating a close relationship of the two substances. We showed that the growth factor was the then unidentified fragment of coenzyme A, and that it was an amide of pantothenic acid with a substance not previously found in nature, β-mercaptoethylamine, now known as cysteamine (**Brown et al., 1950**). We named the -SH form of the synthetic compound pantetheine, and the disulfide form pantethine, by euphoniously combining segments of the names pantothenic acid, β-mercaptoethylamine and cysteine-cystine (**Snell et al., 1950**).

However, we have jumped ahead of events. To return to 1938 and Table

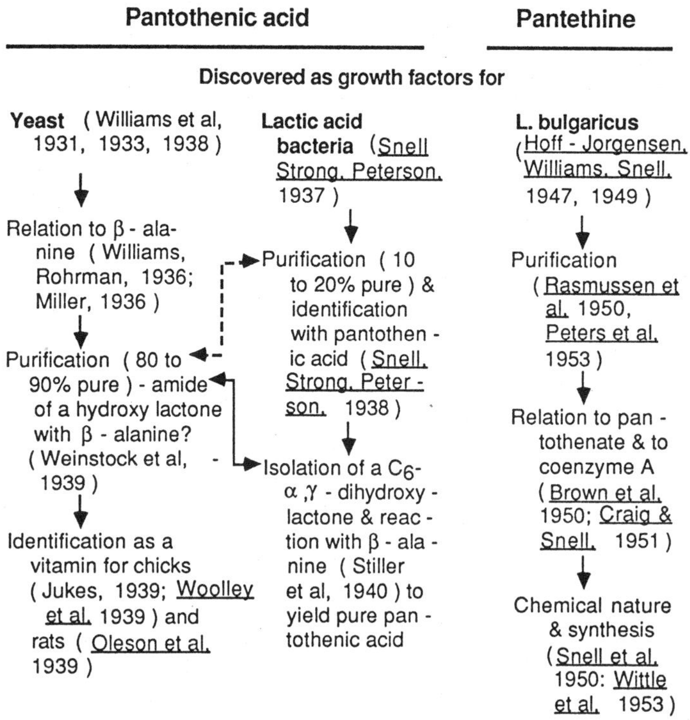

Figure 7.4 A partial chronology of studies leading to synthesis of pantothenic acid and pantetheine.

7.2 we had a medium in which excellent growth responses of *L. casei* to either pantothenate (if riboflavin was added) or to riboflavin (if pantothenate was added) could be obtained. The nutritional significance of riboflavin and its distribution were topics of considerable interest at that time. We felt that use of *L. casei* for its determination would be far superior to other methods then available. Since pantothenic acid was not commercially available, we supplied it (and perhaps several other stimulatory though nonessential growth factors) by adding a yeast extract from which riboflavin had been removed by adsorption on lead sulfide and photolysis (Table 7.3). The growth response of *L. casei* to riboflavin in this medium was excellent (Figure 7.5), whether measured acidimetrically or turbidimetrically. Values obtained for the riboflavin content of a series of standard samples (Table 7.4) were in excellent agreement with those obtained by the much more lengthy, cumbersome, and expensive rat assay. We found later on examining the literature that this was not the first vitamin assay using microorganisms to be suggested (cf. Williams, 1919; Schultz et al., 1937; **West and Wilson, 1938**). It was, however, the first one to be carefully checked against independent methods and to be validated by

Table 7.3 Composition of riboflavin assay medium

	Percent
Alkali-treated peptone	0.5
Photolyzed $Pb(OAc)_2$ filtrate of yeast exract	0.2
Sodium acetate	0.6
Glucose	1.0
Cystine	0.01
Mineral salts (K^+, Mg^{2+}, Mn^{2+}, Fe^{2+}, PO_4^{3-}, SO_4^{2-}, Cl^-)	

Snell and Strong, 1939.
Limiting factor: riboflavin.

careful recovery experiments. In contrast to earlier methods, it was rapidly accepted, widely used, and served as a prototype for similar methods developed later for each of the other B vitamins (review: **Snell, 1950**).

At this point, we were able to begin to analyze the unknown substances supplied by the peptone component of the medium. For this purpose, the sodium hydroxide-treated peptone was replaced by an acid hydrolysate of purified casein (Table 7.5) and riboflavin and a pantothenic acid concentrate were added. Under these conditions a distinct growth response of several lactic acid bacteria to nicotinic acid was obtained. Nicotinic acid (Figure 7.6) was tried because it had been recently implicated in bacterial nutrition by Lwoff and Lwoff (as NAD^+, 1936), by Knight (1937), and by Mueller (1937). I well remember Bob Madden, Wayne Woolley and myself discussing these latter discoveries in a late evening laboratory "bull session," with the result that Madden decided to give a nicotinic acid supplement to his blacktongue-afflicted dogs. Forty-eight hours later the answer was in: it worked!

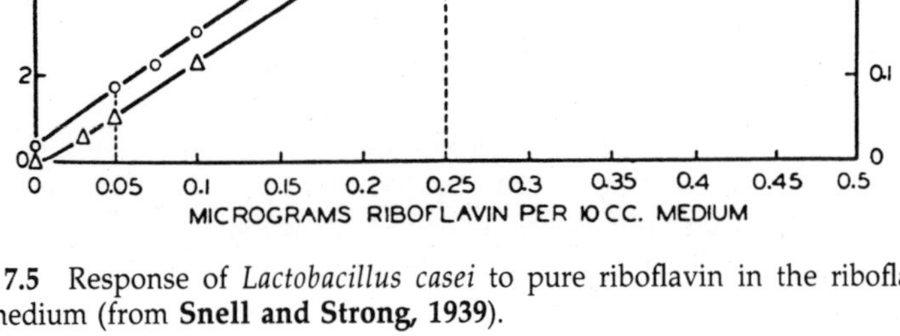

Figure 7.5 Response of *Lactobacillus casei* to pure riboflavin in the riboflavin-free assay medium (from **Snell and Strong, 1939**).

Table 7.4 Comparative riboflavin assays of natural products

	Microbiological assay		Rat assay	
Material	Direct μg/g	Extract μg/g	Direct μg/g	Bourquin-Sherman units
Dried grasses				
Sample I	—	24.4	20.0	—
Sample II	—	24.1	25.0	—
Sample III	—	22.6	22.0	—
Sample IV	—	23.9	20.0	—
Dried oat plant	30.3	31.2	35.0	—
Skim milk powder	17.1	—	17.0	—
Debittered yeast *a*	39.2	38.5	37	—
Debittered yeast *b*	—	36.5	(38.8)*	17.7
Debittered yeast *c*	34.7	33.6	(34.4)	15.7
Yeast extract	149.6	—	(145.6)	66.5
Nondebittered yeast	31.8	32.8	(32.0)	14.6

Snell and Strong, 1939.
*A value of 2.19 μg riboflavin/unit was used for these conversions.

Table 7.5 Composition of Medium III

"Vitamin-free" casein, acid-hydrolyzed	0.5%
Glucose	1.0%
Sodium acetate	0.6%
Cystine	0.01%
Tryptophan	0.01%
Pantothenic acid (concentrate, 5% pure)	0.5 ppm
Riboflavin	0.1 ppm
Mineral salts	

Snell, Strong and Peterson, 1938, 1939.
Limiting factor: Nicotinic acid.

Although, as mentioned earlier, nicotinic acid stimulated growth and acid production on Medium III, this stimulation was not evident upon subculture, apparently due to exhaustion of essential nutrients carried over with the inoculum. We therefore added nicotinic acid to our basal medium (Table 7.6) and examined the additional supplements necessary for growth. Unidentified factors present in both the norite eluate of a liver extract and in the norite filtrate proved essential for growth. The chronology in elucidation of the nature of the norite eluate factor, now known as folic acid or pteroylglutamic acid, is shown (in part) in Figure 7.7. I left Madison for my first job in Texas in 1939. There we continued work on bacterial nutrition, but with another organism, *Streptococcus faecalis* (*S. lactis* R). We very soon realized that the

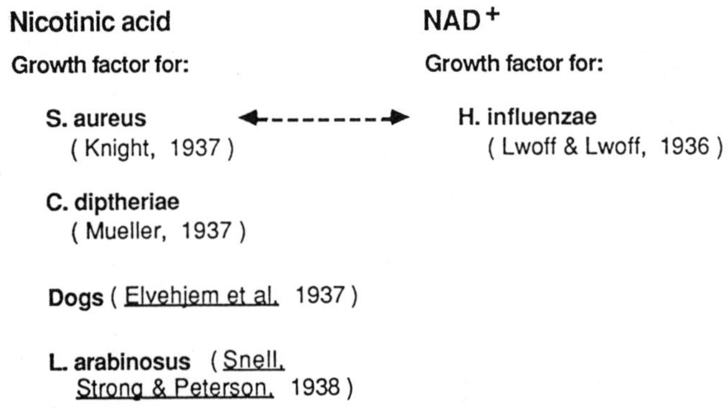

Figure 7.6 Chronology of discovery of nutritional role of nicotinic acid.

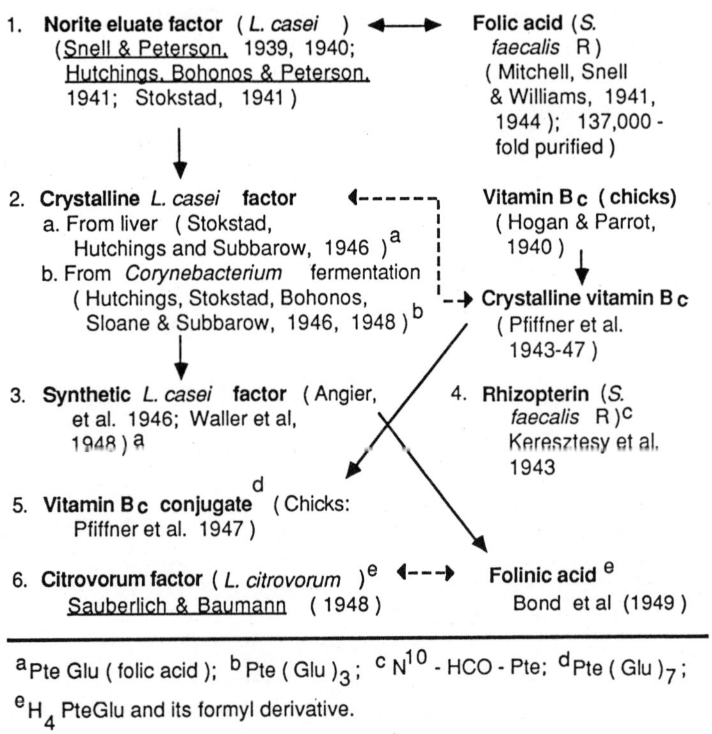

Figure 7.7 Bacterial and chick assays leading to the isolation of folic acid and its congeners. (For a more complete account of the involved early history of these substances, see the reviews by **Peterson and Peterson (1945)**, Robinson (1951), and Wagner and Folkers (1964).

unidentified factor we were fractionating for this organism, which we called folic acid, was the same as the norite eluate factor required by *L. casei*. The two organisms thus introduced have continued to be those most widely used in study of this vitamin and its derivatives. In the meantime, work on the *L. casei* factor continued at Wisconsin in Peterson's laboratory by Hutchings and Bohonos, who later participated in isolation of two crystalline forms of the vitamin, pteroylglutamic acid and pteroyltriglutamate, at Lederle Laboratories. Additional conjugates were isolated by the group under Pfiffner at Parke-Davis and Co., apparently by use of chick assays.

About the time most investigators thought that enough forms of folic acid were already known, **Sauberlich and Baumann (1948)** discovered an essential growth factor for *Leuconostoc citrovorum*. A closely related substance was discovered independently through use of a quite different approach by Shive and his colleagues (Bond et al., 1949) at Texas, identified as a formylated reduced form of folic acid, and shown to play a role in one-carbon metabolism (Flynn et al., 1951).

In addition to folic acid (norite eluate factor), a norite filtrate factor was required for growth of *L. casei* in Medium IV (Table 7.6). Peterson and I showed that pyridoxine, which Moeller (1938) had previously reported as a growth factor for some lactic acid bacteria, partially replaced the norite filtrate but that one or more additional substances were required **(Snell and Peterson, 1940)**. Moeller and Schwarz (1941), in a report which we learned of only later because of war in Europe, had also shown that biotin was required by lactic acid bacteria, and we soon found that it and pyridoxine together completely replaced the norite filtrate.

By further supplementing Medium IV with folic acid and biotin, therefore, we should have a medium (Table 7.7) suitable for quantitative assay of pyridoxine. This medium permitted excellent growth of *L. casei* or *S. faecalis* when supplemented with pyridoxine. Much to our surprise, however, completely

Table 7.6 Composition of Medium IV

Purified casein, acid-hydrolyzed	0.5 g
Sodium acetate	0.6 g
Glucose	1.0 g
Tryptophan	0.01 g
Cystine·HCl	0.01 g
Nicotinic acid	20 μg
Pantothenic acid (5% pure)	50 μg
Riboflavin	10 μg
Mineral salts	
Water to	100 ml

Snell and Peterson, 1939, 1940.
Limiting factors: Norite eluate (folate) and norite filtrate (B_6, biotin).

Table 7.7 Vitamin B_6-free Medium for *S. faecalis* R

Component	mg/l	Component	μg/l
Hydrolyzed casein, charcoal-treated	5000	Thiamine chloride	100
Glucose	10000	Calcium pantothenate	100
Sodium acetate	6000	Nicotinic acid	100
Tryptophan	50	Riboflavin	100
Cystine·HCl	100	Folic acid (conc)	10
Adenine	10	Biotin	10
Guanine	10	Inorganic salts (Mg^{2+}, Mn^{2+}, Fe^{2+}, K^+, Na^+, PO_4^{3+}, SO_4^{2+}, Cl^-)	
Uracil	10		

Snell, Guirard and Williams, 1942.
Limiting factor, vitamin B_6.

unreasonable assay figures resulted and it was evident that something other than pyridoxine was contributing to the growth response of the assay organisms in this medium. Since this work was done in Texas rather than at Wisconsin, I shall not dwell on it; we were able to prove (Snell et al., 1942) that the growth response of *S. faecalis* and *L. Casei* to pyridoxine was due not to pyridoxine per se, but rather to products formed from pyridoxine by rats or other animals. We originally termed this substance "pseudopyridoxine," showed that it resulted from an aldehyde and an amine formed by oxidation or amination of pyridoxine, and eventually, in collaboration with Folkers and his group at Merck and Co., synthesized the responsible compounds, pyridoxal and pyridoxamine. The chronology of events in this field, shown in Table 7.8, illustrates a point made earlier with respect to priority in scientific investigation. If we had not identified pyridoxal in 1944, I am sure that it would have been identified by Strong and his group here in Madison, who, after our first reports on pseudopyridoxine, independently found that partial oxidation of pyridoxine increased its activity for *L. casei*. They postulated correctly that pseudopyridoxine was a partial oxidation product of pyridoxine. Had they by chance failed to identify pyridoxal, Gunsalus et al., who knew that their coenzyme for tyrosine decarboxylase was related to vitamin B_6, also were in an excellent position to identify this substance. Subsequent work, much of it here in Madison, delineated the activity of these new forms of vitamin B_6 in animals, and also resulted in the discovery (1) of the natural occurrence of pyridoxamine 5′-phosphate, and (2) that this substance (and/or pyridoxal 5′-phosphate) was itself an essential growth factor for some lactic acid bacteria (Table 7.8).

From the above discussion, it is apparent that one or another of the lactic acid bacteria require each of the B vitamins required by animals, and that work on bacterial nutrition in this Department and work by its graduates elsewhere played a large role in the discovery of certain of these vitamins

Table 7.8 Vitamin B_6: Identification of pyridoxal (PL) and pyridoxamine (PM) and their coenzyme forms (PLP, PMP)*

Isolation (1938) and synthesis (1939) of PN.
PN essential for *L. plantarum* (Moeller, 1938) and yeast (Schultz et al., 1939; Eakin and Williams, 1939).
"Pseudopyridoxine," a metabolite of PN, is actual growth factor for lactics (Snell, et al., 1942).
PN oxidation → pseudoPN **(Carpenter and Strong, 1944)**
PN oxidation or amination → PL or PM (Snell, 1944a).
Synthesis and activity of PL and PM for various organisms (Harris et al., 1944; Snell, 1944b; Snell and Rannefeld, 1945; **Sarma et al., 1946a, b; Luckey et al., 1945; Linkswiler et al., 1951.**
PLP as coenzyme for Tyr decarboxylase (Gunsalus et al., 1944); synthesis (Heyl et al., 1951).
PMP as bound form of B_6 **(Rabinowitz and Snell, 1947)**; growth factor for *L. helveticus* **(McNutt and Snell, 1948).**

*A more complete discussion of early developments in this area is given by Snell (1981).

and in establishing the identity of others. This is especially true of pantothenic acid, pantetheine, folic acid, tetrahydrofolic acid, and vitamin B_6 (pyridoxal, pyridoxamine, and pyridoxamine phosphate). These bacteria have also provided convenient assay organisms for each of these substances.

But lactic acid bacteria are not the only microorganisms that have played an important role in elucidation of nutritional requirements. I have discussed the use of yeast in the discovery of pantothenate; biotin was also first discovered with this organism. **McDaniel, Woolley, and Peterson (1939)** substantially purified an unidentified factor required for growth of *C. butylicum.* Later that same year, when a sample of Kogl's biotin was made available to us, we showed that it was the active principle for *C. butylicum* (Snell and Williams, 1939). **West and Wilson (1939)** similarly showed that biotin replaced an unidentified factor, coenzyme R, in the nutrition of *Rhizobium trifolii.* The physiological importance of the polyamines, which are currently so stylish in nucleic acid research, and which had long been known to occur naturally, first became apparent when **Herbst and Snell (1948)** isolated putrescine from orange juice as an essential growth factor for *Hemophilus parainfluenzae;* spermidine and spermine also were active. Somehow, the orange juice industry has never used this obviously important finding in their advertising!

I have emphasized the important role played by quantitative studies of microbial growth in discovery of various precursors and active forms of the vitamins. Also important has been the role played by such studies in suggesting metabolic functions of the vitamins. An early instance of this type—the catalytic role of lipoic acid in supplying acetate for growth of lactic acid bacteria—has been discussed. Many additional examples could be cited. The experimental observations here are that the products of the catalyzed reaction (e.g., purine bases and thymine for folic acid; D-alanine and L-amino acids for vitamin B_6; oleic acid and aspartate for biotin) completely replace the need for folic acid vitamin B_6 or biotin, respectively, in a suitable medium, and do not give rise to synthesis of the vitamin by the cell. The enzymatic basis for the effect of aspartate on the biotin requirement was clarified by **Lardy et al. (1947)** when they discovered formation of oxaloacetate from pyruvate and CO_2 in *L. arabinosus* was biotin-dependent. Similarly, the observation that D-alanine replaced vitamin B_6 for *L. casei* in an otherwise complete medium **(Holden and Snell, 1949)** was explained by the findings that (1) D-alanine was an essential component of bacterial cells **(Holden et al., 1949)**; (2) pyridoxal-P is a coenzyme for formation of D-from L-alanine by the alanine racemase in bacterial cells (Wood and Gunsalus, 1951); and (3) D-alanine is an essential component of the cell wall (Snell et al., 1955).

Once all of the vitamins required by the lactic acid bacteria had been identified and became available commercially, it was a simple task to add them to the basal medium, then to replace the hydrolyzed casein of the medium with amino acids to determine which amino acids were required for growth of a given organism. That organism could then be used in a medium deficient in a single essential amino acid for its quantitative determination. As shown in Table 7.9, Wisconsin workers again were prominent in this extension of microbiological methods. Such methods for amino acids were widely used until another graduate of this department, Stanford Moore, developed with W. Stein the chromatographic ion exchange procedures for this purpose that now rule supreme. I have always felt that for repetitive assays of a few selected amino acids, e.g., of essential amino acids in foodstuffs during nutritional survey work, the microbiological methods even now would be simpler, more convenient, sufficiently accurate, and much cheaper (especially in undeveloped areas) than the chromatographic methods.

During study of the amino acid requirements of these bacteria, several

Table 7.9 Microbiological methods for amino acids

Organism	Amino acids	References
L. casei, L. arabinosus	Glu, Leu, Ile, Val, Trp, Phe, Thr, Met	Shankman, 1943 McMahan and Snell, 1944 **Hegsted, 1944**
L. casei, L. arabinosus, S. faecalis, L. mesenteroides	14 amino acids	**Henderson and Snell, 1948**
L. mesenteroides, L. citrovorum	18 amino acids	**Steele et al., 1949**

Table 7.10 Conditions giving rise to peptide "requirements"

1. Destruction of free amino acids: e.g., Tyr vs. Tyr peptides (Tyr decarboxylase in *S. faecalis*)[a]
2. Interference in utilization of free amino acids
   a) By antagonistic amino acids: e.g., D-Ala vs. L-Ala, peptides of L-Ala (in *L. casei*)[b]
   L-Ala vs. L-Ser, peptides of L-Ser (in *L. delbrueckii*)[c]
   Strepogenin (in *L. casei*)[d]
   b) Inefficient uptake of free amino acids, e.g., L-His vs. peptides of L-His (in *L. delbrueckii*)[e]

[a]Kihara et al., 1952; [b]Kihara and Snell, 1952; [c]Prescott et al., 1953; [d]Kihara and Snell, 1960; [e]Peters et al., 1953a, b.

instances were discovered in which peptides of a given amino acid were more active than the free amino acid itself in promoting growth. A number of such instances were analyzed and found to fall into the several categories tabulated in Table 7.10. The significance of these results, if any, for nutrition at the cellular level in animals is unknown. An attractive and untested possibility is that the requirement of animal cell cultures for some of the numerous peptide growth factors now being reported may find its explanation in similar terms.

I have talked too long. I hope, however, to have convinced you of the validity of the thesis with which I began; that is that studies of bacterial growth requirements, in addition to their importance for microbiology itself, have enormously advanced both biochemistry and nutrition by (1) contributing to the discovery of several vitamins and related substances; (2) providing convenient methods for vitamin and amino acid determinations; and (3) contributing to knowledge of the metabolic and enzymatic roles played by these substances. In all of these studies students, former students, and staff of the Department of Biochemistry of the University of Wisconsin have played prominent roles. Some may see a symposium such as this as mere indulgence in self-congratulation; I prefer to consider it an introspective self-examination of past performance aimed at maintaining or improving that performance in the future.

References

Angier, R. B., Boothe, J. H., Hutchings, B. L., Mowat, J. H., Semb, J., Stokstad, E. L. R., Subba Row, Y., Waller, C. W., Cosulich, D. B., Fahrenbach, M. J., Hultquist, M. E., Kuh, E., Northey, E. H., Seeger, D. R., Sickels, J. P., and Smith, J. M., Jr. (1946) *Science* 103:667–669

Bond, T. J., Bardos, T. J., Sibley, M., and Shive, W. (1949) *J. Am. Chem. Soc.* 71: 3852–3853.

Brown, G. M., Craig, J. A., and Snell, E. E. (1950) *Arch. Biochem.* 27: 473–475.

Bullock, M. W., Brockman, J. A., Jr., Patterson, E. L., Pierce, J. V., and Stockstad, E. L. R. (1952) *J. Am. Chem. Soc.* 74: 3455.

Bullock, M. W., Brockman, J. A., Jr., Patterson, E. L., Pierce, J. V., von Salza, M. H., Sanders, F., and Stokstad, E. L. R. (1954) *J. Am. Chem. Soc.* 76: 1828–1832.

Carpenter, L. E. and Strong, F. M. (1944) *Arch. Biochem.* 3: 375–388.

Craig, J. A. and Snell, E. E. (1951) *J. Bact.* 61: 283–291.

Eakin, R. E. and Williams, R. J. (1939) *J. Am. Chem. Soc.* 61: 1932.

Eastcott, E. V. (1928) *J. Phys. Chem.* 32: 1094–1111.

Elvehjem, C. A., Madden, R. J., Strong, F. M., and Woolley, D. W. (1937) *J. Am. Chem. Soc.* 59: 1767–1768.

Flynn, E. H., Bond, T. J., Bardos, T. J., and Shive, W. (1951) *J. Am. Chem. Soc.* 73: 1979–1982.

Goldberger, J. and Lillie, R. D. (1926) *Pub. Health Rep.* 41: 1025–1029.

Guirard, B. M. and Snell, E. E. (1964) *J. Bact.* 87: 370–376.

Guirard, B. M., Snell, E. E., and Williams, R. J. (1946) *Arch. Biochem.* 9: 381–386.

Gunsalus, I. C. (1954) *Fed. Proc.* 13: 715–722.

Gunsalus, I. C., Bellamy, W. D., and Umbreit, W. W. (1944) *J. Biol. Chem.* 155: 685–686.

Harris, S. A., Heyl, D., and Folkers, K. (1944) *J. Biol. Chem.* 154: 315–316.

Hegsted, M. (1944) *J. Biol. Chem.* 152: 193–200.

Henderson, L. M. and Snell, E. E. (1948) *J. Biol. Chem.* 172: 15–29.

Herbst, E. J. and Snell, E. E. (1948) *J. Biol. Chem.* 176: 989–990.

Heyl, D., Luz, E., Harris, S. A., and Folkers, K. (1951) *J. Am. Chem. Soc.* 73: 3430–3433.

Hoff-Jorgensen, E., Williams, W. L., and Snell, E. E. (1947) *J. Biol. Chem.* 168: 773–774.

Hogan, A. G. and Parrott, E. M. (1940) *J. Biol. Chem.* 132: 507–517.

Hogan, A. G. and Richardson, L. R. (1934) *J. Nutrition* 8: 385–396.

Holden, J. T. and Snell, E. E. (1949) *J. Biol. Chem.* 178: 799–809.

Holden, J. T., Furman, C., and Snell, E. E. (1949) *J. Biol. Chem.* 178: 789–797.

Hutchings, B. L., Bohonos, N., and Peterson, W. H. (1941) *J. Biol. Chem.* 141: 521–528.

Hutchings, B. L., Stokstad, E. L. R., Bohonos, N., Sloane, N. H., and Subarow, Y. (1946) *Ann. N. Y. Acad. Sci.* 48: 265–267; (1948) *J. Am. Chem. Soc.* 70: 1–3.

Jansen, B. C. P. and Donath, W. F. (1926) *Mededeel, Dienst. Volksgezondheid Niderland-Indië* 16: 186–199.

Jukes, T. H. (1939) *J. Am. Chem. Soc.* 61: 975–976

Keresztesy, J. C., Rickes, E. L., and Stokes, J. L. (1943) *Science* 97: 465.

Kihara, H. and Snell, E. E. (1952) *J. Biol. Chem.* 197: 791–800.

Kihara, H. and Snell, E. E. (1960) *J. Biol. Chem.* 235: 1409–1414 and 1415–1418.

Kihara, H., Klatt, O. A., and Snell, E. E. (1952) *J. Biol. Chem.* 197: 801–807.

Knight, B. C. J. G. (1937) *Nature* 138: 628.

Langlykke, A. F., Peterson, W. H., and Fred, E. B. (1937) *J. Bacteriol.* 34: 443–453.

Lardy, H. A., Potter, R. L., and Elvehjem, C. A. (1947) *J. Biol. Chem.* 169: 453–454.

Linkswiler, H., Baumann, C. A., and Snell, E. E. (1951) *J. Nutr.* 43: 565–573.

Luckey, T. D., Briggs, G. N., Elvehjem, C. A., and Hart, E. B. (1945) *Proc. Soc. Exp. Biol. Med.* 58: 340–344.

Lwoff, A. and Lwoff, M. (1936) *Compt. Red.* 203: 520–522.

McDaniel, L. E., Woolley, D. W., and Peterson, W. H. (1939) *J. Bact.* 37: 259–268.

McMahan, J. R. and Snell, E. E. (1944) *J. Biol. Chem.* 152: 83–95.

McNutt, W. S. and Snell, E. E. (1948) *J. Biol. Chem.* 173: 801–802; also see (1950) *J. Biol. Chem.* 182: 557–567.

Mitchell, H. K., Snell, E. E., and Williams, R. J. (1941) *J. Am. Chem. Soc.* 63: 2284.
Mitchell, H. K., Snell, E. E., and Williams, R. J. (1944) *J. Am. Chem. Soc.* 66: 267–268.
Moeller, E. F. (1938) *Z. Physiol. Chem.* 254: 285–286.
Moeller, E. F. and Schwarz, K. (1941) *Ber. Deut. Chem. Ges.* 74: 1612–1616.
Mueller, J. H. (1937) *J. Biol. Chem.* 120: 219–224.
O'Kane, D. J. and Gunsalus, I. C. (1947) *J. Bact.* 54: 20–21 (Abst.); also see (1948) *J. Bact.* 56: 499–506.
Oleson, J. J., Woolley, D. W., and Elvehjem, C. A. (1939) *Proc. Soc. Exp. Biol. Med.* 42: 151–153.
Orla-Jensen, S., Otte, N. C., and Snog-Kjaer, A. (1936) *Zentr. Bakt. Parasitenk. Infek. II* 94: 434–447.
Peters, V. J., Brown, G. M., Williams, W. L., and Snell, E. E. (1953a) *J. Am. Chem. Soc.* 75: 1688–1691.
Peters, V. J., Prescott, J. M., and Snell, E. (1953b) *J. Biol. Chem.* 202: 521–532.
Peterson, W. H. and Peterson, M. S. (1945) *Bact. Rev.* 9: 49–109.
Pfiffner, J. J., Binkley, S. B., Bloom, E. S., Brown, R. A., Bird, O. D., Emmett, A. D., Hogan, A. G., and O'Dell, B. L. (1943) *Science* 97: 404–405.
Pfiffner, J. J., Binkley, S. B., Bloom, E. S., and O'Dell, B. L. (1947) *J. Am. Chem. Soc.* 69: 1476–1487.
Prescott, J. M., Peters, V. J., and Snell, E. E. (1953) *J. Biol. Chem.* 202: 533–540.
Rabinowitz, J. C. and Snell, E. E. (1947) *J. Biol. Chem.* 169: 643–650.
Rasmussen, R. A., Smiley, K. L., Anderson, J. G., VanLanen, J. M., Williams, W. L., and Snell, E. E. (1950) *Proc. Soc. Exp. Biol. Med.* 73: 658–660.
Reed, L. J. (1957) *Adv. Enzymol.* 18: 319–347.
Reed, L. J., DeBusk, B. G., Gunsalus, I. C., and Hornberger, C. S., Jr. (1951a) *Science* 114: 93–94.
Reed, L. J., Getzendaner, M. E., DeBusk, B. G., and Johnson, P. M. (1951b) *J. Biol. Chem.* 192: 859–865.
Robinson, F. A. (1951) *The Vitamin B Complex*, John Wiley and Sons, Inc., New York.
Sarma, P. S., Snell, E. E., and Elvehjem, C. A. (1946a) *J. Biol. Chem.* 165: 55–63.
Sarma, P. S., Snell, E. E., and Elvehjem, C. A. (1946b) *Proc. Soc. Exp. Biol. Med.* 63: 284–286.
Sauberlich, H. E. and Baumann, C. A. (1948) *J. Biol. Chem.* 176: 165–173.
Schultz, A., Atkin, L., and Frey, C. N. (1937) *J. Am. Chem. Soc.* 59: 948–949 and 2457–2460.
Schultz, A., Atkin, L., and Frey, C. N. (1939) *J. Am. Chem. Soc.* 61: 1931.
Shankman, S. (1943) *J. Biol. Chem.* 150: 305–310.
Snell, E. E. (1944a) *J. Am. Chem. Soc.* 66: 2082–2088.
Snell, E. E. (1944b) *J. Biol. Chem.* 154: 313–314.
Snell, E. E. (1950) in *Vitamin Methods*, Gyorgy, P., ed., Vol. I, pp. 327–505.
Snell, E. E. (1981) in *Methods in Vitamin B-6 Nutrition*, Leklem, J. E. and Reynolds, R. D., eds., pp. 1–19, Plenum Publishing Corp., New York.
Snell, E. E. and Broquist, H. P. (1949) *Arch. Biochem.* 23: 326–328.
Snell, E. E. and Peterson, W. H. (1939) *J. Biol. Chem.* 128: xciv–xcv.
Snell, E. E. and Peterson, W. H. (1940) *J. Bact.* 39: 273–285.
Snell, E. E. and Rannefeld, A. N. (1945) *J. Biol. Chem.* 157: 475–489.
Snell, E. E. and Strong, F. M. (1939) *Ind. Eng. Chem., Anal. Ed.* 11: 346–350.
Snell, E. E. and Williams, R. J. (1939) *J. Am. Chem. Soc.* 61: 3594.
Snell, E. E., Brown, G. M., Peters, V. J., Craig, J. A., Wittle, E. L., Moore, J. A., McGlohon, V. M., and Bird, O. D. (1950) *J. Am. Chem. Soc.* 72: 5349–5350.
Snell, E. E., Guirard, B. M., and Williams, R. J. (1942) *J. Biol. Chem.* 143: 519–530.
Snell, E. E., Radin, N., and Ikawa, M. (1955) *J. Biol. Chem.* 217: 803–818.
Snell, E. E., Strong, F. M., and Peterson, W. H. (1937) *Biochem. J.* 31: 1789–1799.

Snell, E. E., Strong, F. M., and Peterson, W. H. (1938) *J. Am. Chem. Soc.* 60: 2825; also see (1939) *J. Bact.* 38: 293–308.

Snell, E. E., Tatum, E. L., and Peterson, W. H. (1937) *J. Bact.* 33: 207–225.

Steele, B. F., Sauberlich, H. E., Reynolds, M. S., and Baumann, C. A. (1949) *J. Biol. Chem.* 177: 533–544.

Stiller, E. T., Harris, S. A., Finkelstein, J., Keresztesy, J. C., and Folkers, K. (1940a) *J. Am. Chem. Soc.* 62: 1779–1784.

Stiller, E. T., Harris, S. A., Finkelstein, J., Keresztesy, J. C., and Folkers, K. (1940b) *J. Am. Chem. Soc.* 62: 1785–1790.

Stokstad, E. L. R. (1941) *J. Biol. Chem.* 139: 475–476; also see (1943) *J. Biol. Chem.* 149: 573–574.

Stokstad, E. L. R., Hoffmann, C. E., Regan, M. A., Fordham, D., and Jukes, T. H. (1949) *Arch. Biochem.* 20: 75–82.

Stokstad, E. L. R., Hutchings, B. L., and Subbarow, Y. (1946) *Ann. N. Y. Acad. Sci.* 48: 61; also see (1948) *J. Am. Chem. Soc.* 70: 3–5.

Tatum, E. L., Peterson, W. H., and Fred, E. B. (1935) *J. Bacteriol.* 29: 563–572.

Wagner, A. F. and Folkers, K. (1964) *Vitamins and Coenzymes.* Interscience-Wiley, New York.

Waller, C. W., Hutchings, B. L., Mowat, J. H., Stokstad, E. L. R., Boothe, J. H., Angier, R. B., Semb, J., SubbaRow, Y., Cosulich, D. B., Fahrenbach, M. J., Hultquist, M. E., Kuh, E., Northey, E. H., Seeger, D. R., Sickels, J. P., Smith, J. M., Jr. (1948) *J. Am. Chem. Soc.* 70: 19–22.

Warburg, O. and Christian, W. (1933) *Biochim. Ztschr.* 260: 466–501.

Weinstock, H. H., Jr., Mitchell, H. K., Pratt, E. F., and Williams, R. J. (1939) *J. Am. Chem. Soc.* 61: 1421–1425.

West, P. M. and Wilson, P. W. (1938) *Science* 88: 334–335.

West, P. M. and Wilson, P. W. (1939) *Science* 89: 607–608; also see *J. Bact.* 38: 110–111.

Wildiers, E. (1901) *Cellule* 18: 313.

Williams, R. J. (1919) *J. Biol. Chem.* 38: 465–486.

Williams, R. J. (1941) *Biol. Revs.* 16: 49–80.

Williams, R. J. and Bradway, E. M. (1931) *J. Am. Chem. Soc.* 53: 783–789.

Williams, R. J. and Roehm, R. R. (1930) *J. Biol. Chem.* 87: 581–590.

Williams, R. J. and Rohrmann, E. (1936) *J. Am. Chem. Soc.* 58: 695.

Williams, R. J., Lyman, C. M., Goodyear, G. H., Truesdail, J. H., and Holaday, D. (1933) *J. Am. Chem. Soc.* 55: 2912–2927.

Williams, R. J., Truesdail, J. H., Weinstock, H. H. Jr., Rohrmann, E., Lyman, C. M., and McBurney, C. H. (1938) *J. Am. Chem. Soc.* 60: 2719–2723.

Williams, W. L., Hoff-Jorgensen, E., and Snell, E. E. (1949) *J. Biol. Chem.* 177: 933–940.

Wittle, E. L., Moore, J. A., Stipek, R. W., Peterson, F. E., McGlohon, V. M., Bird, O. D., Brown, G. M., and Snell, E. E. (1953) *J. Am. Chem. Soc.* 75: 1694–1700.

Wood, W. A. and Gunsalus, I. C. (1951) *J. Biol. Chem.* 190: 403–416.

Woolley, D. W., Waisman, H. A., and Elvehjem, C. A. (1939) *J. Am. Chem. Soc.* 61: 977–978.

8

B Vitamin Recollections

LaVell M. Henderson

Department of Biochemistry
University of Minnesota
St. Paul, Minnesota

I extend my warm congratulations to the Department on its 100th birthday. The faculty should be proud of the quality program they have maintained during this century of service to the state and the nation. It has provided great satisfaction to us, the products of your graduate program, that you have maintained your position of eminence in the field. Some of us who have attempted to compete with you realize the challenge you face in keeping the program so outstanding over these many decades.

This occasion is providing many of us with the pleasant experience of returning to the scene of our start in science and serves to remind us of the happy years spent in Madison. In 1939 the faculty membered about ten, counting instructors. The labs were crowded, but it was a stimulating place. Saturday at 8–10 AM was Journal Club, Saturday afternoon the Metropolitan Opera broadcast filled the labs and Sunday morning early, when you came to feed your animals, Dr. Elvehjem gave you all the time you wished.

The history of the B vitamins is a large part of the history of biochemistry at Wisconsin and elsewhere from the middle 30's to the early 50's. Recounting it entails concentrating on the contributions of many of the faculty of that era, including some of the present faculty. It was an exciting subject in an exciting period when discovery and identification of vitamins occupied center stage in biochemistry. A tool had become available which enabled us to define the biosynthetic limitations of animals and microorganisms, and, by taking advantage of these limitations, to produce deficiency states whose characterization led to understanding of vitamin-coenzyme relationships and their metabolic functions. In our recent search for a good way to produce an uncomplicated carnitine deficiency, I have been reminded how useful lack of biosynthetic capacity in animals can be in establishing the function of a substance.

Since Elvehjem and his students made much of the history I will cover, it may duplicate some of Van Potter's remarks of the evening. Connie Elvehjem's scientific career between 1924 and 1959, a period of 35 years, was recorded in his 824 publications involving more than 275 coauthors. He published on the average of two papers a month in those 35 years and in his peak years of the 40's and 50's, more than forty papers per year. Connie was a spokesman for the Department and for the scientific community on vitamins and minerals, their function and metabolism. He and his associates, largely graduate students, had a substantial influence on the development of this new area during two decades beginning about 1935. In today's setting one might expect that such developments were reported in the then new *Journal of Nutrition*. On the contrary, they were more often *Journal of Biological Chemistry* papers. Sixty of Connie's first hundred papers were published in the *Journal of Biological Chemistry*.

Elvehjem and his students appeared on the scene when the complexity of water-soluble B was recognized to be, as he put it, "a full-sized swarm" of separate nutritional factors. The enthusiasm and the understandable confusion which arose when the rations being fed were partially deficient in nutrients other than the one under study, led to many examples of mistaken identity and rediscovery of vitamins. Let me cite one example. Ed Nielson was fractionating a factor which prevented or cured the graying of the hair in black or piebald rats. When it appeared that he had an assay for the gray hair factor, I was asked to examine its distribution in animal tissues, since I was being supported by funds from the National Livestock and Meat Board. On the basis of my experience with rat growth assays for pantothenate, I decided that Ed's ration should contain more of this recently discovered vitamin. There were conflicting reports regarding its efficacy in preventing graying. When the pantothenate was increased in the ration, the gray hair turned black. He had been working with one of the many situations encountered in that era, where the symptom, e.g., feathering, fatty liver, hair loss, anemia, etc., was accentuated by increasing the intake of the limiting nutrient, thereby enhancing growth. Another complication in Nielson's experiments was the lag in observing hair color. Shaving or hair removal with sodium sulfide permitted observation of the color of currently growing hair, thereby limiting the lag. In this case, Nielson's ration contained sufficient pantothenate for nearly optimum growth, but not enough to maintain optimum pigmentation of the hair. With new vitamins expensive and hard to come by, such false leads were to be expected.

A second common cause of confusion was the variation in symptoms of a given deficiency among the species of test animals in common use. Chick pellagra, as the name implies, was the first assumed to be the counterpart of black tongue in dogs and pellagra in humans. Pantothenate cured the chick pellagra and vitamin B_6 cured dermatitis in the rat. As the rations were progressively purified, complications from lack of folate and vitamin B_{12} and on certain rations, biotin, further complicated the picture. In the paper describing the gray hair prevention with pantothenate, we reported (Henderson et al., 1942) maintaining rats through the life cycle for four generations on a ration which contained as water-soluble vitamins only thiamin, pyridoxine, riboflavin, choline, pantothenate, and nicotinic acid (the latter was not required on the 18 percent casein ration). No biotin, folate or vitamin B_{12}, except that present as impurities in the casein, was provided.

This illustrates important features of the experimental work done during the next 7–8 years:

1. Special rations beyond the usual 18 percent casein were needed to complete the unraveling of the B complex, e.g., egg white for biotin and iodinated casein for B_{12} (Betheil and Lardy, 1949), etc.
2. Intestinal synthesis of some B vitamins was a critical feature and was modified by feeding nonabsorbed sulfa drugs (Black et al., 1941).
3. It was necessary to test a number of species to find one in which the deficiency could be produced reliably.

One wonders how this history of discovery would have been altered had the modern methods of separation been available. These methods surely would have hastened the process, but the primary impediments to progress were the biological limitations, chiefly the lack of an uncomplicated assay for the nutrient in question. Esmond Snell has just told us how quickly the isolations occurred when a reliable assay by microbial growth became available. Even with the outmoded techniques for doing the chemistry of the B vitamins, this was accomplished with a great deal more dispatch than was the perfection of assays and the other biological aspects of the experimental work.

A single event in 1937, the isolation and identification of the anti-black tongue factor, reported in a paper by Elvehjem, Madden, Strong and Woolley (1937), gave rise to such excitement that it affected the outlook of graduate students for several years thereafter. Each student conjured up visions of being successful in isolating and identifying a nutrient which would cure some human disease or otherwise benefit humankind so that his or her work would be remembered long after they had left these laboratories. No doubt this discovery motivated students to strive harder and sometimes to make unwarranted assumptions about what their results meant. The identification of the hemorrhagic agent in spoiled sweet clover by Stahmann, Huebner, and Link (1941) had a similar influence on graduate students of that period. There were other such signal contributions which provided added incentive for eager students who had been in school during the depression. Their expectations for financial rewards were modest, but they hoped to achieve success measured in other ways. Unlike some of the other discoveries relating to the B vitamins, the identification of the pellagra preventive factor as nicotinic acid was a contribution not shared with other laboratories and is remembered as a distinct Wisconsin contribution.

During the ensuing 12 years many students pursued what they hoped were new nutrients. With the isolation and recognition of multiple forms and functions of folate and vitamin B_{12}, the varied observations were explained, and the quest for new vitamins tapered off. There remained the hope that germ-free animals, sulfa drugs, antibiotics, thyroxin or an antagonist would limit biosynthesis or otherwise create a deficiency of a new factor. But most of the vitamin seekers turned to other problems and demonstrated that their training had prepared them for biochemical, genetic, pharmacologic and nutritional problems generally.

It might be interesting to list some of the nutritional factors of the discovery period without identifying the discoverers or champions of these factors or even their final identity. They included:

Spectacle eye factor
achromotrichia factor
grass juice factor
chick antidermatitis factor
Factor W
cartilage growth factor
gizzard erosion factor
alkali labile growth factor for dogs
L. casei factor
monkey antianemia factor
guinea pig growth factor
vitamin B_4
vitamin B_{10}
vitamin B_{11}
inositol
p-aminobenzoic acid

At the time these promising nutrients were being pursued, others in the same research groups were extending the classical work on inorganic elements including: iron, copper, manganese, molybdenum, zinc, boron, fluorine, cobalt, arsenic, aluminum, selenium and potassium.

Another important activity which will be dealt with by later speakers is that concerned with the metabolic function of the coenzymes arising from these B vitamins. Leads for experiments in that area were sought and found in the animal and bacterial experiments. Some examples which come to mind are:

1. thiamin sparing action of fat;
2. accentuation of vitamin B_6 deficiency with high protein rations (Conger and Elvehjem, 1941);
3. biotin sparing action of fatty acids and aspartate (Potter and Elvehjem, 1948).

A major effort on the histological changes caused by vitamin deficiencies was mounted by Paul Phillips and his associates. Poliomyelitis, dental caries and other important human diseases were brought under investigation on the basis of hopes that nutritional elements might be crucial.

One characteristic of Connie's program which I appreciated was the freedom to develop your area of research. The broad subject matter of Dr. Elvehjem's publications reflected this freedom he gave to graduate students to follow their interests. I recall I encountered no opposition when I proposed that we test purified amino acid diets, containing only the known vitamins, for the growth of rats and mice. Unfortunately this expensive effort did not lead to the discovery of new micronutrients.

Collaboration was broad within the Department, within the University, and with those in other institutions. We were free to hatch up an arrangement with whomever we wished. A tradition of collaboration with Foods and Nutrition, Animal Husbandry, Dairy Science, Poultry Science, and Microbiology was well established in the 1930's. This collaboration frequently included the use of biochemistry facilities by non-major students, who were so much a part of the scene that it was not always clear what their majors were. The three principal advisors in vitamin research were Hart, Elvehjem and Phillips, but W. H. Peterson became very much involved through his students Strong, Snell, Bohonas, Hutchings and many others. When Karl Link's dicumarol research led him to vitamin K, his group also collaborated with the vitamin research group (Black et al., 1942).

The collaboration with the Hart-Elvehjem-Phillips troika was truly remarkable. These professors talked often, usually in stand-up conferences in the animal rooms, and never became very concerned about where the ideas came from, so long as someone got on with the experimentation. Harry Waisman became a link to the medical microbiology group on the poliomyelitis project. He and Elvehjem's students did much of the first work on the nutritional requirements of primates. Strong, Baumann and Snell frequently had substantial collaboration with one or more members of the B vitamin triumvirate. These arrangements provided strength and greater appreciation and respect among the faculty and graduate students.

In the time remaining, I hope to describe a few highlights of this era in the history of the Department. I shall talk first of the discovery of the cure of black tongue, of the niacin-tryptophan relationship in rats, then of the role of biotin in carboxylation of pyruvate, and finally of the work on folate.

The extensive work of Goldberger and coworkers in 1928 (Goldberger et al., 1928) established black tongue in dogs as the counterpart of human pellagra. It was clear from Koehn and Elvehjem's paper in 1937 that they had not differentiated between the deficiencies causing black tongue and chick pellagra. Both dogs and chicks were used to assay their liver concentrates. Vitamins B_1, B_2 and B_6 were recognized as distinct, but beyond that there was disagreement. It wasn't until nicotinic acid and its amide cured black tongue

(Elvehjem et al., 1937) and was found of no value for chick dermatitis that niacin and pantothenate (Mickelson et al., 1938) were differentiated.

The key observation that nicotinic acid from a bottle from the reagent shelf cured black tongue in dogs was made by R. J. Madden. There are various versions of the events which prompted Madden to test this compound. It is clear that Madden and his colleagues were aware of the discovery of von Euler et al. in 1935 and Warburg and Christian in 1936 that cozymase contained nicotinamide. Others report (Jukes, 1981) that he was aware of the effectiveness of cozymase in replacing factor V for *Hemophilus* growth (Lwoff and Lwoff, 1936) and of the report (Knight, 1937) that nicotinic acid was a growth factor for *Staphylococcus aureus*. With this many clues in the recent literature it seems likely that several of his contemporaries could have made the suggestion. In any case, the cure was dramatic, and nicotinamide was quickly isolated from the concentrates of liver being used as a source of this vitamin.

The sequel to this story came several years later and resulted from the application of Snell and Wright's (1941) microbiological assay for nicotinic acid to a variety of foods (Teply et al., 1942). It became clear that some foods, notably high-protein animal products, prevented black tongue more effectively than they should on the basis of their niacin content. Furthermore, corn, which contains more niacin than milk or eggs, had historically been associated with causation of pellagra and black tongue. This apparent pellagragenic effect of corn on humans and dogs prompted Krehl, Sarma, Teply and Elvehjem (1946) to add corn to a niacin-deficient, purified diet being fed to both dogs and rats. The corn products increased the niacin requirement of the dog and, for the first time, created a recognized requirement for dietary niacin in the weanling rat. The addition of 40 percent corn grits to the 15 percent casein ration reduced the casein to 9 percent of the ration. Krehl and associates reasoned that replacement of casein by grits would reduce the lysine and tryptophan content of the ration, so they added these amino acids to the 40 percent corn grits rations and noted a growth response with the tryptophan comparable to that observed with niacin. We all know that this interchangability of an essential amino acid and a B vitamin was explained by the demonstrated precursor role of tryptophan in niacin biosynthesis a few years later.

At that time the prevailing idea was that the tryptophan was influencing niacin synthesis by intestinal bacteria. This was disproved (Henderson and Hankes, 1949) by showing that rats with their intestinal tracts removed excreted N^1-methyl-nicotinamide, niacin and bound niacin, later identified as quinolinate, in response to tryptophan, almost as well as did the controls. The paper describing this study does not appear in Elvehjem's list of publications. He introduced the paper, but was not a coauthor. I always wondered if (1) he didn't believe the results, or (2) was too committed to the intestinal synthesis hypothesis to detract from it, or (3) felt that he had not contributed enough to the study to be a coauthor. I think it was the latter. I always found him to be generous with credit to his students and colleagues under such circumstances.

The suppression of growth with corn or corn grits below that observed with 9 percent casein plus 0.2 percent cysteine as the sole source of protein implicated the amino acids provided by tryptophan-deficient zein in the induction of the tryptophan-niacin deficiency. The subsequent observation, that replacement of 40 percent corn with 2 percent acid-hydrolyzed casein and finally with 0.078 percent DL-threonine (Hankes et al., 1948), led to many variations of amino acid imbalance effects which occupied much attention during the subsequent 15 years, chiefly in A. E. Harper's group. This particular imbalance appears to result from the dual function of tryptophan for (1) protein synthesis and (2) niacin synthesis such that, when the tryptophan content of the diet is at a critical level (about 0.1 percent of the diet) and the other amino acids are not limiting for growth, niacin synthesis is arrested and a deficiency results. These conditions can be simulated in a niacin-free, amino acid diet (Koeppe and Henderson, 1955).

The explanation for corn's reduction of tryptophan conversion to niacin lies in this imbalance. But the failure of corn to prevent pellagra by virtue of its own niacin content is a continuing puzzle, the explanation for which appears to lie in the presence of an alkali-released form of niacin, which is not available to the animal unless the corn is first exposed to alkali treatment, such as that used in tortilla preparation (see Carpenter, 1981, for review). Little did Krehl, Sarma, Teply and Elvehjem suspect that the rat feeding experiments with corn would expose so many interesting problems.

Some of you will recall a development of 1947 which illustrates collaboration among the staff and how studying a deficiency provided a clue to the molecular basis for vitamin function. R. L. Potter, a student of Elvehjem's, collaborated with Henry Lardy (Lardy et al., 1947) in determining how biotin replaces the aspartate requirement of *L. arabinosus*. They reasoned that biotin was stimulating aspartate synthesis not at the transamination step, as had been suggested, but at the pyruvate carboxylation step. Potter and Elvehjem (1948) demonstrated stimulation of growth by oxaloacetate and oleic acid in a low biotin medium. You all know of the large number of other CO_2 transfer reactions requiring biotin which were discovered soon thereafter.

The history of folic acid's discovery and characterization illustrates nearly all of the problems of vitamin sorting I have discussed. This vitamin was assayed with three different microorganisms and with the monkey and the chick. It took some seven years of intense activity to explain the varied responses observed in the assay organisms to each of the multiple forms. Wisconsin scientists were prominent in these developments. Snell and Peterson (1939) described a new growth factor for *L. casei* in yeast and liver which was adsorbed on norite. After Snell went to Texas, he and his colleagues (Mitchell et al., 1941) isolated folic acid from spinach and showed that it supported chick growth. It was soon clear that this substance was closely related to vitamin M, factor U, factors R and S, vitamin B_c, vitamin B_{10} and B_{11} and possibly other factors. Suboptimum dosage supported slow growth, which accentuated the poor feathering, hence vitamins B_{10} and B_{11}, one for feathering and one for growth. A summary of key observations during these years is presented in Table 8.1.

An outgrowth of this story developed in Carl Baumann's laboratory, where Howerde Sauberlich (Sauberlich and Baumann, 1948) discovered the need for a factor present in refined liver extracts used to treat pernicious anemia, which stimulated the growth of *Leuconostoc citrovorum*. It was not vitamin B_{12} as shown by its migration in an electric field and its stability to alkali. They observed a partial and delayed response to both folate and thymidine. When folate was added to the diet of the rat, citrovorum factor appeared in the urine. The factor was much better than folate in reversing aminopterin inhibition of *Leuconostoc citrovorum* growth. A number of laboratories demonstrated that formylation and reduction of folate yielded citrovorum factor or folinic acid which was characterized as 5-formyl-tetrahydrofolate (5-CHO-H_4PteGlu). This development was useful in unraveling the function of tetrahydrofolate in the transfer of C_1 fragments. The observation earned Howerde the Mead Johnson Award in 1952.

With the multiplicity of folate forms and function it should not be surprising that the discovery phase was complex and confusing. Stokstad and coworkers (Stokstad et al., 1977) have identified 15 folate derivatives in liver. In addition to the various polyglutamates, these derivatives include two oxidation states (oxidized or tetrahydro form) and the absence or presence of the 5-methyl or 5- or 10-formyl substituents. The four major components in

Table 8.1 Historical sequence of folate development

Year	Name	Activity
1935–38	Vitamin M (Day)	Cytopenia in monkeys
1938	Factor U (Stokstad)	Growth of chicks
1938–40	Factors R and S (Norris)	Growth and reproduction in chicks
1939	Norite eluate factor (Snell and Peterson)	*L. casei* growth
1940	Vitamin B_c (Hogan)	Chick anemia
1941	Folic acid (Mitchell et al.)	*S. lactis* R growth
1941–43	Vitamin B_{10} and B_{11} (Mills et al.)	Feathering and growth
1944	Fermentation *L. casei* factor (Hutchings)	*L. casei* not SLR
1945	B_c conjugate (Bird)	Chick anemia not *L. casei*
1948	Citrovorum factor (Sauberlich)	*Leuc. citrovorum* growth

liver were identified as 10-CHO-H_4PteGlu$_5$, 5-CHO-H_4PteGlu$_5$, 5-CH_3-H_4PteGlu$_5$ and H_4PteGlu$_5$.

At this junction in B vitamin identification and isolation, industrial laboratory competition and the lack of a model for human pernicious anemia placed the Wisconsin team at a disadvantage. Nevertheless both Lardy's and Elvehjem's groups played important roles in the early phases of the vitamin B_{12} saga, especially in developing animal assays which were more practical than those employing human pernicious anemia patients. These animal assays would have been crucial, had the microbiological assay of Shorb (1947) not made a timely appearance.

The search for new B vitamins did not end as abruptly as the literature would suggest. For several years stress factors such as thyroxin for the accentuation of vitamin B_{12} deficiency were looked for. Germ-free animals, once deemed a promising approach, did not add to the list of nutrients. Antagonists, useful when you have a known biosynthetic defect, were not helpful in finding new nutritional factors.

The history of discovery of the vitamins ended more than 30 years ago. The development of information regarding vitamin function, metabolism and the implications for optimum nutrition have continued. Since they are dietary essentials, they will continue to be important to nutritionists, and since deficiency states are metabolic tools, they will interest the biochemist. Pyridoxal phosphate is a reagent for the protein and enzyme chemist, and avidin is indispensible to those interested in a wide variety of enzymatic carboxylation reactions.

The availability of most of the B vitamins labeled to high specific activity has made possible studies of absorption, transport, degradation, conversion to coenzymes, etc. under circumstances closely simulating those in the intact organism. The feasibility of such experimentation has stimulated renewed interest in the metabolism of vitamins. A reexamination of older experiments is revealing questions not answered during the stampede of discovery. There is still much unfinished metabolic investigation of the B vitamins, and it can now be done with the improved tools at our disposal.

Future research on the B vitamins will not bring the excitement which characterized the discovery phase which many of us witnessed in this Department, just after the midpoint of the first century of its history. However, there is every reason to believe that equally exciting discoveries will be made in other areas during the next hundred years if you and your successors continue to encourage bright young students to pursue their interests in the natural sciences.

References

Betheil, J. J. and Lardy, H. A. (1949) *J. Nutr.* 37: 495.

Black, S., McKibbin, J. M., and Elvehjem, C. A. (1941) *Proc. Soc. Exp. Biol. Med.* 47: 308.

Black, S., Overman, R. S., Elvehjem, C. A., and Link, K. P. (1942) *J. Biol. Chem.* 145: 137.
Carpenter, K. J. (1981) *Fed. Proc.* 40: 1531.
Conger, T. W. and Elvehjem, C. A. (1941) *J. Biol. Chem.* 138: 555.
Elvehjem, C. A., Madden, R. J., Strong, F. M., and Woolley, D. W. (1937) *J. Am. Chem. Soc.* 59: 1767.
Goldberger, J., Wheeler, G. A., Lillie, R. D., and Rogers, L. M. (1928) *Pub. Health Rep.* 43: 1385.
Hankes, L. V., Henderson, L. M., Brickson, W. L., and Elvehjem, C. A. (1948) *J. Biol. Chem.* 174: 873.
Henderson, L. M. and Hankes, L. V. (1949) *Proc. Soc. Exp. Biol. Med.* 70: 26.
Henderson, L. M., McIntire, J. M., Waisman, H. A., and Elvehjem, C. A. (1942) *J. Nutr.* 23: 47.
Jukes, T. H. (1981) *Fed. Proc.* 40: 1536.
Knight, B. C. J. G. (1937) *Biochem. J.* 31: 966.
Koehn, C. J., Jr. and Elvehjem, C. A. (1937) *J. Biol. Chem.* 118: 693.
Koeppe, O. J. and Henderson, L. M. (1955) *J. Nutr.* 55: 23.
Krehl, W. A., Sarma, P. S., Teply, L. J., and Elvehjem, C. A. (1946) *J. Nutr.* 31: 85.
Lardy, H. A., Potter, R. L., and Elvehjem, C. A. (1947) *J. Biol. Chem.* 169: 451.
Lwoff, A. and Lwoff, M. (1936) *Compt. Rendus.* 203: 520.
Mickelsen, O., Waisman, H. A., and Elvehjem, C. A. (1938) *J. Biol. Chem.* 124: 147.
Mitchell, H. K., Snell, E. E., and Williams, R. J. (1941) *J. Am. Chem. Soc.* 63: 2284.
Potter, R. L. and Elvehjem, C. A. (1948) *J. Biol. Chem.* 172: 531.
Sauberlich, H. E. and Baumann, C. A. (1948) *J. Biol. Chem.* 176: 165.
Shorb, M. S. (1947) *J. Biol. Chem.* 169: 455.
Snell, E. E. and Peterson, W. H. (1939) *J. Biol. Chem.* 128, XCIV. Also see *J. Bact.* 39: 273.
Snell, E. E. and Wright, L. D. (1941) *J. Biol. Chem.* 139: 675.
Stahmann, M. A., Huebner, C. F., and Link, K. P. (1941) *J. Biol. Chem.* 138: 513.
Stokstad, E. L. R., Shin, Y. S., and Tamura, T. (1977) in *Folic Acid, Biochemistry and Physiology in Relation to the Human Nutrition Requirement*, pp. 56–71, Proc. of Workshop. Food and Nutrition Board, NRC, NAS, Washington, DC.
Teply, L. J., Strong, F. M., and Elvehjem, C. A. (1942) *J. Nutr.* 23: 417.
von Euler, H., Albers, H., and Schlenk, F. (1935) *Z. Physiol. Chem.* 237, I.
Warburg, O. and Christian, W. (1936) *Biochem. Z.* 285: 156.

9

Years with Conrad Elvehjem

Van Rensselaer Potter

McArdle Laboratory for Cancer Research
University of Wisconsin Medical School
Madison, Wisconsin

I have chosen the title "Years with Conrad Elvehjem" because it was my privilege to know Dr. Elvehjem rather intimately not only as a graduate student but also as a faculty colleague. I was fortunate in receiving his attention and support for over 20 years on my return to this campus after my postdoctoral years in Stockholm, Sheffield and Chicago as I shall mention. In describing Dr. Elvehjem I hope I will be able to shed some light on the history of the biochemistry department. If I am in error at any point I hope I may be able to make corrections.

Elvehjem's Career Before 1935

Conrad Elvehjem was born May 27, 1901. As he was just 10 years older than I, he was more like an older brother than a father figure. He was born very near Madison, near McFarland, and graduated from the Stoughton High School in 1919. He received all his degrees here at the University of Wisconsin with a B.S. in 1923, M.S. 1924, and Ph.D. in 1927. He married Constance, who, like Conrad, was known as Connie, in 1926. Following the Ph.D. he was immediately appointed as an Instructor. He held this title for two years before beginning the biggest adventure of his life—an experience that was the determining factor in his future career. He and Mrs. Connie went to the University of Cambridge, England, where he was a postdoctoral fellow with National Research Council support for the 1929–1930 academic year. After his sojourn in Cambridge he returned to Madison as an Assistant Professor in 1930.

Although the biochemistry department was extremely inbred as I shall describe, virtually all of her older faculty had done postdoctoral research elsewhere. Dr. Elvehjem fit into that pattern.

The idea of sending promising young Ph.D.'s away from Madison for further training undoubtedly stemmed from the experience of Elvehjem's own mentor, Professor E. B. Hart. After graduating from the University of Michigan, Hart had engaged in full time research at the New York Experiment Station at Geneva (1897–1898) but he felt the need for additional supervised experience. He chose to study in Marburg, Germany, at his own expense under the famous nucleic acid biochemist Albrecht Kossel, who had first reported carbohydrates in nucleic acid in 1891. Kossel and his students isolated the pyrimidines thymine, uracil, and cytosine during the years 1894–1903 (Potter, 1960). After Hart had been in his laboratory for only a year Kossel moved from Marburg to Heidelberg, accompanied by Hart, who was unable to spend the additional time required to obtain a Ph.D. in the new location (Elvehjem, 1953). Professor Hart's experience in Germany must have had an influence on the practice that he evolved in subsequent years, since to a great extent he built the department around students who had completed the Ph.D. in Madison and who had returned after study elsewhere. Harry Steenbock, who received his Ph.D. under Professor Hart's supervision in 1916, also benefitted from the concept by spending a semester at Yale in the nutrition laboratory of Osborne and Mendel and a semester in Berlin with Carl Neuberg during his period as a graduate student with Hart.

The new pattern was set by Karl Paul Link who received all his degrees at U.W., with his Ph.D. in biochemistry under Professor Tottingham. He was an International Education Board postdoctoral fellow at St. Andrews, Graz, and Zurich during the years 1925–1927, returning to Madison as an Assistant Professor in 1927, the year that Elvehjem received his Ph.D. In that same year Marvin Johnson received his B.A. at U.W. and subsequently his Ph.D. with W. H. Peterson in 1932. He left for Europe as a National Research Council Fellow, spending 1932–33 at the German Technical University in Prague. Back at Wisconsin, Marvin unbelievably served as a Research Associate for seven years in the Old Testament tradition before becoming an Assistant Professor in 1940. Carl Baumann also received all his degrees at U.W., receiving his Ph.D. with Steenbock in 1933. Baumann spent 1934–1936 as a General Education Board postdoctoral fellow in Heidelberg, Cambridge, and Copenhagen. He served as a Research Associate for two years and became an Assistant Professor in 1938. Similarly, Frank Strong received his Ph.D. at U.W. (Organic Chemistry, 1932) and studied as a postdoctoral Rockefeller Fellow in Zurich and Utrecht 1935–1936. After serving as a Research Associate with W. H. Peterson during 1936–1938 he also became an Assistant Professor in 1938.

A few years later Bob Burris received his Ph.D. at U.W. in 1940 in Agricultural Bacteriology with Perry Wilson and spent 1940–41 as a National Research Council Fellow at Columbia University with Harold Urey. Henry

Lardy received his Ph.D. with Paul Phillips in 1943 and was a National Research Council Fellow at the Banting Institute in Toronto 1944–1945 with H. O. L. Fischer. Mark Stahmann received his Ph.D. with Link in 1941 and spent three years at Rockefeller Institute and M.I.T. before returning to Madison as an Assistant Professor in 1947.

Thus for 20 years, from 1935–1945, nearly all of the "homegrown" products who were to become biochemistry faculty served as postdoctoral fellows before returning to Madison. The sole exception during this era was Paul Phillips. He obtained his Ph.D. with Professor Hart in 1933, and served as Research Associate 1933–1936 before becoming an Assistant Professor. I believe that although Paul achieved many honors he may have felt at a disadvantage by the lack of research experience outside of the department. However, he had served in Russia as an administrator of Near-East Relief following his B.S. at Washington State.

It may be observed that all the noted examples of the usual pattern occurred at a time when postdoctoral fellowships were extremely rare. As I recall there were less than 20 N.R.C. fellowships per year in all fields of science combined, with only two or three in biological chemistry at most. It would be worthwhile to see a chart showing the number of postdoctoral fellowships in the U.S.A. year by year up to the present.

I have briefly sketched the peer group that Connie entered. At the Ph.D. level he was two years behind Karl Paul Link and five to six years ahead of Strong, Johnson and Baumann. For several years he and Link were the only Assistant Professors in the department and their careers moved more or less in parallel. Connie was identified with vitamins and enzymes while K. P.'s career was greatly advanced by the Dicumarol characterization and isolation.

In retrospect it became clear that when the peer group were students the "Department" was Hart, Tottingham, Peterson and Steenbock. Before then, and blazing their paths, were Babcock and McCollum. The Babcock story is beautifully but all too briefly told in the book *Hunger Fighters* by Paul deKruif (1928). The departure of E. V. McCollum from Wisconsin to Johns Hopkins has never been fully described in print to my knowledge, and no one with first-hand knowledge is now living. I was told that when McCollum departed he emptied all the rats in the animal room from their cages to the floor. The department was divided into segregated compartments that functioned more or less independently of departmental administration. Referring to Professor Hart, Elvehjem wrote, "He wasted no time in his administrative duties. For many years he ran the department without staff meetings." (Elvehjem, 1955.)

Thus as we look at Dr. Elvehjem's record we find that in the six years before Cambridge he published 27 papers as a graduate student and then as an Instructor with Hart and Steenbock, with Hart only, and with Peterson only. Papers with Steenbock and Hart originally involved calcium assimilation. Later their studies branched into the role of iron in nutrition and this work formed the basis of Elvehjem's Ph.D. thesis. A key paper in 1928, opening the crucial role of copper in iron nutrition, was authored by Hart, Steenbock, Waddell and Elvehjem (Hart et al., 1928). By the time Elvehjem returned from Cambridge in 1930 he was fully established as an independent investigator. However, a cordial relationship with Professor Hart, like that of a father and a non-rebellious son, persisted for the life of both, and their jointly sponsored Journal Club became a weekly institution. Elvehjem continued to coauthor many papers with Hart but produced many reports coauthored with only his students or under his sole authorship. An example of the latter was his extensive account of "The biological significance of copper and its relation to iron metabolism" (Elvehjem, 1935) in which the early history of copper in biology from as early as 1917, together with the story of the Wisconsin discovery, is told.

The influence of the Cambridge experience on Dr. Elvehjem was enormous. While there he had contact with Sir Frederick Gowland Hopkins, who has been called the father of the vitamin concept. In particular he worked with David Keilin who had just rediscovered cytochrome C, an iron-containing electron transport protein. Connie's nutritional experience with copper as a trace element prepared him for the generalization that trace elements and vitamins must act by becoming structurally built into one or another enzyme or biocatalyst. This experience prepared him for his contact with Hopkins and with Keilin. Incidentally, I, too, planned to work with Keilin in 1939–1940, but I never even met Keilin at that time because war intervened. In 1958 I entered his office unannounced and said as dramatically as possible, "I'm sorry, Dr. Keilin, I'm 20 years late."

Dr. Elvehjem returned to Madison with several boxes of the now-outmoded Barcroft double respirometers for measuring oxygen uptake and with a micro-spectrophotometer the size of a package of gum for observing the change of cytochrome c from the oxidized to the reduced form. He was determined to find an ecological niche apart from Steenbock, Peterson, and Hart, yes and K. P. Link, too. He was determined to enter the world that embraced enzymes and vitamins, while at the same time going after the isolation of the unidentified vitamins that the Babcock and Hopkins influences had indicated. Was it fate when Professor Hart assigned to him an office with a small adjacent laboratory that bore outside the door a bronze plaque stating simply "The workrooms of Dr. S. M. Babcock 1913–1931"? It was my privilege to share that space with Dr. Elvehjem, 1935–1938, as I shall describe.

Connie's first student who took up the enzyme connection was Frederick J. Stare, who published with Elvehjem in 1933 a paper briefly titled "The Respiration of Animal Tissues." This paper laid out the methodology and use of the Barcroft respirometer and described medium and tissue preparation. Malcolm Dixon and K. A. C. Elliott of Cambridge were quoted to the effect that tissues finely minced with a scissors were as valid as tissues thinly sliced with a razor. Stare and Elvehjem claimed to have confirmed this result but all the data were obtained using the so-called "scissor-mince." No significant differences were seen in comparisons of normal and vitamin-deficient specimens, which must have been a disappointment. This opening paper was followed by a study of tissue respiration in chronic fluorosis and scurvy in the guinea pig, authored by Philllips, Stare and Elvehjem in 1934. Also in 1934, a summer student, Eugene Cohen, and Elvehjem published a paper on the relation of iron and copper to the cytochrome and oxidase content of animal tissues. This paper contained the first hint of copper's role in oxidative metabolism. During 1933 and 1934 fifteen papers were published on nutritional studies, eight including Hart and seven without Hart's name, while only three papers were in the enzyme category. Thus the ratio of nutrition research to enzyme research was 15/3 in Dr. Elvehjem's group in the years 1933 and 1934. Dr. Elvehjem wrote many short reviews or commentaries with sole authorship during his lifetime. These papers were continued right up to the time he became president of the University. The total amounts to about 85 and their titles reveal the breadth of his expertise and interests.

Years as a Graduate Student with Elvehjem, 1935–1938

The element of chance played a role in the most important event in my professional career. I say most important because every subsequent event was related to the fact that Conrad Elvehjem was my major professor. I say chance because I came to Madison intending to interview Professor Steenbock, but without a carefully arranged time schedule.

In 1934 Alvin Moxon and I were working full time in the Experiment Station Chemistry Department at South Dakota State College in Brookings. Moxon had a car, and in late August of 1934 he and I and two friends rather spontaneously decided to go to the World's Fair in Chicago. For my benefit they agreed to let me visit the Agricultural Chemistry Department in Madison and the Biochemistry Department at Purdue University where our former chemistry instructor was working on his Ph.D. under a Professor Corley.

Our first stop on the trip was Madison. I had come to Madison to see Harry Steenbock, the only Wisconsin name that I knew, and when I found that he was out of town I looked around and found Elvehjem in his office and willing to talk to me. I had worked in the Station Lab since 1929, and took five years to get my B.S. degree in 1933. I had a special curriculum approved by a petition signed by my boss, Dr. Kurt W. Franke, and the Dean of Pharmacy, Dr. E. R. Serles. This curriculum appealed to Dr. Elvehjem since I had eliminated all the Education courses. This enabled me to take all the chemistry and biology courses that were available at South Dakota State. By 1934 I had published on toxicity to rats resulting from levels of selenium as low as ten ppm in wheat. The results were of interest to Dr. Elvehjem as a trace metal phenomenon presumably with a connection to some biocatalyst, while I was fascinated with the new field of enzymology in relation to nutrition. Connie liked my curriculum and my transcript and suggested that I might be able to obtain a WARF fellowship since these were being offered for the first time. From the beginning I was impressed with Dr. Elvehjem's open mind and his quick response to new possibilities.

My friends and I went on to Purdue where, through the influence of my former instructor, Ted Tripp, I was offered a fellowship, for which I fortunately postponed a decision. I told Ted I wanted more research experience before embarking on a Ph.D. but he said, "Get your Ph.D. first and then you'll get paid for getting experience." I returned to Madison with the Purdue offer and Connie virtually assured me of the WARF fellowship. But for various reasons I could only begin in the second semester, in February, 1935. I spent that second semester and the following year in the grad houses at Adams and Tripp Hall, where the other WARF fellows were housed. These included Forrest Quackenbush and Howard Schneider, who were WARF fellows with Harry Steenbock.

I came to Wisconsin with several years' experience with rats, hematology and blood chemistry, and I was somewhat appalled at the caging at Wisconsin. My chief in Brookings, Kurt Walter Franke, had invented a new type of cage for individually housing rats (Franke and Franke, 1934). This cage was an absolutely unique and original design, operating like a drawer in a filing cabinet but with a screen bottom. It could easily have been patented but patenting was not even considered. At the very least it should be known today as the Franke cage but unfortunately it is not. The cage design was adopted in Madison after my arrival and they were fabricated by the U.W. shops. Now these cages are made commercially and used all over the world.

Dr. Elvehjem gave me a desk in his office and the two of us shared the space back to back for the whole time I was his student. When he was in his office meeting with another student I always went into the adjacent laboratory and closed the door. This small lab was shared with Bill Sherman, who was my direct link with the Elvehjem program that attempted to define enzyme defects in vitamin deficiency, in this case vitamin B_1 deficiency, using young chicks up to several weeks of age. Brain and kidney were the usual test specimens, and the older chicks were often of frying size, especially the controls. They were frequently so employed, minus brain and kidneys.

Bill Sherman inherited the scissor-mince technique from Fred Stare but I was not enthusiastic about the procedure, and I began to look for other approaches. I wanted to measure individual enzyme activities rather than total oxygen uptake of intact cells. I devised a procedure that came to be called the homogenate technique, which was met with his usual open mind by Dr. Elvehjem. He reported it at a Gordon Conference near Baltimore. K. A. C. Elliott was there and deplored the technique because, as he said, "It's so easy everyone will be using it instead of using tissue slices or minces." Another example of Dr. Elvehjem's progressive attitude was his purchase of the first photoelectric colorimeter (The Evelyn) at my suggestion.

Despite some progress in the enzyme field, the big events in those early years were in nutrition. In 1934 Elvehjem and Carl Koehn reported in *Nature* on the "Non-identity of vitamin B_2 and flavins"; in 1936 Koehn and Elvehjem reported in the *Journal of Nutrition*, "Studies on vitamin G (B_2) and its relation to canine black tongue." The latter symptom in dogs was of course the animal equivalent of pellagra in humans. Elvehjem was now going for the biggest vitamin breakthrough of modern times. Carl Koehn left the scene to take a job in Alabama, with a final paper in the *Journal of Biological Chemistry* in 1937, after his departure, entitled "Further studies on the concentration of the antipellagra factor." Koehn's place was taken by a new graduate student named Robert J. Madden, whose rather inconstant ways led Dr. Elvehjem to remonstrate with him on several occasions. For example, Elvehjem and Hart always held court in the animal room in the attic every morning, seven days a week. Students were encouraged to be there, too, to show rats or chicks with odd symptoms and to elicit comments. Madden was not there as a rule and sometimes his water crocks would be empty. Nevertheless, Madden took up the Koehn concentrates and elicited help from Wayne Woolley, then a graduate student, and Frank Strong, who was a Research Associate. Both were associated with W. H. Peterson. Frank had just assembled a molecular still which could fractionate by sublimation at high vacuum; concentrates prepared with Woolley's help were placed in the apparatus. White crystals were obtained from the material on the collecting surface and proved to rapidly cure black tongue in Madden's dogs. These crystals were identified as nicotinic acid almost overnight, helped by the fact that nicotinic acid was a known compound that was commercially available, although it had no particular use. The question that arises is, how was the identity learned so quickly? Madden, who departed Madison before receiving a Ph.D., wrote in an autobiography that *he* suspected nicotinic acid because Warburg had shown in 1934 in Berlin that it was in an important coenzyme in the form of the amide. But of course Elvehjem and his enzyme students were well aware of the Warburg connection as well as the von Euler connection and Elvehjem surely was eager to test the possibility. At that time, Harold Campbell was a student in K. P. Link's lab and had mastered the micro-techniques for elemental analysis that Link had brought from Zurich. The analyses and melting point were compatible with the formula for nicotinic acid. Moreover, commercial nicotinic acid cured black tongue. A short communication was fired off on August 13, 1937, to the *Journal of the American Chemical Society*, authored by Elvehjem, Madden, Strong and Woolley, with an acknowledgment to Harold Campbell for the analyses and to his chief who was K. P. Link. It appeared in the September number. Dr. Elvehjem left town while the communication was in press. I was alone in his office when the editors called and demanded an answer to their question: "Was Elvehjem aware that nicotinic acid had just been shown (actually August 1937) to be a growth factor for the diptheria bacillus by J. H. Mueller, and would he want to include the reference?" I replied after a moment's reflection that he surely knew of the reference and would have included it if he felt it was appropriate in view of the fact that the communication had essentially no bibliography other than a single reference to Koehn and Elvehjem. This is a story never previously told. The communication indicated an awareness of the coenzyme connection with the words "The fact that nicotinic acid amide is an essential ingredient of the diet is not surprising since it is a component of certain coenzymes." However, no references to the coenzyme literature were cited.

The full publication under the same authorship appeared in the *Journal of Biological Chemistry* in 1938. It provided full documentation to the coenzymes, the bacterial growth requirements, and the biochemical history of nicotinic acid and its amide. The publications by Elvehjem et al. helped me to obtain a postdoctoral fellowship for enzyme experience in the Stockholm laboratory of Hans von Euler, who had extended the Warburg finding to another related coenzyme that was shown to contain nicotinamide.

As soon as he had identified nicotinamide as the factor that cured black tongue in dogs, Elvehjem fired off telegrams to Dr. Tom Spies and to other clinicians who had pellagra patients in hospital wards. Within 72 hours after administration of nicotinamide, the symptoms cleared. In January, 1939, Tom Spies was named scientist of the year, and his picture appeared on the cover of *Time* magazine. I felt that Dr. Elvehjem deserved the credit, and sent an indignant letter to *Time*.

Years with Conrad Elvehjem 1939–1962

Vivian and I returned from abroad about October 1, 1939. We had left Stockholm in August and were in Edinburgh enroute to Hans Krebs' laboratory in Sheffield on September 1 when Hitler invaded Poland. Soon after, we were ordered home by the Rockefeller Foundation's Paris office. We were fortunate to find a haven in Professor Thorfin Hogness' laboratory at the University of Chicago and met with Dr. Elvehjem in Madison quite soon thereafter. Dr. Elvehjem was aware of the impending opening of the McArdle Laboratory and encouraged me to believe I could find a position with Dr. Harold Rusch. I was interviewed by Dr. Rusch and Dean Middleton in October and with Elvehjem's help I began at McArdle as a Jonathan Bowman Fellow on February 1, 1940.

Dr. Elvehjem had never met Professor Hogness and somehow they got together and we began planning for an Enzyme Symposium at Wisconsin, September 11–13, 1941, and a Vitamin Symposium at Chicago, September 15–19. We had acceptances from all the important names of the time—Lipmann, Kalckar, Meyerhof, and many others.

Several years later, in 1945, Ralph Meader sponsored another enzyme symposium at Hershey, Pennsylvania, at the famed Hershey Hotel. I remember having lobster Newburg for the first time.

While there, Dr. Elvehjem and I shared a room for several days. Thus I had a good opportunity to speak to him about my idea of starting an Enzyme Institute at U.W. The rationale was the need for postdoctoral training in the U.S. to replace European laboratories affected by the war. Connie was enthusiastic and promised to talk to the WARF directors. Dean Middleton also got behind the idea and soon it came to pass. I worked closely with Louis Siberz, the architect. Originally, we hoped to build on Henry Mall between Agricultural and Medical School buildings but the Institute ended up by the old Poultry Building. We had a list of candidates topped by Carl Cori. Then Harland Wood, then Severo Ochoa, but each became unavailable. David Green had done excellent work with Dixon at Cambridge and, after some months of negotiations, he became the first team leader at the Enzyme Institute, followed by Henry Lardy and Gobind Khorana.

Professor Elvehjem had become chairman of Biochemistry in 1944. In 1946 he became Dean of the Graduate School. It would seem like an unsurmountable task to hold both of these offices at a time when he was publishing between 40 and 50 papers a year. However, the old idea that the department ran itself might have remained in his mind. While he had many collaborators, he still continued to publish frequent reviews and state-of-the-art commentaries under his sole authorship.

In addition, Dr. Elvehjem served on many national scientific panels and in particular the Food and Nutrition Board. I am indebted to Ms. Jean R. St. Clair, Archivist for the National Academy of Sciences, for the following account, which indicates the high regard that his colleagues at the national level had for Dr. Elvehjem. Ms. St. Clair was a staff member of the Board in 1958 and attended the spring and fall meetings. She wrote (August 17, 1983):

> For the March 1958 meeting, Dr. Elvehjem, Chairman of the Board, had sent word that he could not attend the Friday meeting, but that he hoped to attend the dinner and the Saturday morning session. The appointment of Dr. Elvehjem to the Presidency of the University of Wisconsin, effective July 1, had been announced at the Friday meeting, as had his decision that, under that circumstance, he would be unable to continue as Chairman of the Board.
>
> The speaker for the dinner was George McGovern, Congressman from South Dakota [who] had decided, via his membership on the House Committee on Education and Labor to develop a guide to inform the American people on what to eat to be healthy. . . . The Board's plan was to listen to what he had to say, and then tactfully offer the Congressman its assistance.
>
> In Dr. Elvehjem's absence, Dr. Grace Goldsmith, Vice Chairman of the Board, had just introduced McGovern who had delivered a sentence or two of his speech, when Dr. Elvehjem entered the room. Applause broke out and McGovern said, "Well, I can't compete with that," and sat down. A standing ovation followed for "Connie" Elvehjem, not so much for his new assignment at Wisconsin as for himself as a person. It was a remarkable show of affection and respect.
>
> Dr. Elvehjem succeeded in restoring quiet. He reintroduced McGovern who,

after his speech, had to leave for another appointment. It was just as well, because it turned out to be the wrong night for an outsider. I don't think the planners of the dinner meeting had anticipated what would happen if Dr. Elvehjem were to appear in the middle of the program, and it was evident that Elvehjem was surprised as well.

At the close of the dinner session, Dr. Elvehjem was presented with a signed scroll which read: "On the occasion of his movement into a new orbit, members and friends of the Food and Nutrition Board join in expressing to Conrad A. Elvehjem their appreciation of his scientific leadership, his sustained wisdom, his common sense, and his good humor as a member of the Board from its beginning and as its Chairman from 1955 to 1958."

In July, 1958, Conrad Elvehjem became President of the University of Wisconsin. Only four years later his life ended suddenly while he was at his desk in July, 1962. It is relevant to state that blood cholesterol has been proved to be raised to high levels by stress. Dr. Elvehjem had been under great stress by the tasks that he had accepted. Finally, the administrative problems, especially with the Medical School—problems for which he had no training—were too much for him. At the height of the medical school turmoil he called me for a private meeting in his car. He picked me up in front of Ag. Hall and we parked hear Babcock Hall. He asked me to become Dean of the Medical School. I was aghast, torn between my great love for him to whom I owed everything and, on the other hand, by my conviction that I was not suited for the Deanship. I said that I felt I should not accept because I lacked the M.D. degree. He said, "Oh I thought you got an M.D." He quietly let me off near McArdle and went on his way alone and isolated. Only two years later I learned of his death. I mourned the loss of the man who supported me in every way but whom I could not help when it was too late for help.

I hope I may be forgiven, and that all who hear my story may ponder the lessons to be learned from the life of this great man who would surely have received the Nobel Prize had he continued as a scientist.

One more anecdote may shed some light on the forces that played upon the life of Conrad Elvehjem.

Sometime during the year 1937 Dr. Elvehjem was being considered for the Chairmanship of the Department of Biochemistry at the Cornell University Medical School in New York City. I doubt very much that he applied for the job but he did agree to go to New York for an interview. I can well imagine the contrast between the décor of the chairman's office in New York and the décor of his own office in Madison. In New York the floor was probably carpeted and the furnishings would include at least a luxurious desk, a sofa, and a couple of easy chairs. In Madison, the walls needed painting and the floor of his office was covered with a weather-beaten maroon linoleum that was never subjected to soap and water by the one elderly janitor who served the entire building. [Here it may be noted that in his account of "Thirty-two years' association with E. B. Hart" (Elvehjem, 1955), Elvehjem was to write in 1955 that from the day he first entered Professor Hart's office it "did not change one iota from that day until he moved out of the office 20 years later. Not one drop of paint had been used in that room."] Dr. Elvehjem returned from New York on a Friday night. On Saturday morning he came into the office with a scrub bucket, got down on his hands and knees and began to scrub the floor. I came in through the door from the laboratory, took one look, and fled the building in utter embarrassment. Dr. Elvehjem evidently had decided that his duty lay with the University of Wisconsin. This was the final act of humility—this was "Christ washing the feet of the disciples." Connie never mentioned the interview or whether he had been offered the job at Cornell or not. The facts are that Dr. Vincent du Vigneaud, who in 1937 was Chairman of Biochemistry at George Washington University, occupied the Chair at Cornell from 1938 until his retirement in 1966.

In my opinion, Dr. Elvehjem did not serve the University of Wisconsin as Chairman of Biochemistry, as Dean of the Graduate School, and as President through egotism or personal ambition. I believe he had developed a life-long habit of doing what he was asked to do. Finally the burden became too great. In a very real sense he gave his life for this great university. Ironically his inaugural address contained the lines, "Give us hills to climb, and the strength to climb them."

References

Cohen, E. and Elvehjem, C. A. (1934) *J. Biol. Chem.* 107: 97–105.

deKruif, P. (1928) *Hunger Fighters,* Harcourt Brace and Co., New York.

Elvehjem, C. A. (1935) *Physiol. Reviews* 15: 471–507.

Elvehjem, C. A. (1953) *J. Nutrition* 51: 1–14.

Elvehjem, C. A. (1955) *Food Tech.* 9: 4–7.

Elvehjem, C. A. and Koehn, C. J., Jr. (1934) *Nature* 134: 1007–1008. Also see Koehn, C. J., Jr. and Elvehjem, C. A. (1936) *J. Nutrition* 11: 67–76.

Elvehjem, C. A., Madden, R. J., Strong, F. M., and Woolley, D. W. (1937) *J. Am. Chem. Soc.* 59: 1767–1768. Also see Elvehjem, C. A., Madden, R. J., Strong, F. M., and Woolley, D. W. (1938) *J. Biol. Chem.* 123: 137–149.

Franke, K. W. and Franke, W. R. (1934) *J. Lab. Clin. Med.* 19: 669–671.

Hart, E. B., Steenbock, H., Waddell, J., and Elvehjem, C. A. (1928) *J. Biol. Chem.* 77: 797–812.

Koehn, C. J., Jr. and Elvehjem, C. A. (1937) *J. Biol. Chem.* 118: 693–699.

Mueller, J. H. (1937) *J. Biol. Chem.* 120: 219–224.

Phillips, P.H., Stare, F. J., and Elvehjem, C. A. (1934) *J. Biol. Chem.* 106: 41–61.

Potter, V. R. (1960) *Nucleic Acid Outlines,* pp. 10–11, Burgess Publishing Co., Minneapolis.

Stare, F. J. and Elvehjem, C. A. (1933) *Am. J. Physiol.* 105: 655–664.

10

The Discovery of Dicumarol and Its Sequels*

Karl Paul Link†

Department of Biochemistry
University of Wisconsin
Madison, Wisconsin

The story of Dicumarol** has been told several times by me in the past 17 years, and often by others. Like any good story it need not be told in exactly the same manner each time. In Wisconsin it has become a kind of legend. I shall consider only the high water marks of certain chapters.

Fortunately the basic scientific facts on the discovery and development have already been thoroughly recorded (Link, 1943–44; 1945a; 1945b; Shapiro and Weiner, 1949; Anon., 1950; Marple and Wright, 1950) so that little new information on Dicumarol and its sequels needs to be revealed here. However, when I do introduce new material it will be restricted to that which is documented or sustainable via memoranda or letters.

The story begins some 36 years ago on the prairies of North Dakota and in Alberta, Canada. In the 1920's a new malady of cattle involving fatal bleeding showed up almost simultaneously in these areas. The veterinarians, Schofield and Roderick, were forced to conclude that the cause of the disease was neither a pathogenic organism nor a nutritional deficiency. The origin of the new malady was traced to stacks of sweet clover hay mysteriously gone bad. Hence the disease became known in veterinary practice as "sweet clover disease" and it was found that it was caused *only* by improperly cured hay

*Given February 25, 1958 at the New York Academy of Medicine under the auspices of the Section of Medicine and the New York Heart Association, on the program, "The Historical and Physiological Aspects of Anticoagulants." Reprinted from *Circulation* 19: 97–107 (1959).

†Deceased.

**Dicumarol is the trademark for 3,3′-methylene-bis (4-hydroxycoumarin). The anticoagulant was made available in 1940 and 1941 for clinical use by the cooperative efforts of the Wisconsin Alumni Research Foundation, Madison, Wis., the Abbott Laboratories, North Chicago, Ill., Eli Lilly and Company, Indianapolis, Ind., and E. R. Squibb and Company, New Brunswick, N.J. The official U.S.P. name is bishydroxycoumarin.

made from the common varieties of sweet clover. When first observed, this disease was in a sense without parallel in animal pathology or human medicine. When cattle or sheep ate the spoiled hay, the disease slowly became manifest by a progressive diminution in the clotting power of the blood (about 15 days) and resultant internal hemorrhage which usually became fatal in about 30 to 50 days.

It was recognized by Schofield and Roderick that the disease was reversible. It could be controlled in cattle by the withdrawal of the spoiled hay from the diet and by transfusion of blood freshly drawn from normal cattle, provided the hemorrhagic extravasation had not proceeded too far. Indeed, they showed that even in desperate cases, recovery could be hopefully anticipated after transfusion and change in diet (good hay).

In a comprehensive and thorough study of the pathology and physiology of the disease Roderick in 1931 emphasized that the delayed or abolished coagulability of the blood was due to a "prothrombin" deficit. Indeed he showed that the severity of the hemorrhagic condition paralleled the reduction in "prothrombin content or activity." He did this by using the technique developed by that great American pioneer of blood coagulation, the late Professor W. H. Howell. Solutions of what was then called "prothrombin" prepared by precipitation of normal bovine plasma with Howell's acetone method when added to the "sweet clover blood" promoted coagulation. In contrast, preparations of "prothrombin" made in a parallel manner from "sweet clover blood" did not produce coagulation. The other constituents for the maintenance of normal coagulability known at that time (fibrinogen, calcium, platelets, and inhibitory substance) appeared to be unaffected.

I first learned about the hemorrhagic sweet clover disease of cattle in December, 1932, through the late Ross A. Gortner, who then headed the Biochemistry Department of the University of Minnesota. He had offered me a post and I had come to St. Paul to consider it. Since the "sweet clover disease" was also a problem in Minnesota it was one of the projects open for study if I chose to accept. It was Gortner who supplied me with the original publications of Roderick. Some attempts had been made in Gortner's department to extract the hemorrhagic agent but they, like those of Roderick and others, had failed.

Curiously, the "official" start of *our* work in January, 1933, in cooperation with Professor R. A. Brink and W. K. Smith of our Genetics department was on a different aspect of the sweet clover problem. They sought to develop a strain of sweet clover suitable for Wisconsin climatic conditions low in, or free from, coumarin. Though coumarin smells sweet (the characteristic smell of new mown hay is due to its presence) it tastes bitter, and it was known that the bitter taste of green sweet clover plants, *Melilotus alba* and *M. officianalis*, paralleled the total coumarin content. In actual practice it was observable that cattle (or rabbits) would eat the less bitter plants first.

Tragedy out on the Farm

Quite apart from the "official" start, concerned primarily with the palatability question, my laboratory had a direct catalytic hit from agricultural practice.

Indeed, on a Saturday afternoon in February, 1933, following the first conferences with Brink, while a blizzard was howling and the mercury was hovering near zero, a farmer from the vicinity of Deer Park, Wisconsin, some 190 miles from Madison, appeared with what the late Professor A. J. Carlson might have called "the evidence." Curiously the farmer's name was Ed Carlson. The hemorrhagic sweet clover disease of cattle was rampant on his farm. He had fed sweet clover hay for years previously without encountering any difficulties and he doubted the veterinarian's diagnosis. Accordingly, he was advised to go to the Agricultural Experiment Station authorities to get the facts. The office of the State Veterinarian had closed and pure chance had brought him to the Biochemistry Building.

Farmer Carlson's multiple evidence was a dead heifer, a milk can containing blood completely destitute of clotting capacity, and about 100 pounds of spoiled sweet clover—the only hay he had to feed his cattle.

His account of the over-all course of the disease coincided perfectly with the classical "sweet clover poisoning" picture. Late in December he had lost two young heifers. In January, one of his favorite old cows had developed a massive hematoma on a thigh, and following a skin puncture, fatal bleeding set in rapidly. Finally, two young cows had died on Friday and the bull was oozing blood from the nose. So he took off for Madison in a blizzard.

I immediately had to tell farmer Carlson that we could do no more at this time than to recommend the teachings of Roderick and Schofield. He had to stop feeding *that* hay, and possibly transfuse those desperately sick cattle, if he wanted to save them. Eventually it might become possible to make some usable recommendations to avoid such disasters, but not now.

I can still see him take off for home about 4:00 p.m. Those 190 miles of drifted roads between our laboratory and his barn must have appeared to him like a treacherous and somber ocean.

I cannot take the time to tell all the details of this slice of the Dicumarol story, but I can assure you its impact on me was immense. I will relate a part of it exactly as I did in my first lecture on Dicumarol given at the Mayo Clinic on March 12, 1942.

When farmer Carlson came to see us, my senior student and old man Friday was Eugen Wilhelm Schoeffel, a volatile Schwabian who came to the U.S. in 1926 with a diploma in Agricultural Chemistry. After serving a 2-year apprenticeship in the Chicago Stock Yards he began to study with me in 1929. Schoeffel is interesting, energetic, and loyal. He was then and still is, somewhat of a mystic and inclined in ordinary conversation to quote freely from Goethe's *Faust*, Shakespeare, and the Bible, as well as other primary sources. In 1933 his spoken English was not only strongly gutteral, but also very earthy, punctuated frequently with Schwabian German. After farmer Carlson left, Schoeffel stormed back and forth in the laboratory shouting, "Vat da Hell, a farmer shtruggles nearly 200 miles in dis Sau-wetter, driven by a shpectre and den has to go home vit promises dat might come true in five, ten, fifteen years, maybe never. Who knows? 'Get some good hay—transfuse.' Ach!! Gott, how can you do dat ven you haf no money?" he snarled.

He dipped his hands into the milk can repeatedly and while rubbing them muttered, "Dere's no clot in dat blut! BLUT, BLUT, VERFLUCHTES BLUT. 'Die Menschen dauern mich in ihren Jammertagen.' " (*Faust* Prolog. line 297) and then, "Vat will he find ven he gets home? Sicker cows. And ven he and his good voman go to church tomorrow and pray and pray and pray, vat will dey haf on Monday? MORE DEAD COWS!! He has no udder hay to feed—he can't buy any. And if he loses de bull he loses his seed. Mein Gott!! Mein Gott!! Vy didn't ve anti-shi-pate dis? Ya, ve should haf anti-shi-pated dis."

We took the blood and hay and played about with them until about 7:00 p.m. when I headed for home. As I left the laboratory, Schoeffel grabbed me by the shoulders, looked me squarely in the face and said, "Before you go let me tell you something. Der is a desh-tiny dat shapes our ends, it shapes our ends I tell you! I vill clean up and gif you a document on Monday morning."

Development of the Bioassay

Two fundamental issues confronted us. First there were no *chemical* criteria available to establish the presence of the hemorrhagic agent. Therefore, a *bioassay* involving a small experimental animal (rabbits) offered the only practical means of appraising the anticoagulant activity of test hays and extracts prepared therefrom.

It was clear from the pioneer papers of Roderick that the sweet clover disease *was completely reversible*. The eating of spoiled hay, even over long periods, caused no permanent functional change, no demonstrable morphologic change, and no detectable pathologic change of the liver, the assumed primary site of prothrombin synthesis. Nevertheless, the immediate prospects of developing a reliable and simple bioassay were not bright; indeed they were dark, "dark like the inside of a cow." We had not had previous experience with that complex problem—blood coagulation.

Schoeffel and Roberts first showed that the Howell method for estimating prothrombin activity did not have the precision required. Smith and Roberts showed that the whole blood coagulation time was too variable, and that the Quick 1-stage method using whole plasma left much to be desired. Smith also showed that there was a wide variation in the response of individual rabbits to the standard dose of 50 g of the spoiled hay. So Campbell and Smith bred and reared a susceptible rabbit colony specifically for the assay.

At that time, 1935–1938, a bloody and amusing polemic raged among the coagulation specialists on how to estimate "prothrombin concentration or activity"—whether it should be done by the 1-stage method of Quick* or the 2-stage method of H. P. Smith and co-workers (Link, 1943–44; 1945a; 1945b; 1948). We tried to keep out of that brawl. In 1938 Campbell finally got over the chief obstacles. He adapted the Quick 1-stage method to our conditions, primarily by relying on the clottability of diluted plasma within the concentration range 12.5 to 8.34 percent. He eliminated some of the inherent daily variations by fasting the assay rabbit 24 to 36 hours before feeding any preparation under test, by making the plasma clottability tests promptly after drawing the blood, and by comparing the test plasma against the normal plasma of each rabbit.

Through the use of individually standardized rabbits (the standard response being that induced by the anticoagulant in 50 g of spoiled hay) and by having the assay on a strictly differential basis, the everpresent problem of biologic variation was greatly reduced.

Some side observations were made by Campbell on the plasma of rabbits fed the spoiled hay or fractions thereof that were later reported by others. A plasma factor beyond that needed by the classical blood coagulation expression of Morawitz-Field and Spero was hinted at in one of Campbell's reports. But these hares were not hunted. Our goal was to make a substance that abolished the clottability of cattle blood in agricultural practice. To use the vernacular, the bioassay using the 1-stage plasma clottability was altered so that "it worked," and few of the valuable assay rabbits were lost in the process. One of them, known as Bess Campbell, was used for about 200 individual assays, over a period of 5 years.

Isolation, Crystallization, Identification, and Synthesis of Dicumarol

Between that fateful Saturday in February, 1933, and June, 1939, a long and arduous trail was followed by Smith, Roberts, and especially Campbell, to lay the anticoagulant out on the bench. I would like to detail some of the chemical extraction, separation, and isolation problems that the spoiled sweet

*The intricacies of the blood coagulation phenomenon are outside the scope of this discourse. Suffice it to state that it is now accepted by most "coagulationists" that a prolonged Quick 1-stage "prothrombin time" (when the fibrinogen is normal) induced by Dicumarol and the like is a primary deficiency in factor VII and prothrombin. See *British Medical Bulletin*, vol. II, no. 1, Blood Coagulation and Thrombosis, Medical Department, The British Council, London, November (1955), and the lectures by Owren, P. A. on Coagulation of Blood, etc., *Northwest Medicine*, January, pp. 31–39, February, pp. 159–166, and March, pp. 298–307, 1957.

clover hay presented. This hay was indeed a kind of biochemical grab-bag and yielded many inactive products, some new, most of them old. But suffice it to state that many a seething and simmering hope did not become reality. At times the hemorrhagic agent appeared to hover before us like thistle down only to elude us like the will-of-the-wisp. At one time it was thought to be a porphyrin-like substance, a pheophytin resulting from the degradation of the chlorophyll in the spoiling process.

Finally in the dimness of dawn on June 28, 1939, after working all night, Campbell saw on a microscope slide what turned out to be crystalline Dicumarol. Two hours later he had collected about 6.0 mg of it.

When I reached the laboratory that morning Campbell was asleep on the laboratory couch; the door to the room was guarded by one Chet Boyles, a soldier of fortune on the W.P.A. relief roles, who assisted Campbell with the bioassays. Boyles was an excellent handler of animals for he had served two years as helper to a veterinarian before he came to us.

As I walked into the room, Boyles was taking a nip from the contents of a bottle whose bottom layer consisted of carpet tacks, the upper layer of 95 percent ethanol. Without the flicker of an eyelash Boyles said to me, "I'm celebrating, Doc. Campy has hit the jack-pot." (As though I didn't know that he had been hitting that bottle for months.)

But Boyles' surmise was correct this time. Campbell did have Dicumarol and the first bioassay to establish its anticoagulant potency was already in process!

Campbell avoided me for two days—until the results of the assay were available—and then he came in to report.

There is a bedrock of matter-of-fact common sense in Campbell's makeup. He was not inclined to show his emotions, but it was apparent that he was secretly as happy as a boy who had just caught his first big fish. He passed the vial to me and said, "This is H. A.!" (H. A. was the laboratory code for hemorrhagic agent.) I did not disclose that Boyles had given me the tip-off. I told Campbell that I knew a couple of lines of German poetry that fitted the occasion, and I recited to him,

> "So halt 'ich's endlich denn in meinem Händen
> Und nenn' es in gwissen Sinne mein."

We sent a short wire to Schoeffel, who was then in the control laboratory of the American Medical Association in Chicago. He responded at once with a 200-word reply wherein he expressed his *complete* confidence in Nature, Fate, and us.

Mass isolation was started at once, and a stock of about 1,800 mg of the crystalline anticoagulant was accumulated (Stahmann). The problem of determining its structure fell to the sensitive, brilliant, and deft C. F. Huebner, who with some assistance from his lively imagination made the correct structural diagnosis as 3,3′-methylene-bis (4-hydroxycoumarin). He set the sights for the synthesis, which was achieved on April Fool's day, 1940. The synthetic and the natural product were shown to be chemically identical. Subsequently, Overman and Sullivan, through carefully conducted tests on the rabbit, rat, guinea pig, mouse, and dog, hallmarked the natural and synthetic products as biological equals.

The determination of the structure of the anticoagulant as a 3-substituted derivative of 4-hydroxycoumarin makes it appear that both of the undesirable aspects of the common sweet clovers—their unpalatability (bitterness) in the green state and the tendency of the hays to cause hemorrhage when improperly cured—have a common basis in the coumarin molecule. The biological synthesis during spoilage can be rationalized as an oxidation of coumarin to 4-hydroxycoumarin which upon coupling with formaldehyde leads to Dicumarol (Figure 10.1).

Physiologic Action of Dicumarol

After synthetic Dicumarol became available in quantity, the essentials of its physiologic action were quickly established. It was shown that there is a lag in response, a variation in the intensity and duration of the hypoprothrombinemia (plasma prothrombin clotting time), depending on the size of the dose. In each species tested a certain single dose level gives the most efficient response. Below this level the efficiency of action is decreased by a threshold effect and at high levels by incomplete absorption of the drug.

Due to the latent or lag period of 12 to 24 hours before the drug's action becomes apparent, there is a cumulative effect following repeated administration. Thus it was anticipated that in clinical practice this action will vary with the individual and because of this variation optimal therapeutic effects without hemorrhage would be obtained only when the dosage is individualized.

A brief summary (Link, 1943–44; 1945a; 1945b) of the details follows:

1. There is a wide species difference in the response induced in the rabbit, rat, guinea pig, mouse, dog, cat, and chicken, and this varies with the age and sensitivity of each individual. Broadly speaking,

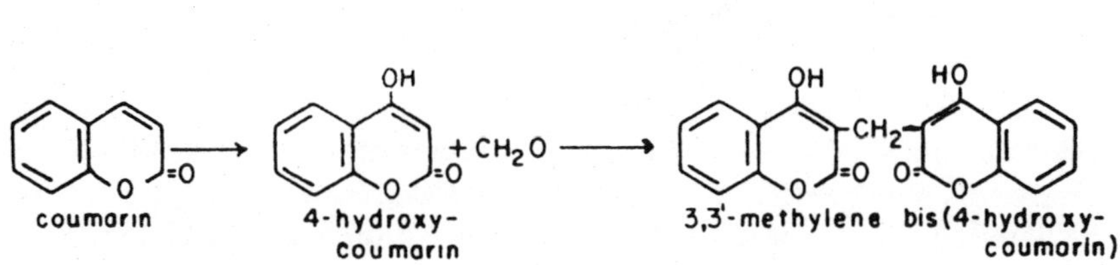

Figure 10.1 The biological synthesis of Dicumarol during spoilage.

the rat and mouse are the most sensitive, and cat and dog intermediate, and the rabbit, the cow, and the chicken the least sensitive.

2. The vitamin K and C levels in the diet affect not only the intensity but also the duration of the anticoagulant action. I propose to elaborate on this later.
3. The nutritional status of the animal affects the anticoagulant response—fasting generally enhances it in all species.
4. Any pre-existing hypoprothrombinemia like that inducible by the salicylates (aspirin), the sulfa drugs, or mild chloroform anesthesia *augmented* the response.
5. The hepatic and renal function influences both the intensity and duration of the response.
6. The presence of drugs that affect the total functioning capacity of the liver, like the methylxanthines (theophyllin) and the digitalis drugs, have a mild but definitely detectable counteraction.
7. Pregnant or lactating females show a slight resistance to the drug's anticoagulant action.

These observations did not exhaust the conditions that can influence Dicumarol's action but they cover the essential points. Finally, it should be added that in Dr. Best's department at Toronto, Dale and Jaques (1942) first and later Meyer and co-workers (Thill et al., 1943) at Wisconsin General Hospital, and others (Richards and Cortell, 1942; 1943; Bollman and Preston, 1942) were able to show that a primary relationship exists between thrombus formation and the clotting mechanism of the blood. These studies established for the first time that an effective reduction of extravascular and intravascular thrombus formation parallels the diminished hypocoagulability induced by Dicumarol. It was also shown by Spooner and Meyer (1944) that, when Dicumarol is given to dogs in safely usable therapeutic doses, it definitely decreases platelet adhesiveness; at the same time Quick showed that it also reduced platelet agglutinability (Boronofsky and Quick, 1943). Thus the clinical use of the anticoagulant as a prophylactic agent for (against) thrombosis rested on a sound experimental basis.

Breaking the Bonds of the Usual Pattern of Thought

When we turned Dicumarol over to the clinicians in the years 1940 to 1942, one significant point, clearly established by our work, was at first missed, in fact denied (Marple and Wright, 1950). I have reference to the capital fact that vitamin K (all forms—some better than others) can counteract the action of Dicumarol.* I emphasized this in letters, personal conversations, and in my first lecture on Dicumarol at the Mayo Clinic and at Wisconsin General Hospital. In spite of these efforts the first clinical reports carried the statement that "vitamin K has no effect as an antidote to the administration of Dicumarol." The editorial and annotation writers for the medical journals, those who only "think" but "don't try," innocently reiterated this statement (Link, 1943–44; 1945a). While in error, the clinicians were in good company, for an authority on blood coagulation (Quick, 1937) had written in 1937, and again in his book published in 1942, that "vitamin K will not restore the prothrombin concentration" depleted by Dicumarol (Quick, 1942).

Originally these denials made me very unhappy. The misfortune of being accused of error was not the primary basis for the unhappiness, for we were certain that the antidotal capacity of vitamin K would in time be sustained in the clinic. What did disturb me was the needless induction of the hemorrhagic "sweet clover disease" in man and the stigma temporarily attached to Dicumarol, that it was a dangerous drug (Biggs and McFarlane, 1949). *And this did happen.*

A feature of science that has always appealed to me is that sooner or later, and usually sooner, "the truth will conquer."

Dr. Shepard Shapiro in New York City was the first clinician (February, 1942) to sustain our claims that vitamin K can counteract the anticoagulant action of Dicumarol in man when liver function is adequate (Shapiro et al., 1943; Shapiro and Weiner, 1949). Subsequently he was independently supported by Townsend and Mills (1942) in Canada, Lehman (1943) in Sweden, and finally by Cromer and Barker (1944) at the Mayo Clinic, as well as others. Today it is accepted that the water-soluble forms of vitamin K or vitamin K_1 given orally can successfully antidote overdosing with Dicumarol, provided they be employed when reversal is still possible.

Let us briefly examine why the error arose. The clinicians did not use a 1-stage prothrombin assay as sensitive as the one Campbell developed for our experimental animals. They were originally conditioned to the low levels of vitamin K effective in obstructive jaundice, biliary fistula, cholemic bleeding, etc. It was also thought that the menadione form of vitamin K might be toxic. Over 10 years were required to wipe out this error from clinical practice.

To summarize, surmise, faulty thinking, and not enough trying kept vi-

*For an account of how, in January, 1939, a bull desperately sick from eating spoiled sweet clover hay (he was "down," blood was oozing from the nose, and a massive hematoma adorned the right thigh), was rescued from the clutches of death via a vitamin K_1 concentrate prepared from alfalfa hay, see Link, 1945a). Originally not even this "bull story" could break the bonds of the usual pattern of thought. The equation is a completely reversible reaction:

$$\text{Normal bull} \xrightleftharpoons[\text{Vitamin K}_1\text{ concentrate}]{\text{Spoiled sweet clover hay}} \text{Bleeding bull}$$

tamin K from being the corner building stone in Dicumarol therapy that it deserved to be from the outset.

In 1950 Marple and Wright (1950, pages 149 and 181) wrote, "When bleeding occurred from the clinical use of Dicumarol the fault rested with the physician who administered the drug."

Enthusiasm—Muddle—Consolidation

Within two years after Dicumarol was synthesized, over 100 related 3-substituted 4-hydroxycoumarins were prepared in my laboratory. Synthesis ran substantially ahead of biochemical appraisal. Accordingly, when I gave the Harvey Society lecture on "The Anticoagulant from Spoiled Sweet Clover Hay" in January 1944 (Link, 1943–44) it was indicated that

> . . . it would not be valid to conclude from the relative appraisals on activity made with the rabbit—that Dicumarol is the most desirable compound for clinical use . . . In the course of the routine appraisal of the many compounds tested it was learned that some of them exhibited a slower but more sustained hypoprothrombinemic action, while the action of others is of shorter duration. It will take some time before final judgment can be passed on this subject. From the experience gained with other pharmacological agents it is abundantly clear that the final test is the action in man under a variety of conditions. The unpredictable can be surprising, so, as we see it, we might now be at the beginning of things and not at the end in this field of study.

Being an agriculturist I have little confidence in predictions, including my own. The situation can now be appraised in the light of wisdom after the event. Bear in mind that the statement quoted was made less than four years after Dicumarol became known to us and before extensive clinical information on the response in man was available. About 50 reports on the clinical use of Dicumarol had appeared between 1941 to 1944 (Shapiro and Weiner, 1949; Marple and Wright, 1950).

The appearance of any new drug creates an interesting cycle of events, and Dicumarol went through that cycle quite rapidly. The first preliminary reports indicated that an atmosphere of optimism prevailed. They evoked prompt favorable editorial comment in the *Lancet* (September 13, 1941) under the title, "Heparin and a Rival." Then came the second period—a period of muddle. Enthusiasts and skeptics for anticoagulant therapy with Dicumarol were created, and it can be stated that some of the skeptics condemned the drug in no uncertain terms, though they were largely armed with surmise, faulty, or no prothrombin clotting time determinations and they used the antidote vitamin K inadequately. Then came the third period of consolidation, from which it can now be concluded that a better anticoagulant of the Dicumarol type was desired.

Since Dr. Wright asked for aspects of human interest, let me add another slice from the Dicumarol story. Early in September, 1945, I was fed up with laboratory work, etc., and I went off on a canoe trip with my family. On this trip we were caught in a cold rain storm. I got soaked and overexhausted. Two weeks later I came down with what I had had once before—after a similar heavy physical bout, as a student in Switzerland—wet pleurisy. At first my doctor thought I had pneumonia; then I told him about my previous bouts of tuberculosis; so the diagnosis was changed to reactived pulmonary tuberculosis. I spent two months at Wisconsin General Hospital and then was transferred to Lakeview Sanatorium headed by the double cross of Lorraine. Here I was supposed to vegetate like a topped carrot. I did rest there, physically, for 6 months, took nothing stronger than cod liver oil and 3 bottles of beer a day, but kept the aged tuberculosis out of my mind by studying laboratory records and reading the history of rodent control from ancient to modern times (Link, 1950; deKruif, 1951).

A "Janus" in the Coumarin Family

Now brace yourselves, for I propose to shift from a "cow poison" that had become a drug of substantial clinical usefulness, to a "rat poison" converted to a drug, which has, I believe, most of the desirable features that can be expected from an anticoagulant to be given primarily via the oral route.

The many coumarins synthesized between 1940 and 1944 were listed by numbers in logical groups based on their chemical structure (Link, 1943–44). While I was in the sanatorium in 1945–1946 the laboratory work was practically at a standstill. There were few students available, since most of them were still in the armed forces. So I had ample time to reexamine all the chemical and bioassay data available. Upon the return of L. D. Scheel from service in the spring of 1946, he was assigned to the task of reappraising the anticoagulant activity of the compounds numbered from 40 to 65. They were made by Ikawa in 1942–43. Instead of using only rabbits for the bioassays Scheel also used rats, mice, and dogs. In 1946–1948 he defined coumarin numbers 42 and 63 as being much more potent than Dicumarol in the rat and dog, as capable of producing a more uniform anticoagulant response, and as having the quality of maintaining a more severe state of hypothrombinemia without inducing visible bleeding. Certain chemical properties were also considered: the degree of purity readily attainable (absence of taste and odor), the cost of making them, and the property of being convertible to stable water-soluble salts.

Back in 1940 to 1942, Overman, Field, and my colleague, C. A. Baumann, had studied extensively the action of Dicumarol in the laboratory rat, and the effect of diet on the response, specifically the influence of vitamin K and foods rich in it. Later in 1942 I personally, with the help of good old Schoeffel, set up field trials to ascertain the suitability of Dicumarol for rodenticidal purpose. It was concluded that the activity of Dicumarol in the rat was not high enough to make it practical for rodent control. This was found to be largely due to the vitamin K content of mature grains and the availability of green foods with a high vitamin K content. It was shown that rats could tolerate a daily intake of 2.0 mg of Dicumarol for 60 or more days due to the vitamin K content of the natural foods available. On a semisynthetic diet essentially free from vitamin K the survival time was about 15 to 23 days. When 5 mg of vitamin K per day were added to the artificial diet, the rats also tolerated 2.0 mg of Dieumarol daily for over 60 days.

Early in 1945 I told Scheel and Dorothy Wu that I wanted to propose no. 42 for rodenticidal use (Link, 1950; deKruif, 1951). This proposal shook the laboratory. I can sum up by stating the consensus of opinion "the boss has really gone off the deep end this time." Scheel favored no. 63 for clinical purposes. They are chemically closely related, no. 63 being a direct derivative of no. 42. To make a long story very short, early in 1948 no. 42 was promoted for rodent control under the auspices of the Wisconsin Alumni Research Foundation through the able, enthusiastic, and public-spirited Ward Ross, General Manager of this organization. Within a short time this effort revolutionized the art of rodent control (multiple doses as opposed to the single dose of the highly toxic poisons), and warfarin rapidly became and still is the leader in the rodenticide field.* The name Warfarin was coined by me by combining the first letters of the Wisconsin Alumni Research Foundation with the "arin" from coumarin—and it is now a household word throughout the world.**

Between 1948 and 1952 Dicumarol was, so to speak, being squeezed by chemical kin stemming primarily from European studies (Koller and Mertz, 1954). "Imitation is the sincerest form of flattery." Curiously, one of them, a derivative of Dicumarol, trade-named Tromexan, was not seriously considered by us as early as 1940 (Figure 10.2). Though it acted somewhat faster than Dicumarol, it required substantially larger doses to get the equivalent anticoagulant action.† The second, Marcumar, a close kin to warfarin, was also

---

*Just as cattle eat hemorrhagic sweet clover hay until they die without visible sensory responses, the rat eats warfarinized cereal grain bait until fatal hemorrhage sets in. Neither bait refusal nor bait shyness develops. Indeed, the rodent's departure is biblical "death without sting." In Maxwell Anderson's drama, Elizabeth the Queen says, "To the end of time it will be so . . . the rats inherit the earth." Since warfarin has become available, this need not be so. Furthermore, via the water-soluble warfarin sodium the rat can drink unto death.

**Warfarin is the safest rodenticide known. Up to now, in the United States there is no recorded case of a warfarin-induced fatality in man, although over 140,000,000 pounds of warfarin-containing bait (0.025 percent) has been distributed since 1950.

†The clinical promotion of Tromexan (bis-3,3′-4-oxycoumarinyl) ethyl acetate, referred to as B.O.E.A. in the article by Burt, C. C., Wright, H. P., and Kubik, M. (1949) *Brit. M. J.* 2: 1250, precipitated interesting editorial comment under the heading, *Dangers of Dicumarol* (pp. 1279–1280). In this editorial it was suggested that since Tromexan seemed to be superior to Dicumarol "owing to its shorter-lived action . . ." and in view of recent reports of the drug's (Dicumarol's) efficiency as a rat poison, it may be that Dicumarol will ultimately be more useful for that purpose."

Unfortunately the significance of our paper on the action of Dicumarol in the rat dealing specifically with the effect of diet and vitamin K on the anticoagulant action (*J. Nutrition* 23: 589–602, 1942) was not appreciated by O'Conner, J. A. (1948) *Research* 1: 334, who suggested the use

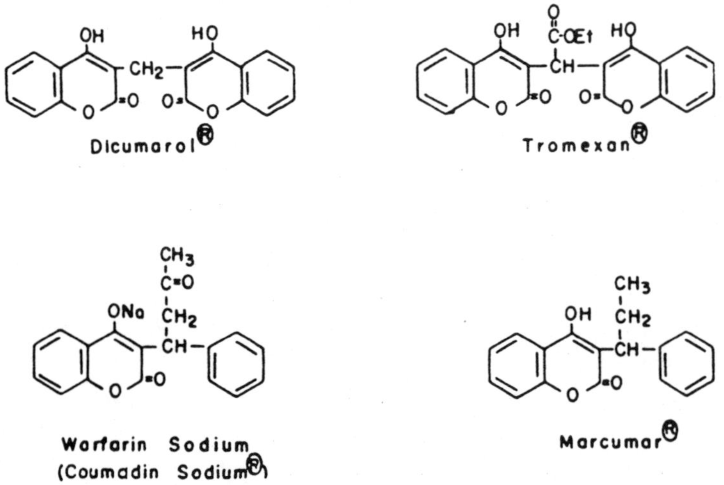

Figure 10.2 Structure of Dicumarol, Tromexan, Warfarin, and Marcumar.

passed by us, since its water-soluble sodium salt is less stable than warfarin sodium. Milligram for milligram, Marcumar is more active than warfarin and its action is also more prolonged. But as a result of the claims made about Tromexan and Marcumar, I took another good look at the mass data in the light of what clinicians were seeking, namely an anticoagulant that could be used via any route with the retention of the virtues of Dicumarol but without its limitations.

Based in part on the supposition that the response of the rat to warfarin (no. 42) was a very reliable index of how humans would respond, late in 1950 I told Dr. S. Shapiro and Dr. O. O. Meyer that the water-soluble sodium salt of warfarin should be tried on humans. In 1941 the clinicians had literally

---

of Dicumarol for rodent control. Had O'Conner read our paper carefully, he would not have made this suggestion. The critical issue is that Dicumarol's anticoagulant action in the rat subsisting on natural grain foods is too slow to be practical. The level of Dicumarol in the bait has to be set so high that other animals (cat, dog) and children (accidental ingestion) would be vulnerable.

It was the inefficiency and slowness of Dicumarol to kill rats under practical field conditions that caused me not to suggest its use as a rodenticide in 1941–1943 (letter, Link, K. P., to the National Defense Research Council, Washington, D.C., dated March 10, 1943, and confidential disclosures, 1942–1943, to the late Professor Homer Adkins and Professor H. Gilman, official investigators and project leaders of N.D.R.C. and O.S.R.D., confirmatory letter of Gilman to Link, June 11, 1952). Instead of Dicumarol the much more potent and efficient (no. 42) warfarin was recommended. Nevertheless, O'Conner's paper served a useful purpose in rodenticide control circles, and he must be accredited with being the first one to stimulate, via the printed page, the backward pest control workers by pointing out the potentials of anticoagulants (Link, K. P., letter December 6, 1948, to U.S.D.I. Fish and Wildlife Service, Denver, Colo.). I had attempted to create an interest in warfarin via letters and memoranda, which at first failed to reach the objective (see Link, 1950, and particularly deKruif, 1951). A complete history of the warfarin development based on 10 years of practical field experience is in the process of being prepared.

snatched the "cow poison" from us, but the transition to a substance originally promoted to exterminate rats and mice was a bit more than they could accept with real enthusiasm. Then on April 5, 1951, we were informed by Captain J. Love (MC) in the U.S.N. at Philadelphia that an army inductee was admitted to the Naval Hospital who had taken over a period of five days a concentrate of warfarin designed for rodent control (Holmes and Love, 1952). The package contained 567 mg of warfarin corn starch. The inductee had followed the multiple dose directions on the package. It became clear to him that warfarin was not an efficient agent "to shuffle off" this "mortal coil." It allowed too much time for thinking—so he went to the hospital with a fully developed case of hemorrhagic "sweet clover disease." He was treated per the directions—blood transfusion and large doses of vitamin K—and made an uneventful recovery (Kellum, 1952).

This incident acted as a catalyst. Shapiro (1953) and Meyer (Clatanoff et al., 1954) both concluded from their carefully done work with warfarin sodium that it did possess certain properties not inherent in Dicumarol or the other anticoagulants they had tried. After Collin Schroeder perfected the process of making warfarin sodium, I induced my long-standing friend, Dr. S. M. Gordon, of the Endo Laboratories, Richmond Hill, N.Y., to make it available for clinical use. This he did, under the trade name Coumadin Sodium. Today it would appear, from the 15 to 20 clinical papers on warfarin sodium that have been published (see Shapiro and Ciferri, 1957), that most of the drawbacks of Dicumarol have been overcome. Warfarin sodium is at least five and possibly ten times more potent than Dicumarol. It is the only synthetic anticoagulant available today for therapeutic anticoagulation that can be given orally, intravenously, intramuscularly, or rectally (Shapiro and Ciferri, 1957.) The rate of absorption is almost the same, irrespective of how it is administered. No other anticoagulant of the Dicumarol type has all these virtues. Of course, an overdosage can be readily corrected via vitamin K. It acts faster than Dicumarol, and fewer prothrombin times are required in its routine use. To use the words of both Shapiro and Meyer, "It is easier to handle clinically." It is my firm belief that in time it will replace Dicumarol on the basis of its performance over a wide variety of conditions and that other anticoagulants of the Dicumarol type will not be superior.

It always seems appropriate to me to visualize successful anticoagulant therapy with the Dicumarol-type drugs as being shaped like a triangle with accurate "prothrombin assays" at one corner, vitamin K at another, and sound clinical judgment at the third (Figure 10.3).

Each corner is linked to the other by way of the connecting sides. There should be no separation; each is vitally dependent on the other two. Though the clinical judgment be good and the "prothrombin time" accurate, the vitamin K corner might still have to be evoked, since each individual patient is essentially "an unstandardized biologic entity," errors in dosage can be

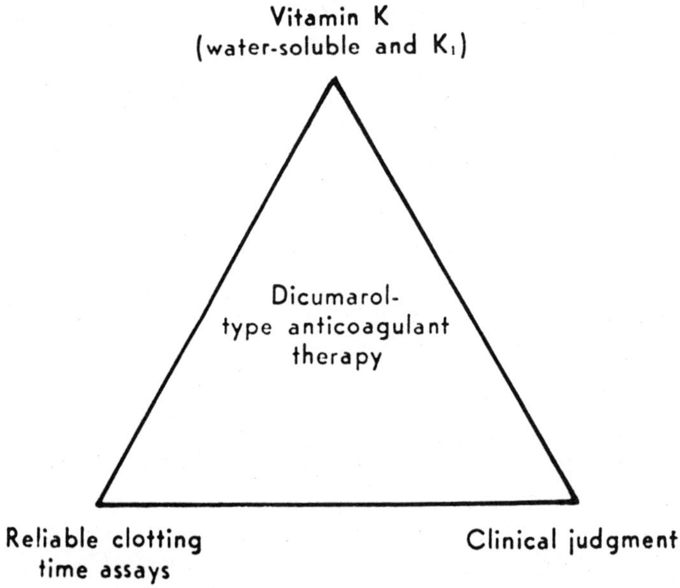

Figure 10.3 The triangle of anticoagulant therapy.

made by the hospital service, the patient might have a silent ulcer, or the functioning of the *liver* or *kidney* might unknowingly be penumbral.

On September 29, 1955, I got a card from a former Wisconsinite working in Fitzsimmons Army Hospital in Denver, Colorado, which read, "The President is getting one of your drugs and it's not Dicumarol." A day later press secretary J. C. Hagerty announced (Hagerty, 1955), "The heparin which was used initially as the anticoagulant has been replaced by a drug of the Dicumarol type. The present prothrombin level has been well maintained." I knew of Colonel Pollock's paper, "Clinical experience with warfarin (coumadin sodium a new anticoagulant)," read before the first annual meeting of the American College of Angiology, Atlantic City, N.J., on June 4, 1955, and I surmised that the most important man in the world today was being anticoagulated via warfarin sodium (Pollock, 1955). This surmise proved to be correct and since then it is an open secret that warfarin sodium was being used. *"The unpredictable can be surprising."*

In closing I wish to indicate that what my laboratory has achieved in the past 2½ decades represents the combined effort of many students. It is fun to be the reporter or narrator of this highly successful adventure. To use the words of the late Allan Gregg (1951), my students represented much "emergent ability." I think the secret of their success is three-pronged: they never ceased to wonder, they kept on trying, and they were on a project directed toward doing mankind *some good* instead of trying to *destroy it*.

References

Anonymous. (1950) New York Heart Association Quarterly, vol. 2, pp. 6–11, New York.

Biggs, R. and McFarlane, R. J. (1949) Estimation of prothrombin in dicoumarin therapy. *Brit. M. J.* 296.

Bollman, J. L. and Preston, F. W. (1942) Effects of experimental administration of dicoumarin 3,3′-methylene-bis (4-hydroxycoumarin). *J. A. M. A.* 120: 1021.

Boronofsky, I. D. and Quick, A. J. (1943) Heparin and agglutination of platelets in vitro. *Proc. Soc. Exper. Biol. and Med.* 53: 173.

Clatanoff, D. V., Triggs, P. O., and Meyer, O. O. (1954) Clinical experience with coumarin anticoagulants Warfarin and Warfarin sodium. *Arch. Inst. Med.* 94: 213.

Cromer, H. E., Jr. and Barker, N. W. (1944) Effect of large doses of menadione bisulfite (synthetic vitamin K) on excessive hypoprothrombinemia induced by dicoumarol. *Proc. Staff Meet. Mayo Clin.* 19: 217.

Dale, D. U. and Jaques, L. B. (1942) Prevention of experimental thrombosis by coumarin. *Canad. M. A. J.* 46: 546.

deKruif, P. (1951) Sure Death to Rats, in *Reader's Digest*, March, 1951, Reader's Digest Association, Pleasantville, New York.

Gregg, A. (1951) Emergent ability. *The Pharos of Alpha Omega Alpha* 15: 3.

Hagerty, J. C. (1955) Associated Press release as published in the *Capital Times*, Madison, WI, September 30, 1955.

Holmes, R. W. and Love, J. (1952) *J. A. M. A.* 148: 935.

Kellum, J. M. (1952) Warfarin for suicide. *J. A. M. A.* 148: 1443.

Koller, T. and Merz, W. R. (1954) Eds., *I. International Conference on Thrombosis and Embolism*. B. Schwabe, Basel, Switzerland.

Lehmann, J. (1943) Thrombosis: Treatment and prevention with methylene-bis-(hydroxycoumarin). *Lancet* 1: 611.

Link, K. P. (1943–44) The anticoagulant from spoiled sweet clover hay. *Harvey Lecture Series* 34: 162.

Link, K. P. (1945a) The anticoagulant Dicumarol. *Proc. Inst. Med. Chicago* 15: 370.

Link, K. P. (1945b) The anticoagulant 3,3′-methylene-bis (4-hydroxycoumarin). *Fed. Proc.* 4: 176.

Link, K. P. (1948) *Dicumarol and the Estimation of Prothrombin. Transactions of the First Conference on Blood Clotting and Allied Problems*, p. 148. Josiah Macy, Jr. Foundation, New York.

Link, K. P. (1950) The Rat. Paper read before the Madison Literary Club, Madison, Wisconsin, January 9, 1950, and filed in the Wisconsin Historial Library, Madison, Wisconsin.

Marple, C. D. and Wright, I. S. (1950) *Thromboembolic Conditions and Their Treatment with Anticoagulants*. Charles C. Thomas, Springfield, IL.

Pollock, B. E. (1955) Clinical experience with Warfarin (coumadin) sodium, a new anticoagulant. *J. A. M. A.* 159: 1094.

Quick, A. J. (1937) Coagulation defect in sweet clover disease and in hemorrhagic chick disease of dietary origin; consideration of source of prothrombin. *Am. J. Physiol.* 118: 260.

Quick, A. J. (1942) *The Hemorrhagic Disease and the Physiology of Hemostatis*. Charles C. Thomas, Springfield, IL.

Richards, R. K. and Cortell, R. (1942) *Proc. Soc. Exper. Biol. and Med.* 50: 237.

Richards, R. K. and Cortell, R. (1943) *Proc. Soc. Exper. Biol. and Med.* 52: 358.

Shapiro, S. (1953) Warfarin sodium derivative (coumadin sodium): Intravenous hypoprothrombinemia-inducing agent. *Angiology* 4: 380.

Shapiro, S. and Ciferri, F. E. (1957) Intramuscular administration of the anticoagulant Warfarin (coumadin) sodium. *J. A. M. A.* 1655: 1377.

Shapiro, S., Redish, M. H., and Campbell, H. A. (1943) *Proc. Soc. Exper. Biol. and Med.* 52: 12.

Shapiro, S. and Weiner, M. (1949) *Coagulation, Thrombosis and Dicumarol.* Brooklyn Medical Press, New York.

Spooner, M. and Meyer, O. O. (1944) Effect of Dicumarol [3,3′-methylene-bis (4-hydroxycoumarin)] on platelet adhesiveness. *Am. J. Physiol.* 142: 279.

Thill, C. J., Stafford, W. T., Spooner, M., and Meyer, O. O. (1943) Hemorrhagic agent 3,3′-methylene-bis (4-hydroxycoumarin); its effect in prevention of experimental thrombosis. *Proc. Soc. Exper. Biol. and Med.* 54: 333.

Townsend, S. R. and Mills, E. S. (1942) Effect of synthetic haemorrhagic agent 3,3′-methylene-bis (4-hydroxycoumarin) in prolonging coagulation and prothrombin time in human subject. *Cand. M. A. J.* 46: 214.

11

Carbohydrate Chemistry and Anticoagulants—an Historical Link

Clinton E. Ballou

Department of Biochemistry
University of California
Berkeley, California

I am pleased to be here at this Centennial Celebration of the Wisconsin Biochemistry Department and to bring you congratulations from my own department at Berkeley. It is appropriate that I should do so for, until recently when Esmond Snell moved from Berkeley to Austin, Texas, four of our 16 faculty were graduates of this department and three of the four have served as chairman at Berkeley.

I have been asked to speak about "carbohydrate chemistry and anticoagulants." When the topic was proposed, I immediately thought of some lines from "The Walrus and the Carpenter" written by Lewis Carroll (Bartlett, 1968).

> "The time has come," the Walrus said, "to talk of many things:
> of shoes—and ships—and sealing wax—of cabbages—and kings—"

"Carbohydrate chemistry and anticoagulants"—are they really as unrelated as "cabbages and kings"? Well, certainly not for a Wisconsin biochemist, for the link between them is a very real and human one, namely the Karl Paul Link.

When we talk of carbohydrate chemistry and anticoagulants, we focus on a period in the history of this department that spans 50 of those 100 years—certainly a highly interesting, exciting and important 50 years in biochemistry. It was in 1922 that Karl Link entered graduate school in Agricultural Chemistry at Madison, it was in 1930 that he became the first faculty member at Madison to hold the title "Professor of Biochemistry," and it was in 1971 that he retired from the university (Behm, 1979).

Link began research in graduate school under the supervision of W. E. Tottingham, a graduate of Hopkins, who joined the Wisconsin faculty about 1910 (Cattell, 1944). Tottingham was interested in "climatic effects on the nutrition and metabolism of plants," and he served as president of the Society of Plant Physiologists in 1931. From 1923 to 1929, Link was concerned in his research mainly with analyzing the constituents of plant tissues, a rather unimaginative project compared to the high-powered research of today, that probably represented the interests of Tottingham more than Link (Link and Tottingham, 1923). During these same years, and unknown to Link, in Canada and North Dakota a new hemorrhagic disease of cattle caused by spoiled sweet clover was being described by Schofield (1924) and Roderick (1929a). They demonstrated that the disease could be reversed by transfusion or by change of diet, and Roderick (1929b) also found that the disease symptoms resulted from a reduction in the blood prothrombin level.

Link says that he became aware of the "sweet clover disease" in 1932, but that he didn't take up the active search for the hemorrhagic agent until 1934. The work moved slowly until Harold Campbell developed a good bioassay, and he obtained the first crystals of the hemorrhagic agent in 1939 (Link, 1959). The isolation, the characterization as 3,3′-methylene-bis(4-hydroxycoumarin), and the synthesis of the hemorrhagic agent were reported simultaneously in 1941 (Campbell et al., 1941; Campbell and Link, 1941; Stahmann et al., 1941). First called dicoumarin, the anticoagulant was given the trade name Dicumarol when it was released for clinical testing in 1941.

(Actually, dicoumarin had been synthesized and reported in the literature by Anschütz in 1903 (Anschütz, 1903). Because 4-hydroxycoumarin has an activated methylene group, it reacts spontaneously with aldehydes to give the bis derivatives. This facile reaction undoubtedly explains dicoumarin formation in spoiled sweet clover, probably as a consequence of microbial growth that leads to oxidation of coumarin to 4-hydroxycoumarin and the reaction of the latter with formaldehyde.)

In 1944, the synthesis and anticoagulant activities of a large number of Dicumarol analogs were reported from the Link laboratory (Overman et al., 1944). In a way, this publication represented the end of an era and the culmination of some 10 years of hard but exciting work by a crew of dedicated graduate students. There was, of course, Harold Campbell, who went on to a career in medical research at the McArdle Laboratory; William Sullivan, who became Director of Research at Hoffmann-La Roche; Charlie Huebner, research director at Ciba Pharmaceuticals; Mike Ikawa, now Professor of Biochemistry at the University of New Hampshire and an expert on fungal toxins; Dave Doherty, a biochemist at Oak Ridge National Laboratory; Ed Popenoe, a biochemist at Brookhaven National Laboratory; and Mark Stahmann, whom you all know as a professor in this department since 1947 (Cattell, 1971). It was Mark Stahmann whose large-scale isolation of Dicumarol was carried out according to Link "with relentless zeal" and made the structural work possible (Link, 1943–1944). It was also Mark who, I remember, helped to hold the laboratory together during those long months when Link's health kept him away from the campus.

I came to the Link laboratory in 1946. It was a bustling place full of hungry ex-GIs who were anxious to complete a doctoral thesis and find a job. My first interview with Link was held at Lake View sanatorium, where he was recovering from an attack of tuberculosis. There we outlined my master's thesis project, which was to develop a method for the direct estimation of Dicumarol in blood so that its metabolism could be studied. The results of my study were not remarkable, but they did produce a thesis, and later Link allowed me to switch to a carbohydrate problem for my doctoral dissertation.

An event occurred in 1948 that stirred the lab to new activity. This was the publication by O'Connor (1948) in England of a study on the use of Dicumarol as a rodenticide. Link claims (Link, 1959), based on a letter to the Research Council and "confidential disclosures" to collaborators, that he had long been thinking of the design of an anticoagulant rodenticide. In what seems today to be a rather ungenerous reference to O'Connor's interesting report, Link commented "the significance of our paper on the action of Dicumarol in the rat dealing specifically with the effect of diet and vitamin K on the anticoagulant action was not appreciated by O'Connor." Link's point was that, although Dicumarol was an effective anticoagulant in the rat, its effect was readily reversed by vitamin K. Of course, Link was also aware that the lab had a large number of Dicumarol analogs on the shelf and that one of these might be much better than Dicumarol for this purpose.

My own recollection of this period is somewhat different, for I remember Link storming into the lab one morning shouting "We can't let them make Dicumarol into a rat poison. The M.D.'s will never touch it again." Then, he set Lester Scheel and Dorothy Wu to work to select one of the analogs that had high anticoagulant activity in the rat and that also enhanced internal hemorrhage by increasing the fragility of the blood capillaries. This search led to compound #42, which Link named Warfarin after discarding all of the suggestions we students in the lab made. It is a fact, as Link has commented, that the rodent-control industry did not take readily to the idea of a slow-acting poison, and he devoted a great deal of time selling the idea. The worldwide success of Warfarin is now well-known. It is ironic, perhaps, that in 1951 Warfarin, under the new name of Coumadin, was adopted for clinical use and was administered to President Eisenhower after his heart attack in 1955 (Link, 1959). Thus, a research project that set out to protect Dicumarol for clinical use led, eventually, to its replacement.

Although Link continued to work and publish in the coumarin field, notably with Collin Schroeder and Mark Hermodson, a new development changed the focus of interest in blood anticoagulants. This started in 1960 with the report by Boyle (1960) of the occurrence of Warfarin-resistant rats, which was followed in 1967 by evidence (Greaves and Ayres, 1967) that the resistance was controlled by a single dominant gene. [Link had written in 1959, when extolling the virtues of Warfarin as a rodenticide, that one no longer need assume that rats would inherit the earth (Link, 1959). In hindsight, we now can see that when you "try to fool Mother Nature" you need to be more clever than most people.] Attention began to center on the Warfarin-vitamin K antagonism as a tool for gaining an understanding of the biochemical function of the vitamin. This attention was well rewarded, and rapid advances followed (Stenflo and Suttie, 1977). Ganrot (Ganrot and Nilèhn, 1968) observed that Dicumarol causes the accumulation in the serum of an abnormal prothrombin-like protein, and Suttie (Shah and Suttie, 1974) found that vitamin K functions to modify this apparent precursor molecule to form active prothrombin.

Clearly, Dicumarol and Warfarin inhibited some vitamin K-dependent reaction by which an inactive prothrombin precursor was converted to the activated molecule. The nature of this activation process was demonstrated by Stenflo (Stenflo et al., 1974) in 1974 to be the γ-carboxylation of glutamate residues in the protein. This new amino acid, γ-carboxyglutamic acid (Gla), was synthesized by Howard Morris in 1975 and its structure was confirmed by mass spectrometry (Morris et al., 1975). In the same year, Suttie demonstrated the *in vitro* vitamin K-dependent carboxylation of the prothrombin precursor (Esmon et al., 1975).

The story does not end here, of course, for there was and still is much to learn about the mechanism of vitamin K. In 1976, Suttie (Sadowski et al., 1976) found that the carboxylation reaction requires reduced vitamin K (vitamin KH_2), and later it was shown by Fasco (Fasco and Principe, 1982) that Warfarin inhibits the vitamin K reductases. Fasco and Suttie (Fasco et al., 1982) have presented evidence that the reductases of Warfarin-resistant rats have a reduced sensitivity to the anticoagulant. Finally, a whole new area of research is opening up in the study of the effect of Warfarin on the synthesis of other carboxyglutamic acid-containing proteins, such as the "bone Gla protein." Price (Price and Sloper, 1983) has shown that Warfarin interferes with mineralization and calcium mobilization in bone, and he has traced this effect to an inhibition of the carboxylation of the bone Gla protein.

I know that it brought a great sense of satisfaction to Karl Link to see some of the secrets of vitamin K function revealed before his death in 1978. It was a subject of continuing interest to him, and he once speculated that vitamin K might function to "activate prothrombin" (Overman et al., 1942). That major advances in this field were and are being made here at Madison in the laboratory of Professor John Suttie is particularly appropriate. I only hope that my attempts to review these developments have not dealt unfairly with any of the participants.

Now I turn to carbohydrate chemistry, the other cylinder that helped to propel the Linkmobile. This interest may have been instilled by Tottingham, as I suggested earlier, but as it grew Link shifted from the analysis of plant constituents (Link, 1929) to the chemistry of hexuronic acids (Link, 1932), and his publications in this area still serve as primary references in the field.

Still, carbohydrate chemistry was and remained a kind of hobby with Link, although it performed as an important vehicle for training graduate students in the rigors of doing good science (Link, 1936). This apprenticeship was the sole training for men such as Carl Niemann (Niemann and Link, 1934), who became Professor of Chemistry at the California Institute of Technology; Sam Morell (Morell and Link, 1935), Professor of Biochemistry, Medical College of Wisconsin; Harold Sell (Sell and Link, 1938), emeritus Professor of Biochemistry, Michigan State University; and Stanford Moore (Moore and Link, 1940), past Professor of Biochemistry, Rockefeller University; each of whom had a distinguished career in other fields (Link, 1943–1944). Clearly, the values instilled by this department and by the Link carbohydrate laboratory served them well.

Did Link's carbohydrate research influence developments in the field outside this University? I think it did (Ballou, 1981). Not one to attend national meetings, Link did regularly go to the Starch Round Table conferences sponsored by the Corn Industries Research Foundation, and his sharp wit and good sense of humor added weight to his scientific opinions. Then, there was the period in the 1940's when several of Link's students (Robert Dimler, Roland Lohmar, and Ivan Wolff, for example) went to staff the Regional Research Laboratories of the U.S. Department of Agriculture where work was heavily concentrated on plant polysaccharides (Link, 1943–1944). Later, during my own time at Madison, I watched with admiration as Link tamed Saul Roseman and molded him into a carbohydrate chemist who would later contribute so imaginatively to glycoconjugate research (Roseman, 1970) and, singlehandedly, be responsible for discovery of the important sugar transport process known as the phosphoenolpyruvate phosphotransferase system (Roseman, 1972). Roseman, now Professor of Biology at Johns Hopkins, has assured perpetuation of the Link tradition by the many excellent young scientists he has trained and sent on to important positions in biochemistry.

For my own part, the interest I gained in carbohydrates at Madison has remained dominant throughout my career. With Link, I learned the complexities (and heartbreaks) of synthetic carbohydrate chemistry as I studied the relationship of a glycon structure to alkali-lability of the glycosidic linkage (Ballou, 1954). Much of my thesis dealt with glycosides of β-hydroxycarbonyl compounds and I am embarrassed to admit that I never had the leap of imagination that could have led me to see the relevance of my studies to the carbohydrate-serine linkage in some glycoproteins. Perhaps the block had something to do with overcoming a prevalent view that most carbohydrate found in proteins was a contaminant. It remained for Elvin Kabat, in a paper on the alkaline-degradation of blood-group glycoproteins published in 1966 (Lloyd et al., 1966), to draw attention to the studies I carried out with Link as being a primary reference for rationalizing his observations.

Link supervised the studies of 43 master's students and of 55 who were awarded the doctorate in Biochemistry. In addition, his lectures on carbohydrate chemistry influenced a whole generation of graduate students at Madison. Although he was not an official Link student, I am certain that Professor Laurens Anderson was strongly influenced by these lectures and that they played a part in his decision to specialize in the organic chemistry of carbohydrates. He has represented this field very ably at Madison and is at the forefront of research on the synthesis of complex oligosaccharides (Nashed and Anderson, 1983).

It has been an unusual privilege for me to have this chance to review 50 years in the life of a great man that paralleled half a century in the life of a great institution. It is a measure of the quality of this Department that it provided Karl Link with the freedom to grow and the facilities to work, and it is a measure of the man that he left a standard in teaching, research and public service that will not often be surpassed.

References

Anschütz, R. (1903) *Ber.* 36: 465.
Ballou, C. E. (1954) *Adv. Carbohydr. Chem.* 9: 59.
Ballou, C. E. (1981) *Adv. Carbohydr. Chem. Biochem.* 39: 1.
Bartlett, J. (1968) in *Bartlett's Familiar Quotations*, 14th edition, p. 745, Little Brown and Co., Boston.
Behm, D. (1979) in *Karl Paul Link: A Biographical Memoir.*
Boyle, C. M. (1960) *Nature* 188: 517.
Campbell, H. A. and Link, K. P. (1941) *J. Biol. Chem.* 138: 21.
Campbell, H. A., Smith, W., Roberts, W. L., and Link, K. P. (1941) *J. Biol. Chem.* 138: 1
Cattell, J. (1944) *American Men of Science*, 7th edition, The Science Press, Lancaster, Pennsylvania.
Cattell, J. (1971) *American Men and Women of Science*, Cattell-Bowker Press, New York.
Esmon, D. T., Sadowski, J. A., and Suttie, J. W. (1975) *J. Biol. Chem.* 250: 4744.
Fasco, M. J., Hildebrandt, E. F., and Suttie, J. W. (1982) *J. Biol. Chem.* 257: 11210.
Fasco, M. J. and Principe, L. M. (1982) *J. Biol. Chem.* 257: 4894.
Ganrot, P. O. and Nilèhn, J.-E. (1968) *Scand. J. Clin. Lab. Invest.* 22: 23.
Greaves, J. H. and Ayres, P. (1967) *Nature* 215: 877.
Link, K. P. (1929) *J. Am. Chem. Soc.* 51: 2516.
Link, K. P. (1932) *Science* 28: 386.
Link, K. P. (1936) in his horse-radish and vinegar speech, presented at research discussion hour, October 16, in his biochemistry research laboratory.
Link, K. P. (1943–44) *The Harvey Lecture Series* 39: 162.
Link, K. P. (1959) *Circulation* 19: 97.
Link, K. P. and Tottingham, W. E. (1923) *J. Am. Chem. Soc.* 45: 439.
Lloyd, K. O., Kabat, E. A., Layng, E. J., and Gruezo, F. (1966) *Biochemistry* 5: 1489.
Moore, S. and Link, K. P. (1940) *J. Biol. Chem.* 133: 293.
Morell, S. and Link, K. P. (1935) *J. Biol. Chem.* 108: 763.
Morris, H. R., Thompson, R. M., and Dell, A. (1975) *Biochem. Biophys. Res. Commun.* 62: 856.
Nashed, M. A. and Anderson, L. (1983) *Carbohydr. Res.* 114: 43.
Niemann, C. and Link, K. P. (1934) *J. Biol. Chem.* 104: 743.

O'Connor, J. A. (1948) *Research (London)* 1: 334.

Overman, R. S., Field, J. B., Baumann, C. B., and Link, K. P. (1942) *J. Nutrition* 23: 589.

Overman, R. S., Stahmann, M. A., Huebner, C. F., Sullivan, W. R., Spero, L., Doherty, D. G., Ikawa, M., Graf, L., Roseman, S., and Link, K. P. (1944) *J. Biol. Chem.* 153: 5.

Price, P. A. and Sloper, S. A. (1983) *J. Biol. Chem.* 258: 6004.

Roderick, L. M. (1929a) *J. Am. Vet. Med. Assn.* 74: 314.

Roderick, L. M. (1929b) *Am. J. Physiol.* 96: 413.

Roseman, S. (1970) *Chem. Phys. Lipids* 5: 270.

Roseman, S. (1972) *Metabolic Pathways*, p. 41, Vol. 4, Academic Press, New York.

Sadowski, J. A., Esmon, C. T., and Suttie, J. W. (1976) *J. Biol. Chem.* 251: 2770.

Schofield, F. W. (1924) *J. Am. Vet. Med. Assn.* 64: 553.

Sell, H. and Link, K. P. (1938) *J. Am. Chem. Soc.* 60: 1813.

Shah, D. V. and Suttie, J. W. (1974) *Biochem. Biophys. Res. Commun.* 60: 1397.

Stahmann, M. A., Heubner, C. F., and Link, K. P. (1941) *J. Biol. Chem.* 138: 513.

Stenflo, J., Fernlund, P., Egan, W., and Roepstorff, P. (1974) *Proc. Natl. Acad. Sci. USA* 71: 2730.

Stenflo, J. and Suttie, J. W. (1977) *Annu. Rev. Biochem.* 46: 157.

12

Blood Plasma Proteins: A Unique Mixture

Robert E. Feeney

Department of Food Science and Technology
University of California
Davis, California

In its 100th year in existence, The Department of Biochemistry at the University of Wisconsin, Madison, stands forth in excellence for its many accomplishments, particularly those accomplishments related to highly important substances in biological materials. Biochemistry at Wisconsin has ranged from a study of function of vitamins and coenzymes to the composition and properties of biological fluids, as well as to the study of helpful and harmful substances in plant and animal materials. When one thinks of blood and Wisconsin, a universally accepted tribute is inevitably given to the studies of Professors Karl Paul Link, Mark A. Stahmann and co-workers on Dicumarol and eventually on the warfarins, which started over four decades ago. Indeed, these two names, the dicumarols and their derivatives and the warfarins, are almost household words throughout the world. But this is only a small part of the contributions of Wisconsin.

The Department of Biochemistry at Wisconsin has excelled in the training of biochemists as well as in doing fine research. My major professor was Frank Morgan Strong, who had obtained his Ph.D. in organic chemistry and supplemented this with postdoctoral work on natural products chemistry in the European laboratories of Paul Karrer and Fritz Kögl. When Frank took me as his first Ph.D. student in 1938, he had just been appointed as an assistant professor and the Depression was strongly in progress. Consequently, we initially suffered from a near absence of supplies, equipment and space (my *laboratory* for the first two years was a section of a bench in the basement teaching laboratory). But Frank more than compensated for these deficiencies in teaching me how to improvise by modifying procedures, by blowing glass and by constructing equipment. He gave me breadth of training by splitting my research between the development of microbial vitamin assays and the characterization of the structure of a mold pigment by microchemical techniques (World War II caused this research to remain unpublished). He instilled such a strong regard for methodology and the study of natural products that most of my career has been spent in such studies (Feeney and Allison, 1969; Means and Feeney, 1971). It was his training for independent work and his teachings on improvisation that supplied a valuable background for later researches where facilities were greatly restricted (Feeney, 1947; 1974a; 1982).

Blood—A Unique Mixture

The history of the chemistry and the biochemistry of blood reads like the history of the inception and progress of biochemistry itself. This intense interest has not only been due to the obvious overall importance of blood, or even to the halo of mystery early associated with it, but perhaps more to the early recognition of its unique properties. The following is a list of some of the major functions or properties of the plasma and plasma proteins:

1. The transport of cellular constituents such as erythrocytes.
2. Transport of many organic and inorganic compounds such as lipids, iron, copper, endocrines, and vitamins.
3. Prevention of toxic reactions by formation of complexes with undesirable concentrations of substances. Here the serum albumin serves as a biochemical buffer, but other substances such as the haptoglobins are highly important for forming specific complexes, such as with hemoglobin. Still others, like the transferrins, have a main function of iron transport, but also prevent free iron from being toxic.
4. Defensive actions. The most notable substances here are the antibodies found in the γ-globulin fractions.
5. Probably the most dramatic property of blood plasma is its sophisticated system of clotting and dissolution of clots once formed. Perhaps as many as two dozen components are, or can be, involved in these reactions.
6. Blood plasma, of course, must maintain the pH within a narrow range, a viscosity within narrow limits, and the osmotic pressures within a definite range. It is this last requirement, the maintenance of an osmotic pressure within an acceptable range, that is responsible for polar fish that live within the ice-laden areas of normal-salinity ocean waters having antifreeze proteins in their blood (Feeney, 1974b; Feeney and Yeh, 1978). The proteins function additively with the colligatively acting substances (which affect the osmotic pressure) but function noncolligatively, and therefore do not adversely increase the osmotic pressure.

Properties of Some of the Principal Blood Plasma Proteins

A very abbreviated summation of properties of some of the human blood plasma proteins is listed in Table 12.1. Although it is now a few years old, the series *The Plasma Proteins,* edited by F. W. Putnam (1975), should be consulted for general information. In addition to the proteins listed in Table 12.1, there are a myriad of proteins and enzymes and polypeptides present in relatively small amounts, many of which vary with the physiology or the pathology of the individual.

Some major Wisconsin contributions Wisconsin's contributions to the understanding of the properties of blood started many years ago. Starting from the nutritional side, Conrad A. Elvehjem and Edwin B. Hart reported on values of iron and copper for proper levels of hemoglobin (Elvehjem, 1935; Hart and Elvehjem, 1936). In the last half a century many notable and creditable advances have been made, either coming directly from Wisconsin or based on initial studies there. The major area of their contributions on

Table 12.1 Some major proteins of human blood plasma*

Protein	Approx. conc. (mg/100 ml)	Function
Albumin	3500–4500	Osmotic regulation, transport of fatty acid, anions, bilirubin, etc.
α_1-Globulins		
Lipoprotein	300–800	Transport of triacylglycerols, steroids, phospholipids
α_1-Antitrypsin	200–400	Inhibits proteinases
α_2-Globulins		
Lipoprotein	200	Triacylglycerol transport
Ceruloplasmin	15–60	Copper transport
Haptoglobulins	100–300	Transport of hemoglobin from destroyed red cells
Prothrombin	10	Precursor of thrombin
β-Globulins		
β_1-Lipoprotein	400–1000	Lipid transport
Transferrin	200–320	Iron transport
Plasminogen	30	Precursor of fibrinolysin
Fibrinogen	300	Blood clotting
α_2-Macroglobulin	150–400	Inhibits proteinases
Complement components	10–150	Complement fixation
γ-Globulins	700–1500	Antibodies
Blood group globulins	800	Antigen-antibody reactions

*Adapted in part from Putnam (1975).

proteins has been related to blood clotting, but in addition, several workers have pursued outstanding careers in proteins other than those related to blood clotting. For the sake of brevity, only two of those nonblood-clotting researchers will be mentioned.

Amongst the several names of Wisconsin chemists and biochemists studying the nonblood-clotting proteins is the name of Harold F. Deutsch of the Department of Physiological Chemistry of the Medical School. A partial listing of Deutsch's earlier distinguished publications (from 1946 to 1965) includes: separation and purification of a new globulin (Deutsch et al., 1946a); recovery of γ-globulin (Deutsch et al., 1946b); immunological properties of γ_1-globulin (Deutsch et al., 1947); animal γ-globulin (Hess and Deutsch, 1948); relation of Bence-Jones proteins to normal serum proteins (Deutsch et al., 1955); crystalline macroglobulin (γ_1) (Kratochvil and Deutsch, 1956); preparation of crystalline ceruloplasmin (Deutsch, 1960); and preparation of crystalline human superoxide dismutase (Stansell and Deutsch, 1965).

Deutsch was also one of the first individuals to see the value of comparative biochemistry in protein chemistry and its use as a tool to study structure-function relationships. During this earlier period he consequently extended his work to a series of in-depth programs on chicken egg white proteins and comparative studies on other egg white proteins. Most of these have been reviewed elsewhere (Feeney and Allison, 1969). Today it is well recognized that there are important homologies between some of the blood plasma proteins, the egg white proteins, and milk proteins, and the pioneer studies of Deutsch stimulated such studies by many laboratories.

The elegant studies of Oliver Smithies on the genetics of human transferrins and haptoglobins came a little later (Smithies, 1959; Smithies et al., 1962). His studies bridged the theretofore serious void between biochemistry and genetics. In addition, his development and exploitation of the technique of starch gel electrophoresis was a boon to many areas of biochemistry and genetics.

Wisconsin's Famous History on the Study of Blood Clotting

Wisconsin's ascendancy to stardom in blood clotting research had its beginning with the studies of Karl Paul Link and his students and colleagues approximately half a century ago. This, of course, was initially the isolation, identification and structural characterization of Dicumarol (Figure 12.1) (Peterson and Strong, 1953) and the soon-following now world-widely used anticoagulant rodenticide warfarin (Figure 12.2) (Peterson and Strong, 1953). These and other derivatives developed as anticoagulants for human use have had one of the greatest impacts on human health of our times.

A few years later the studies of Professor John D. Ferry of the Department

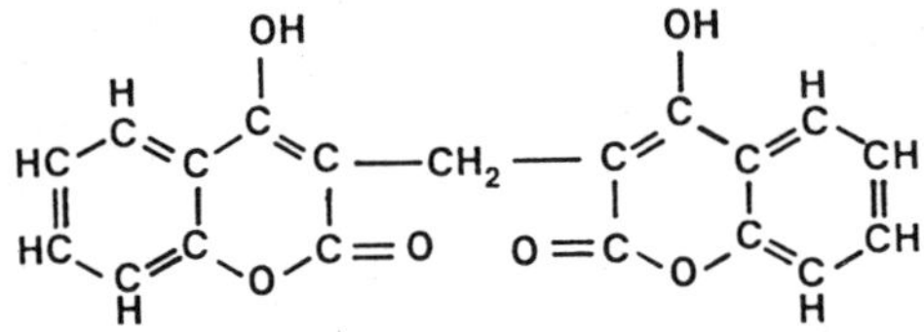

Figure 12.1 The structure of Dicumarol. From Peterson and Strong (1953).

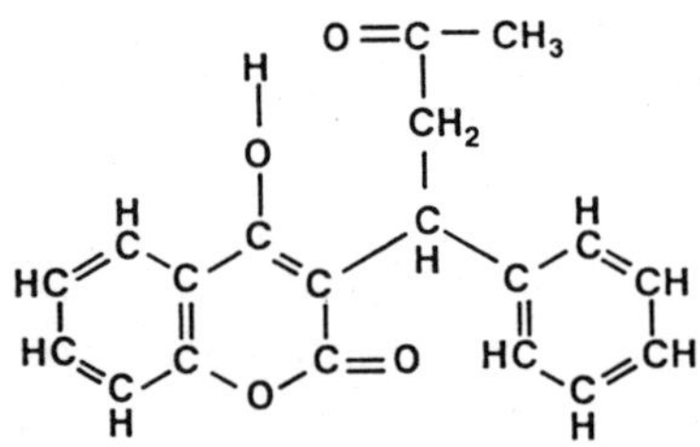

Figure 12.2 The structure of "warfarin." From Peterson and Strong (1953).

of Chemistry on the physical chemistry of the polymerization of fibrinogen began to appear. It was due to the careful physical chemical studies occurring during the polymerization of fibrinogen to fibrin that the groundwork was laid for the eventual characterizations of the enzymatic processes and the final characterization of the nature of the cross-linkages. In a review of his major works (Ferry, 1954), Ferry stated:

> Our task is now to try to discover why the altered molecule has the faculty of polymerizing in such a highly specific manner. There is a variety of evidence that the polymerization itself involves no further changes in configuration or shape; the thrombin-activated fibrinogen molecule is simply a building block for construction of the network strands which compose the fibrin clot. What, then, is the geometrical pattern of construction, and what are the forces that guide the blocks into the proper alignment and hold them in place?

Ferry's laboratory attacked the fibrinogen to fibrin problem with most of the physical techniques then available. One of the initial problems he studied was the remarkable property occurring at pH 6 in which the intermediate polymers formed in the presence of hexamethylene glycol dissociated when the solution was diluted, indicating a reversible equilibrium with rather weak binding (Ferry et al., 1953). In other attempts to characterize the conversion sequence, a study was made of the effect of pH at different ionic strengths on clotting times (Shulman and Ferry, 1950) (Figure 12.3).

During the last decade the name of John W. Suttie, Professor, Department of Biochemistry, University of Wisconsin, has been at the top of most lists of

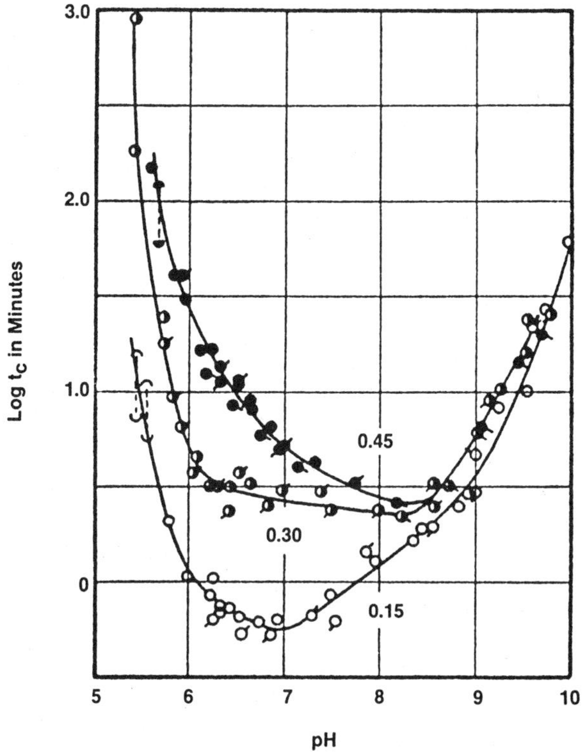

Figure 12.3 Logarithm of clotting time plotted against pH, for bovine fibrinogen (5 gm/l) and thrombin (1 unit/ml), at three values of ionic strength as indicated. From Shulman and Ferry (1950).

significant references related to the enzymatic conversions involved in the blood clotting sequence. These studies have involved vitamin K-dependent enzymes as well as the mechanisms of action of inhibitors such as the coumarins and warfarin. Suttie has summarized much of this program in a current review article, "Synthesis and Secretion of Vitamin K-Dependent Clotting Proteins" (Suttie, 1983). Much of Suttie's work has been concerned with the vitamin K-dependent clotting factors in coagulation, and these are outlined in Figure 12.4 (Suttie, 1983). All of these vitamin K-dependent clotting factors are now known to contain γ-carboxyglutamic acid (Gla), and there are now at least six or seven of these known (Suttie, 1983). The main interest in Suttie's laboratory has been in the biosynthesis of these factors as well as having a good hand in identifying the presence of Gla in some of them. Figure 12.5 (Suttie, 1983) summarizes Suttie's current concepts of the role of vitamin K and its inhibitors in the synthesis of Gla in the vitamin K-dependent proteins.

One of the more recent contributions of the Suttie laboratory has been

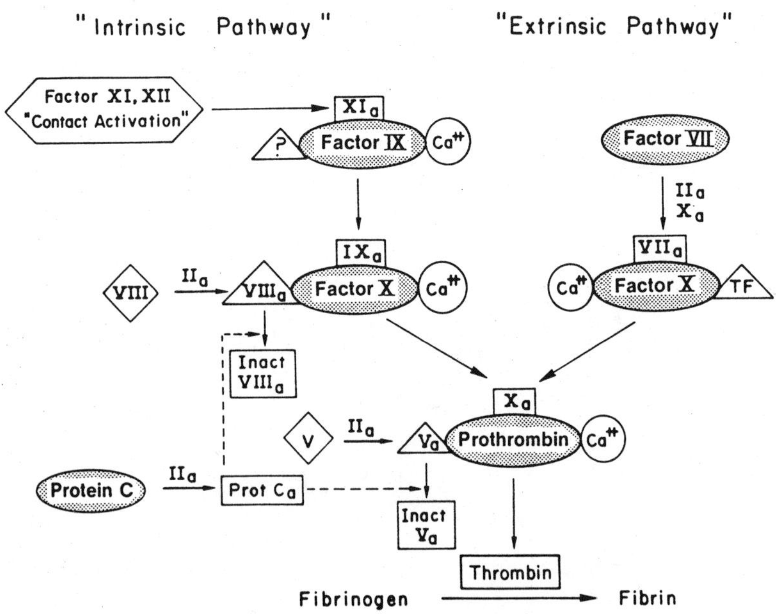

Figure 12.4 Involvement of vitamin K-dependent clotting factors in coagulation. From Suttie (1983).

the study of the vitamin K epoxide reductase from rat liver microsomes (Hildebrandt and Suttie, 1982; Fasco et al., 1983). Suttie's group has used the coumarin-type inhibitors both as a tool in studying the mechanisms of biosynthesis as well as in understanding the mechanism of these important inhibitors. Recently they have studied the relative effects of the anticoagulants difenacoum and warfarin on different enzymes involved in the vitamin K synthesis system using warfarin-resistant and normal strains of rats. They found that the only enzyme that demonstrated a significant differential inhibition by the two coumarins in warfarin-resistant strains of rats was the vitamin K epoxide reductase (Hildebrandt and Suttie, 1982). Carrying these studies still further (Fasco et al., 1983), they have demonstrated the formation of hydroxyvitamin K [2(3)-hydroxy-2,3-dihydro-2-methyl,3-phytyl-1,4-naphthoquinone] by vitamin K epoxide reductase of warfarin-resistant rats. This metabolite was not found as a product in hepatic microsomes in normal rats. The hydroxyvitamin K is neither a substrate nor an inhibitor of epoxide reduction. The authors go on to say, "although there is as yet no direct evidence that coumarin anticoagulants bind to the active site of vitamin K epoxide reductase, formation of hydroxyvitamin K may well be mechanistically related to warfarin resistance."

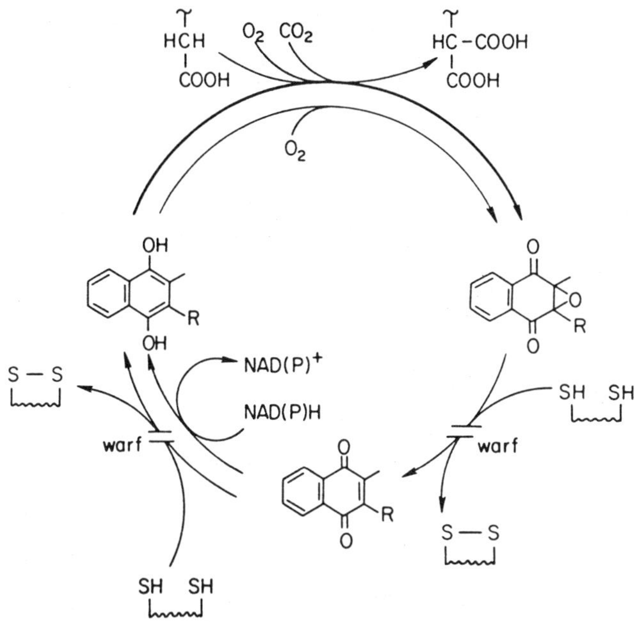

Figure 12.5 Vitamin K-related metabolic activities in rat liver microsomes. The pathways crossed and indicated "warf" are those that are sensitive to the action of the coumarin anticoagulants. The figure indicates that vitamin K epoxide is reduced by a warfarin-sensitive pathway "vitamin K epoxide reductase" that uses dithiothreitol (DTT) as a reducing agent, and that the quinone form of the vitamin can be reduced to the hydroquinone by either a warfarin-sensitive DTT-driven pathway or by a nicotinamide nucleotide-linked dehydrogenase. From Suttie (1983).

Antifreeze Glycoproteins from Polar Fish Blood Sera

Antifreeze substances such as salts and low molecular weight organic compounds like those found in certain insects, are commonly found in the fluids of most biological systems. Higher animals, however, have difficulties in adding large amounts of salt or organic compounds to blood. An adjustment would be to add something that does not affect the osmotic pressure and yet significantly influences the freezing process. Certain polar fishes and northern fishes appear to have achieved this adaptation by having in their blood proteins that function noncolligatively and lower the freezing temperature without affecting the melting temperature (Scholander et al., 1957; DeVries and Wohlschlag, 1969; DeVries et al., 1970; Feeney, 1974a; Yeh and Feeney, 1978; Feeney and Yeh, 1978; Feeney et al., 1980). Approximately one-third of the lowering of the freezing temperature in the blood sera of several Antarctic fish, such as *Pagothenia borchgrevinski*, and the named antifreeze glycoproteins (AFGP) and are numbered 1–8 (AFGP 1–8) according to their electrophoretic mobilities in gels. The mobility differences are almost entirely due to differences in size and not to structure. AFGP 1–5, with molecular weights of 32,000 to 10,500, have similar antifreeze activities and differ only by the length of the polypeptide chain, which consists of the repeating tripeptide unit Ala-Ala-Thr, with threonine glycosidically linked to the disaccharide galactosyl-N-acetyl-galactosamine (Figure 12.6). The shorter chain components (AFGP 6–8) are very similar in structure but contain proline and have low antifreeze activity when tested alone (Table 12.2). For an extensive review of the older literature and some general theoretical discussions see Feeney and Yeh (1978).

While working on penguin proteins in Antarctica in 1964, I observed the studies of Donald E. Wohlschlag on the general metabolism of fish living among the ice crystals in Antarctic waters at less than −1.8°C (Wohlschlag, 1964). Our studies on antifreeze proteins stemmed from these observations, and the next year we initiated a biochemical study of the muscle proteins (Greene and Feeney, 1970) and blood plasma proteins (Komatsu et al., 1970b) of Antarctic fish. Later, Per Scholander at Scripps Institute of Oceanography told me about his earlier studies showing that northern polar fishes had substances in the blood that abnormally lowered its freezing temperature (Scholander et al., 1953, 1957; Gordon et al., 1962), and thus my interests were entrenched. When Arthur L. DeVries joined our laboratory in 1968 for three years of postdoctoral study, the AFGP became the most prominent area of study in my group. DeVries, a former assistant to Wohlschlag at Stanford,

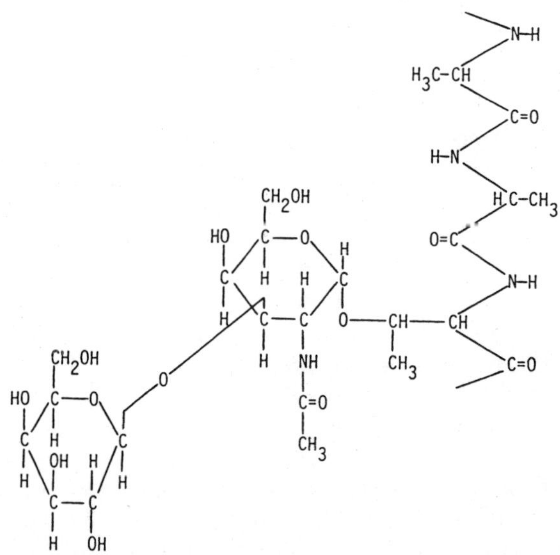

Figure 12.6 Polymer unit of the active antifreeze glycoprotein. From Feeney et al. (1981).

Table 12.2 Properties of different components of antifreeze glycoproteins (AFGP)[a]

AFGP component	Relative amount in serum (%)	Approximate number of glycotripeptides[b] (Ala-Ala-Thr)	Approximate molecular weight (g)	Relative activity
1	1	?	?	Strong
2		35	25,000	Strong
3	2	30	21,000	Strong
4	4	24	17,000	Strong
5	4	16	11,000	Strong
6	4	7–10[c]	6,000[d]	Weak[e]
7	85	5[c,f]	3,288	Weak[e]
8		4[c,f]	2,646	Weak[e]

[a]Adapted from Feeney et al. (1982).
[b]Complete sequence has two additional terminal residues, usually Ala.
[c]The small AFGP 6 to 8 have Pro substituting for some Ala following the Thr.
[d]Average of several components.
[e]The small AFGP 6 to 8 have approximately 25% the antifreeze activity of the large AFGP 1–5 on a weight basis. AFGP 8 is approximately 5% as active as the large AFGP 4 or AFGP 3 on a molar basis.
[f]Based on sequence.

had worked on the AFGP for his thesis and completed a preliminary chemical characterization as a part of our laboratory's program (DeVries and Wohlschlag, 1969). More definitive characterization soon followed (DeVries et al., 1970; Komatsu et al., 1970a; DeVries et al., 1971).

Physical and chemical properties of antifreeze glycoproteins (AFGP)
Most of the studies have been done with the AFGP that are strong antifreeze agents. Nearly identical structures have been found for the antifreeze in *P. borchgrevinki* and *Dissostichus mawsoni*, fishes from the southern polar regions, and in *B. saida*, from the northern polar regions.

The effect of increases in concentrations of *P. borchgrevinki* AFGP and comparisons with the effects of other substances show that AFGP is nearly twice as active as sodium chloride at concentrations of 3–4 mg/ml (Figure 12.7). On a molar basis, it is therefore about 500 times more effective in lowering the freezing temperature than calculations from its molecular weight would suggest. The activity is additive with that of sodium chloride in solutions of their mixtures. At higher AFGP concentrations the plots tend to plateau.

Many structural studies have been done on the effects of chemical modifications and degradations of the different AFGP fractions (Feeney and Yeh, 1978). Acetylation of 30% of the total hydroxyl groups caused greater than

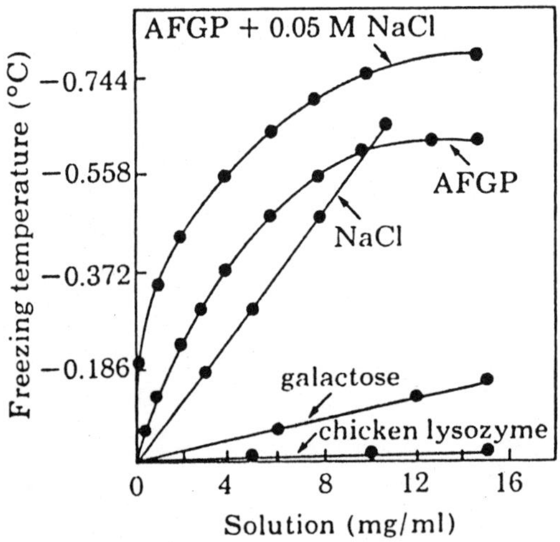

Figure 12.7 Freezing temperatures of solutions of NaCl, galactose, lysozyme, and a mixture of antifreeze glycoproteins (AFGP 3, 4 and 5 from *P. borchgrevinki*), and a mixture of glycoprotein and NaCl. Adapted from DeVries et al. (1970).

80% inactivation, while deacetylation with hydroxylamine gave a complete recovery of activity. Oxidation with periodate destroyed activity, while oxidation of the C-6 carboxyl group of the sugar side chains to aldehydes caused no loss of activity. Formation of complexes of this C-6 aldehyde with sulfite or further oxidation of the aldehyde groups to carboxyl groups, however, destroyed the activity. The borate complexes of the native AFGP were likewise inactive. Modifications of the small AFGP polypeptide AFGP-8 (with only 14 amino acids) by methylation or acetylation of the lone amino (terminal) group or amidation of the lone carboxyl (terminal) group with ethanolamine caused no losses in activities (Geoghegan et al., 1980). With Edman degradations, however, even the removal of a single amino-terminal alanine resulted in losses in activity. On synthesis of a fraction like the larger, more active, AFGP-5 (16 triglycopeptides) by joining two of the smaller fractions of AFGP-6 (7 glycotripeptides), the activity of the dimer was only similar to that of the original monomer on a weight basis (Burcham et al., 1984).

Physical measurements have been by viscosity, ultracentrifugation, osmotic measurements, circular dichroism, Raman, laser and nuclear magnetic resonance spectroscopy (Feeney and Yeh, 1978; Franks and Morris, 1978; Berman et al., 1980; Bush et al., 1981). There are definite discrepancies, but a more recent interpretation is that the structure may have a 3-fold left-handed helix as based on circular dichroism experiments (Bush et al., 1981).

A different type of antifreeze protein (AFP) has been found in the Newfoundland winter flounder (*Pseudopleuronectes americanus*) (Duman and DeVries, 1976; Ananthanarayanan and Hew, 1977). This is a nonglycoprotein but it has the same high percentage of alanine as do the antifreeze glycoproteins (i.e., nearly two-thirds of the amino acids are alanine). This nonglycoprotein antifreeze, however, only has about 40% of the activity as the AFGP. Mixtures of AFGP and AFP function additively, indicating a similar function (Osuga et al., 1980). Other proteins with abnormal effects on lowering the freezing temperature have also been reported. Those from the beetle larvae, *Tenebrio molitor* (Schneppenheim and Theede, 1979) and the sea raven (a fish), *Hemitripterus americanus* (Slaughter et al., 1981), have much lower contents of alanine. Carbohydrate thus does not appear essential, although it is helpful, for antifreeze function. A possible requirement is hydrophobic and hydrophilic areas.

Freezing characteristics Several of the more important characteristics related to antifreeze activities of the glycoproteins are given in the following list:

1. Depresses the freezing temperature without lowering the melting point.
2. Does not lower the melting point but gives an easily obtainable sharp melting point as obtained with H_2O.
3. Depresses the freezing temperature greater than 500 times as based on molecular weight.
4. Depresses the freezing temperature additively with salt.
5. There is no evidence for interaction with H_2O.
6. Ice is normal by x-ray.
7. Is easily entrapped in bulk ice phase.
8. Functions in presence of ice.
9. Growing edge of ice crystals is disrupted.
10. When the activities of the smaller AFGP 6–8 are tested by observing inhibition of growth of ice crystals, these smaller AFGP 6–8 are inactive with ice made by nucleating H_2O at $-6°C$; with ice made at $-1°C$ they are active. The larger AFGP 1–5 function with either ice.
11. A cooperative potentiation is observed in that when some of the larger AFGP 1–5 is mixed with the smaller AFGP 6–8, the smaller ones now function with $-6°C$ ice.

Many of these observations have been previously reviewed (Yeh and Feeney, 1978; Feeney and Yeh, 1978). The melting and freezing characteristics have been recently described by Knauf et al. (1982) and the ice-crystal effects by Burcham et al. (1982, 1983) (Table 12.3).

Table 12.3 Effect of ice crystalline structure on AFGP activity*

	Freezing temperature of solution in the presence of ice made at:	
Sample	−1°C	−6°C
AFGP 1–4 (4 mg/ml)	−0.39°C ± 0.02	−0.36°C ± 0.03
AFGP 7–8 (10 mg/ml)	−30°C ± 0.01	−0.05°C ± 0.02
AFGP 1–4 (4 mg/ml) plus AFGP 7–8 (10 mg/ml)	−0.51°C ± 0.02	−0.51°C ± 0.02

*From Burcham et al. (1982).

Mechanism of antifreeze function Extensive discussions and presentations of possible mechanisms have been presented in several reviews (Yeh and Feeney, 1978; Feeney and Yeh, 1978; Ananthanarayanan and Hew, 1980; Raymond and DeVries, 1977). Currently there appears to be reasonable agreement that the mechanism does not involve interaction with the bulk H_2O or with the ice crystal structure once formed, but more likely involves a function at the ice-solution interfacial region. Indeed, a suggestion was once made that the function could include the formation of structures resembling clathrates on the surfaces of ice crystals (and developing ice nuclei) (Vandenheede et al., 1972). Studies in our laboratories are currently projected toward examining possible interactions at the surfaces of growing ice crystals and at the interfacial regions between the ice and the solution. Thermodynamic, kinetic, and statistical mechanical aspects have been reported (Feeney and Yeh, 1978; Yeh and Feeney, 1978).

A Wisconsin ending With glycoproteins as large as these and existing in as many polymeric sizes, there is always the possibility of some structural characteristics being unnoticed. In fact, the very monotony of the structure has detracted from confidence in its characterization. For this reason continued studies on the structures are necessary, but a very important advance would be the chemical synthesis of glycopeptides with antifreeze activity. Although a synthesis of an active polypeptide has not been achieved as yet, a major advance toward the synthesis has been achieved by the laboratory of Professor Laurens Anderson (Anderson and Anisuzzaman, 1983) of the Biochemistry Department, University of Wisconsin. The synthesis of a glycotripeptide unit is shown in Figure 12.8.

Thus, the contributions of Wisconsin biochemistry to blood-protein chemistry are continuing and have been extended to include this inhibitor of ice-crystal development.

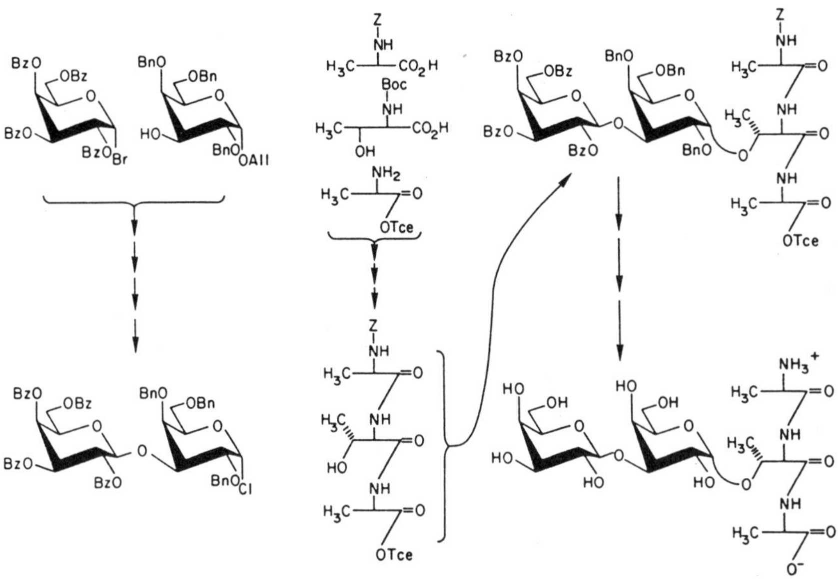

All = allyl Bn = benzyl Boc = tert-butyloxycarbonyl

Bz = benzoyl Tce = 2,2,2-trichloroethyl Z = benzyloxycarbonyl

Figure 12.8 Synthesis of the Gal-Gal analog of the repeating unit of AFGP from derivatives of D-galactose, L-alanine, and L-threonine. The number of arrows at each stage shows the number of chemical steps required for that stage. [Figure supplied by L. Anderson. From presentation at American Chemical Society Meeting, Washington, D.C., Aug. 28–Sept. 2, 1983 (Anderson and Anisuzzaman, 1983).]

References

Ananthanarayanan, V. S. and Hew, C. L. (1980) Structural studies on a freezing-point-depressing protein and its synthetic analogue, in *Biomolecular Structure, Conformation, Function and Evolution* (Srinivasan, R., ed.), Vol. 2, pp. 191–198. Pergamon Press, New York.

Ananthanarayanan, V. S. and Hew, C. L. (1977) Structural studies on the freezing-point-depressing protein of the winter flounder, *Pseudopleuronectes americanus. Biochem. Biophys. Res. Commun.* 74: 685–689.

Anderson, L. and Anisuzzaman, A. K. M. (1983) Synthesis of a close analog of the repeating unit of the antifreeze glycoproteins of Antarctic fish. Paper presented at the 186th National Meeting of the American Chemical Society, Division of Carbohydrate Chemistry, Aug. 27–Sept. 2, 1983, Washington, D.C.

Berman, E., Allerhand, A., and DeVries, A. L. (1980) Natural abundance carbon 13 nuclear magnetic resonance spectroscopy of antifreeze glycoproteins. *J. Biol. Chem.* 255: 4407–4410.

Burcham, T. S., Knauf, M. J., Osuga, D. T., Feeney, R. E., and Yeh, Y. (1984) Antifreeze glycoproteins: influence of polymer length and ice crystal habit on activity. *Biopolymers* 23: 1379–1395.

Burcham, T. S., Osuga, D. T., Yeh, Y., and Feeney, R. E. (1982) Antifreeze glycoprotein activity as a function of ice crystalline habit. *Cryo-Letters* 3: 221–226.

Bush, C. A., Feeney, R. E., Osuga, D. T., Ralapati, S., and Yeh, Y. (1981) Antifreeze glycoprotein. Conformational model based on vacuum ultraviolet circular dichroism data. *Int. J. Peptide Protein Res.* 17: 125–129.

Deutsch, H. F. (1960) The preparation of crystalline ceruloplasmin from human plasma. *Arch. Biochem. Biophys.* 89: 225–229.

Deutsch, H. F., Alberty, R. A., and Gosting, L. J. (1946a) Biophysical studies of blood plasma proteins. IV. Separation and purification of a new globulin from normal human plasma. *J. Biol. Chem.* 165: 21–35.

Deutsch, H. F., Alberty, R. A., Gosting, L. J., and Williams, J. W. (1947) Biophysical studies of blood plasma proteins. VI. Immunological properties of γ_1-globulin from the plasma of normal humans. *J. Immunol.* 56: 183–194.

Deutsch, H. F., Gosting, L. J., Alberty, R. A., and Williams, J. W. (1946b) Biophysical studies of blood plasma proteins. III. Recovery of γ-globulin from human blood protein mixtures. *J. Biol. Chem.* 164: 109–118.

Deutsch, H. F., Kratochvil, C. H., and Reif, A. E. (1955) Immunochemical relation of Bence-Jones proteins to normal serum proteins. *J. Biol. Chem.* 216: 103–111.

DeVries, A. L., Komatsu, S. K., and Feeney, R. E. (1970) Chemical and physical properties of freezing point-depressing glycoproteins from Antarctic fishes. *J. Biol. Chem.* 245: 2901–2908.

DeVries, A. L., Vandenheede, J., and Feeney, R. E. (1971) Primary structure of freezing point-depressing glycoproteins. *J. Biol. Chem.* 246: 305–308.

DeVries, A. L. and Wohlschlag, D. E. (1969) Freezing resistance in some Antarctic fishes. *Science* 163: 1073–1075.

Duman, J. G. and DeVries, A. L. (1976) Isolation, characterization and physical properties of protein antifreezes from the winter flounder, *Pseudopleuronectes americanus. Comp. Biochem. Physiol.* 54B: 375–380.

Elvehjem, C. A. (1935) The biological significance of copper and its relation to iron metabolism. *Physiol. Rev.* 15: 471–507.

Fasco, M. J., Preusch, P. C., Hildebrandt, E., and Suttie, J. W. (1983) Formation of hydroxyvitamin K by vitamin K epoxide reductase of warfarin-resistant rats. *J. Biol. Chem.* 258: 4372–4380.

Feeney, R. E. (1947) The isolation and identification of *Clostridia* in the field. *The Military Surgeon* 101: 50–56.

Feeney, R. E. (1974a) *Professor on the Ice.* Pacific Portals, Davis, California.

Feeney, R. E. (1974b) A biological antifreeze. A glycoprotein in the blood of polar fishes lowers the freezing temperature. *American Scientist* 62: 712–719.

Feeney, R. E. (1982) Penguin egg-white and polar fish blood-serum proteins. *Int. J. Peptide Protein Res.* 19: 215–232.

Feeney, R. E. and Allison, R. G. (1969) *Evolutionary Biochemistry of Proteins: Homologous and Analogous Proteins from Avian Egg Whites, Blood Sera, Milk, and Other Substances.* John Wiley and Sons, New York.

Feeney, R. E., Osuga, D. T., Ahmed, A. I., and Yeh, Y. (1981) Antifreeze proteins from fish bloods. Relationship of the function to structure, solvent, and freezing conditions. *Lebensm.-Wiss. u.-Technol.* 14: 171–175.

Feeney, R. E., Osuga, D. T., Reid, D. S., and Yeh, Y. (1982) Techniques for studying the mechanism of action of antifreeze proteins. *Protein, Nucleic Acid and Enzyme* 27: 1645–1657.

Feeney, R. E., Osuga, D. T., and Yeh, Y. (1980) Anomalous depression of the freezing temperature by blood-serum proteins of fishes, in *Proteins at Low Temperatures* (Fennema, O., ed.), Adv. Chem. Series 180, pp. 83–107. American Chemical Society, Washington, D.C.

Feeney, R. E. and Yeh, Y. (1978) Antifreeze proteins from fish bloods. *Adv. Protein Chem.* 32: 191–282.

Ferry, J. D. (1954) Polymerization of fibrinogen. *Physiol. Reviews* 34: 753–760.

Ferry, J. D., Shulman, S., Gutfreund, K., and Katz, S. (1953) The conversion of fibrinogen to fibrin. XI. Light scattering studies on clotting systems inhibited by hexamethylene glycol. *J. Amer. Chem. Soc.* 74: 5709–5715.

Franks, F. and Morris, E. R. (1978) Blood glycoprotein from Antarctic fish. Possible conformational origin of antifreeze activity. *Biochim. Biophys. Acta* 540: 346–356.

Geoghegan, K. F., Osuga, D. T., Ahmed, A. I., Yeh, Y., and Feeney, R. E. (1980) Antifreeze glycoproteins from polar fish. Structural requirements for function of glycopeptide 8. *J. Biol. Chem.* 255: 663–667.

Gordon, M. S., Amdur, B. H., and Scholander, P. F. (1962) Freezing resistance in some Northern fishes. *Biol. Bull.* 122: 52–62.

Greene, F. C. and Feeney, R. E. (1970) Properties of muscle glyceraldehyde-3-phosphate dehydrogenase from the cold-adapted Antarctic fish *Dissostichus mawsoni. Biochim. Biophys. Acta* 220: 430–442.

Hart, E. B. and Elvehjem, C. A. (1936) Mineral metabolism. *Annu. Rev. Biochem.* 5: 271–294.

Hess, E. L. and Deutsch, H. F. (1948) Biophysical studies of blood plasma proteins. VIII. Separation and properties of the γ-globulins of the sera of normal cows. *J. Amer. Chem. Soc.* 70: 84–88.

Hildebrandt, E. F. and Suttie, J. W. (1982) Mechanism of coumarin action: sensitivity of vitamin K metabolizing enzymes of normal and warfarin-resistant rat liver. *Biochemistry* 21: 2406–2411.

Knauf, M. J., Burcham, T. S., Osuga, D. T., Ahmed, A. I., and Feeney, R. E. (1982) Comparisons of the freezing and melting characteristics of antifreeze glycoproteins and other substances. *Cryo-Letters* 3: 221–226.

Komatsu, S. K., DeVries, A. L., and Feeney, R. E. (1970a) Studies of the structure of freezing point-depressing glycoproteins from an Antarctic fish. *J. Biol. Chem.* 245: 2909–2913.

Komatsu, S. K., Miller, H. T., DeVries, A. L., Osuga, D. T., and Feeney, R. E. (1970b) Blood plasma proteins of cold-adapted Antarctic fishes. *Comp. Biochem. Physiol.* 32: 519–527.

Kratochvil, C. H. and Deutsch, H. F. (1956) A crystalline macroglobulin from human serum. *J. Biol. Chem.* 222: 31–37.

Means, G. E. and Feeney, R. E. (1971) *Chemical Modification of Proteins.* Holden-Day, San Francisco.

Osuga, D. T., Feeney, R. E., Yeh, Y., and Hew, C.-L. (1980) Co-functional activities of two different antifreeze proteins: the antifreeze glycoprotein from polar fish and the nonglycoprotein from a Newfoundland fish. *Comp. Biochem. Physiol.* 65B: 403–406.

Peterson, W. H. and Strong, F. M. (1953) *General Biochemistry.* Prentice-Hall, New York.

Putnam, F., ed. (1975) *The Plasma Proteins,* 2nd ed., Vol. I. Academic Press, New York.

Raymond, J. A. and DeVries, A. L. (1977) Adsorption inhibition as a mechanism of freezing resistance in polar fishes. *Proc. Natl. Acad. Sci. USA* 74: 2589–2593.

Schneppenheim, R. and Theede, H. (1979) New results on occurrence and function of antifreeze proteins, in *Animals and Environmental Fitness* (Abstracts of the First European Society of Comparative Physiology and Biochemistry Conference) (Gilles, R., ed.), pp. 97–98. Pergamon Press, New York.

Scholander, P. F., Flagg, W., Walters, V., and Irving, L. (1953) Climatic adaptation in Arctic and tropic poikilotherms. *Physiol. Zool.* 26: 67–92.

Scholander, P. F., Van Dam, L., Kanwisher, J. W., Hammel, H. T., and Gordon, M. S.

(1957) Supercooling and osmoregulation in Arctic fish. *J. Cell Comp. Physiol.* 49: 5–24.

Shulman, S. and Ferry, J. D. (1950) The conversion of fibrinogen to fibrin. II. Influence of pH and ionic strength on clotting time and clot opacity. *J. Phys. Colloid Chem.* 54: 66–79.

Slaughter, D., Fletcher, G. L., Ananthanarayanan, V. S., and Hew, C. L. (1981) Antifreeze proteins from sea raven, *Hemitripterus americanus*. Further evidence for diversity among fish polypeptide antifreezes. *J. Biol. Chem.* 256: 2022–2026.

Smithies, O. (1959) Zone electrophoresis in starch gels and its application to studies of serum proteins. *Adv. Protein Chem.* 14: 65–113.

Smithies, O., Connel, G. E., and Dixon, G. H. (1962) Chromosomal rearrangements and the evolution of haptoglobulin genes. *Nature* 196: 232–236.

Stansell, M. J. and Deutsch, H. F. (1965) Preparation of crystalline erythrocuprein and catalase from human erythrocytes. *J. Biol. Chem.* 240: 4299–4305.

Suttie, J. W. (1983) Synthesis and secretion of vitamin K-dependent clotting proteins, in *Plasma Protein Secretion by the Liver* (Glaumann, H., Peters, T., Jr., and Redman, C., eds.). Academic Press, New York.

Vandenheede, J. R., Ahmed, A. I., and Feeney, R. E. (1972) Structure and role of carbohydrate in freezing point-depressing glycoproteins from an Antarctic fish. *J. Biol. Chem.* 247: 7885–7889.

Wohlschlag, D. E. (1964) *Antarctic Res. Series* 1: 33–62.

Yeh, Y. and Feeney, R. E. (1978) Anomalous depression of the freezing temperature in a biological system. *Acct. Chem. Res.* 11: 129–135.

13

One Hundred Years of Research on Photorespiration and Biological Nitrogen Fixation at Wisconsin

Israel Zelitch
Connecticut Agricultural Experiment Station
New Haven, Connecticut

Agricultural Chemistry and the Experiment Station Movement in the Nineteenth Century

Before discussing these scientific subjects it seems appropriate to first outline how Experiment Stations were started in America and to say something about the scientific and historical events that led to the appointment of a Professor of Agricultural Chemistry at the Wisconsin Agricultural Experiment Station in 1883. I shall therefore comment briefly about some interesting people and events that made such an appointment possible, and then present a perspective of the scientific knowledge and controversies that introduced biochemistry research to Wisconsin. Later I will summarize some research achievements in photorespiration and biological nitrogen fixation during the past one hundred years attributable to work at Wisconsin.

After 1830 in the eastern United States there was an overall decline of agriculture because of poor soil and the emergence of western competition (Rossiter, 1975). It is therefore perhaps not surprising that the movement in America that led to the establishment of the first Experiment Station was centered in New Haven, Connecticut. The idea was imported from Germany by American students who had observed there the *Versuchsstationen* ("experiment station"). Eastern farmers had to specialize in certain crops and adopt new techniques, especially the use of chemical fertilizers, and many of these being sold were of dubious quality. Agricultural chemistry in this country in the middle of the last century became the professional speciality of a very small group of men; of these, John Pitkin Norton (1822–1852) (Gillispie, 1981) and Samuel W. Johnson (1830–1909) (Osborne, 1911) were the most active, and they were centered around Yale University in New Haven.

In 1840 Justus Liebig (Figure 13.1), then in Giessen, Germany, had published his book *Organic Chemistry in Its Applications to Agriculture and Physiology* in which he tried to explain for the first time the scientific basis of certain farming practices. This book had a tremendous impact in Europe and the United States because it came from a respected organic chemist. With Wöhler he had discovered the benzoyl radical in 1832, and had extended the scope of his interests to include a new perspective on agricultural chemistry based on speculations concerning soil analysis.

John Pitkin Norton, son of a Connecticut farmer, studied chemistry with Benjamin Silliman at Yale in 1840–1841, and then went to Scotland to study with James F. W. Johnston. There he carried out an analysis of the chemical changes in five varieties of oats during development, and on trips to the continent met the famous European agricultural chemists of his time. As his studies were finishing abroad, two professorships in agricultural chemistry were established at Yale College in 1846. One was held by Benjamin Silliman, Jr. and the other was given to Norton. These posts were established "for the purpose of practical instruction in the applications of science to the arts and agriculture" (Jenkins, 1925). While abroad Norton had acquired misgivings about Liebig's views of the role of soil analysis. He lectured on agricultural chemistry and vegetable and animal physiology, and gave 33 lectures in a

Figure 13.1 Portrait of Justus Liebig.

course that included such subjects as the chemical constituents of plants, soil chemistry, and nutrition. He also began to lecture on agricultural chemistry among the farmers of Connecticut and to train teachers for the new agricultural colleges, but the political and intellectual climate was such that the Connecticut legislature was indifferent to the idea of "book farming" (Rossiter, 1975). He was also involved in soil and fertilizer analysis, but his brilliant career came to a close in 1852 when he died at the age of 30.

Samuel W. Johnson was born in 1830 and raised on a farm in upstate New York (Figure 13.2). He studied with both Norton and Liebig, and more than any other person in America developed the new rationale for scientific agriculture in the mid-1850's that led it beyond Liebig with his soil analysis and into the twentieth century (Rossiter, 1975). Johnson entered Yale in 1850 to study chemistry after meeting and being greatly impressed by Norton. In 1851 as a college sophomore Johnson wrote an article for the *Albany Cultivator* proposing that "farmers set up an agricultural institute . . . to afford greater facilities for experimental agriculture . . . and to provide men and means for striking out into the path of discovery, for increasing as well as diffusing knowledge. . . ." (Horsfall, 1969). He left for further study in Germany in 1853, spending his first year with Erdmann, the editor of *Journal für praktische*

Figure 13.2 Portrait of Samuel W. Johnson.

Chemie, where he visited the first *Versuchsstation* that had two well qualified chemists on its staff. He then worked with Liebig, who had recently moved to Munich. Liebig was so prominent in his day that Johnson was able to purchase a bust of him in Munich (Figure 13.3). Johnson returned to Yale in 1855 as Professor of Agricultural Chemistry. There he lectured and wrote about the importance of applying to the art of agriculture the teachings of natural science and the need to establish an experiment station to carry out such investigations. In the 1870's he led the drive that culminated in the creation of the nation's first agricultural experiment station in Connecticut in 1875, the institution that has been located primarily in New Haven since 1877, the year he became its Director. This Station is still functioning according to Johnson's concepts (Figure 13.4).

Johnson is probably best known for his two books (Figures 13.5 and 13.6): *How Crops Grow. A Treatise on the Chemical Composition, Structure, and Life of the Plant* (1868, second edition 1890), and *How Crops Feed. A Treatise on the Atmosphere and the Soil* (1870). The latter book is essentially a text on plant physiology. Both books resulted from his courses at Yale in agricultural chemistry and plant physiology. They became the standard textbooks in agricultural colleges for the next forty years (Rossiter, 1975) and were translated into German, Russian, Italian, Japanese, and Swedish. J. B. Lawes, founder of the Experimental Station in Rothamsted, England, wrote Johnson in 1870 after receiving a copy of *How Crops Grow* that . . . "it is by far the best summary of information on this subject which has yet been published . . ." (Osborne, 1913). I will later use these books and what Johnson has to say about photosynthesis, respiration, and nitrogen metabolism in plants, because they pro-

Figure 13.3 Bust of Liebig purchased by Samuel W. Johnson in Munich in 1854.

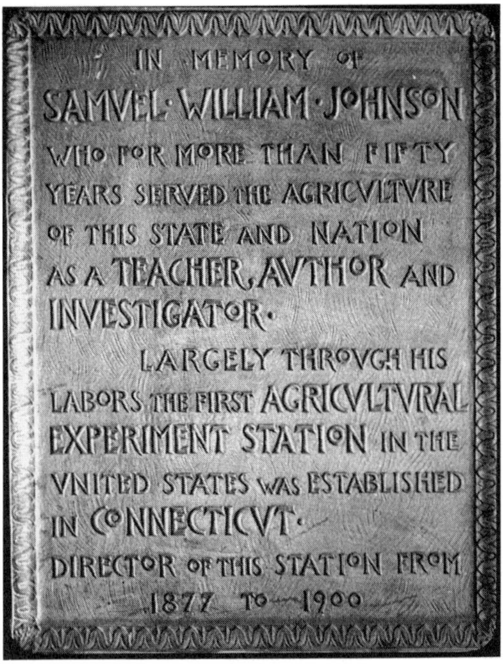

Figure 13.4 Bronze plaque in The Johnson Laboratory at the Connecticut Agricultural Experiment Station, New Haven.

vide a scholarly and vividly written analysis of what was known and argued about 100 years ago. Since his books were so respected and widely read, they provide a basis for comparison with more recent work. On a personal note, I am pleased to say that I became the proud owner of each of these famous books by Samuel Johnson signed by "the author" when I quickly retrieved them from a pile of surplus agricultural books that were discarded by the Yale University library 15 years ago.

In 1877, six years before the start of the Wisconsin Station, the staff of the Connecticut Agricultural Experiment Station consisted of Samuel W. Johnson (Director) and two chemists: Edward H. Jenkins (who became Director when Johnson retired in 1900) and Henry Prentiss Armsby. Armsby (1853–1921) (Ivins and Winship, 1925) was born in Massachusetts, and studied chemistry with Johnson at Yale. In 1876, together with his friend Edward H. Jenkins of Yale, he went to the University of Leipzig, Germany, for further study. There he had contact with professors at the University as well as the director of the *Versuchsstation* at Möckern, near Leipzig, where digestion experiments with cattle were being conducted (Ivins and Winship, 1925). Soon after returning to the United States he was appointed chemist at Johnson's new Connecticut Agricultural Experiment Station in New Haven, where he

HOW CROPS GROW.

A TREATISE ON THE

CHEMICAL COMPOSITION, STRUCTURE,
AND LIFE OF THE PLANT,

FOR

ALL STUDENTS OF AGRICULTURE.

WITH NUMEROUS ILLUSTRATIONS AND TABLES OF ANALYSES.

BY

SAMUEL W. JOHNSON, M. A.,

PROFESSOR OF ANALYTICAL AND AGRICULTURAL CHEMISTRY IN THE SHEFFIELD SCIENTIFIC SCHOOL OF YALE COLLEGE; CHEMIST TO THE CONNECTICUT STATE AGRICULTURAL SOCIETY; MEMBER OF THE NATIONAL ACADEMY OF SCIENCES.

NEW YORK:

ORANGE JUDD & COMPANY,

245 BROADWAY.

Figure 13.5 Title page of *How Crops Grow* (1868).

remained from 1877–1881. While there he began a translation of Wolff's *Landwirtschaftliche Fütterungslehre,* but it evolved into a new book when it appeared in 1880 under the title *Manual of Cattle Feeding*. In 1879, he and Jenkins received their Ph.D.'s from Yale University. Armsby appeared frequently before scientific organizations and farm groups and lectured on the value of silage and other topics of interest to farmers. In 1881–1883 he was professor of Agricultural Chemistry at Storrs Agricultural School, now the University of Connecticut, and in 1883 he became Professor of Agricultural Chemistry at the University of Wisconsin and Associate Director of the new Agricultural Experiment Station. It is this appointment of Armsby that we now celebrate 100 years later. Indeed, the present Department of Biochemistry was called "Agricultural Chemistry" until the 1940's. Armsby left Wisconsin in 1877 to become Director of the Pennsylvania Agricultural Experiment Sta-

HOW CROPS FEED.

A TREATISE ON THE

ATMOSPHERE AND THE SOIL

AS RELATED TO THE

Nutrition of Agricultural Plants.

WITH ILLUSTRATIONS.

BY

SAMUEL W. JOHNSON, M.A.,

PROFESSOR OF ANALYTICAL AND AGRICULTURAL CHEMISTRY IN THE SHEFFIELD SCIENTIFIC SCHOOL OF YALE COLLEGE; CHEMIST TO THE CONNECTICUT STATE AGRICULTURAL SOCIETY; MEMBER OF THE NATIONAL ACADEMY OF SCIENCES.

NEW YORK:
ORANGE JUDD AND COMPANY,
245 BROADWAY.

Figure 13.6 Title page of *How Crops Feed* (1870).

tion, where he later became famous for his digestion studies with farm animals, including the construction of a calorimeter large enough to accommodate mature cattle. In this respect his career was not unusual, because Yale and the Connecticut Agricultural Experiment Station continued to be the training grounds of a number of distinguished scientists who staffed the Experiment Stations throughout the United States, including E. H. Farrington in Dairy Husbandry at Wisconsin.

The Agricultural Experiment Station was organized at the University of Wisconsin on October 1, 1883, as experiment stations were being organized in a number of states. The First Annual Report out of Madison at the end of 1883 is most interesting because it covered only a period of three months and was therefore long in philosophy if sparse in research results. The introduction, by Hiram Smith, Chairman of the Farm Committee of the Board of Regents, describes his pride in the three professors who staffed the Station: W. A. Henry, Agriculture; W. Trelease, Botany and Horticulture; and H. P. Armsby, Agricultural Chemistry. Smith expressed optimism about the knowledge to be gained at the new Station and its importance to the farmers of Wisconsin:

> The knowledge here gained will be of vastly more importance to the citizens of this state in the next decade than to learn how to manipulate the puts and calls and option deals of grain gamblers and stock jobbers (that bring such periodical financial ruin to the country), or the highest attainments of criminal lawyers, whose greatest ambition too often seems to be to corrupt juries and befool judges with technicalities considered more sacred than justice or the rights of the community.

The First Annual Report (1883) contains a statement by the Officers of the Station that although the Station is in a sense the successor of the experimental farm,

> . . . its sole purpose is the discovery of new truths and laws which may be of benefit to agriculture. . . . The Station is provided with a chemical laboratory, a botanical laboratory, and an herbarium, which together with its office are situated in South Hall on the University grounds. In addition, the University farm affords the Station opportunities for the conduct of field, feeding and dairy experiments, supplying all necessary materials, space and labor. *By this union of farm and laboratory facilities, the Station is enabled to do far more useful work than it could with either alone* (italics mine).

This statement made in Wisconsin, emphasizing the opportunities as well as the concern for both the theoretical and the practical and having them worked at in close proximity, comes directly from the teachings of Samuel Johnson in Connecticut who believed that the practice and the theory ought to go together. This idea became the backbone of the Experiment Station concept in America. By titling his book *How Crops Grow* rather than *How to Grow Crops* Johnson sensed that if scientists know how crops grow, they could and would provide the farmer with the information for growing vastly better crops, so that society would be plentifully fed (Horsfall, 1969).

Photosynthesis and Respiration Studies in the Nineteenth Century

The word "photosynthesis" does not appear in either of Johnson's books (Johnson, 1868, 1870), but he writes of "absorption" or "fixation" of CO_2 ("carbonic acid"), and of the release of O_2 by leaves during photosynthesis as the "exhalation." He recognized that respiration (O_2 uptake and CO_2 release) was essential for plant growth, and in *How Crops Grow* he states that the absorption of O_2 and the corresponding production of CO_2 is ". . . chemically analogous with the breathing of animals, and may hence be designated as *respiration*. We may speak of the other process [the absorption of CO_2 and evolution of O_2] as the *fixation of carbon*. These opposite changes obviously cannot take place at the same points, but must proceed in different organs or cells, or in different parts of the same cells." It was known that chlorophyll was concentrated in "grains" in leaf cells, but long before knowledge of the biochemical role of chloroplasts, mitochondria, and peroxisomes, it is a rather prophetic statement. Johnson stated further that since in sunlight the "absorption of CO_2 and evolution of oxygen proceed so much more rapidly [than respiration] that when we experiment with the entire plant . . . [the respiration] is completely masked . . . we can, in fact, only measure the preponderance of the [photosynthesis] over the [respiration]." In *How Crops Grow* the role of the stomata, the microscopic pores of leaves, in regulating the diffusion of CO_2 into leaves is also clearly presented. "The epidermis is further provided with vast number of curious 'breathing pores,' or *stomata*, by means of which the intercellular spaces in the interior of the leaf may be brought into direct communication with the outer atmosphere. Each of these stomata consists usually of two curved cells, which are disposed toward each other nearly like the two sides of the letter o, or like the halves of an elliptical carriage-spring."

How Crops Feed contains a rather vague statement about the biochemistry of the photosynthetic assimilation of carbon by plants, and there is barely mention of a connection with chlorophyll. The average composition of CO_2 in the atmosphere is given as 0.04% by volume, although it was surely no greater than 0.03% at that time. The section of the book on photosynthesis concludes with a statement that under the influence of sunlight, CO_2 is absorbed by leaves and is ". . . decomposed within the plant, its carbon being retained, and in an unknown manner becoming a part of the plant itself, while the oxygen is exhaled into the atmosphere in the free state." Liebig had earlier speculated that there must be a stepwise reduction of CO_2 assimilated during photosynthesis through such common organic acids as oxalate, tartrate, and malate. Organic acids are the first intermediates in CO_2 reduction, though not the ones he proposed. In 1870 A. von Baeyer suggested that formaldehyde was the first product of CO_2 assimilation, because it is a one-carbon compound at the same oxidation state as carbohydrate, and it condenses to produce hexoses in the presence of alkali. This hypothesis persisted, and in 1931 Kostychev, a noted plant physiologist, could write about the formaldehyde hypothesis that ". . . this theory is now more than 50 years old and still no directly contradictory facts have been uncovered." Experiments with radioactive carbon as tracer and the isolation and separation of enzymes provided definitive data about the path of carbon during photosynthesis and led to an understanding of the importance of photorespiration.

Photorespiration Research and its Connection with Wisconsin

In 1920 Otto Warburg (Warburg, 1920) observed that the rate of photosynthesis in *Chlorella* was inhibited with increasing oxygen concentration. Perry Wilson (1935), as discussed later under nitrogen fixation, also observed the inhibitory effect of 40% and 60% oxygen on the growth of clover plants. It is now known that in all plants with C_3-type photosynthesis (the first product of CO_2 assimilation is 3-phospho-D-glyceric acid) there is a 33% to 50% inhibition of photosynthesis in 21% oxygen compared with 1–3% oxygen. This inhibition is often called the Warburg effect, and I have recently termed it "oxygen stress" (Zelitch, 1982) to call attention to this inhibition that occurs in all outdoor environments. Oxygen stress is caused mainly by the process of photorespiration in which early products of photosynthesis, especially glycolate, are oxidized to CO_2. This results in a marked decrease in net CO_2 assimilation. The history of the discovery of photorespiration and its importance in the carbon metabolism of plants has been described previously (Zelitch, 1971, 1979) and will be mentioned only briefly here.

Rabinowitch in his 1945 scholarly treatise on photosynthesis (Rabinowitch, 1945) wrote that the possibility of the existence of photorespiration ". . . is a nightmare oppressing all who are concerned with the exact measurement of [gross] photosynthesis and various attempts have been made to bring it to light . . . Unfortunately, the best that can be stated is that no evidence of such a phenomenon has yet been found." These earlier attempts were unknowingly made under conditions that inhibited or obscured photorespiration (Zelitch, 1971), thereby prolonging its discovery.

Work at Wisconsin by Tottingham and his students (Tottingham et al., 1943) had been concerned with the role of polyphenol oxidase in the blackening of boiled potatoes, and when Burris took over the laboratory in 1944 upon Tottingham's sudden death, his curiosity about terminal oxidases in plants broadened the studies to the role of organic acids in respiratory systems of leaves. The first paper by Clagett, Tolbert, and Burris (Clagett et al., 1949) published in 1949 states that their attention was drawn to glycolic acid "because it supported very active respiration [in soybean leaf extracts] and had been studied little." Crude extracts of leaves of many species were capable of oxidizing glycolate and lactate. The enzyme activity was purified 10-fold from tobacco-leaf extracts, and they failed to find that hydrogen peroxide was formed in the reaction. Since the enzyme activity was found only in green tissues, they suggested that it might function in photosynthetic reactions. A second paper published later that year by Tolbert, Clagett, and Burris (Tolbert et al., 1949) described further purification of glycolate oxidase from tobacco leaves and demonstrated that glyoxylate was formed from glycolate oxidation and pyruvate from lactate. They also showed that formate was produced from the C-2 of glyoxylate, and concluded the paper by suggesting that glycolate or its oxidation products may be intermediates in the photosynthetic reactions. These papers initiated biochemical investigations on photorespiration that are still active today in a number of different aspects throughout the world.

The stoichiometry of the glycolate oxidase reaction was firmly established later when I isolated the enzyme in highly purified form in Severo Ochoa's laboratory (Zelitch and Ochoa, 1953) and showed that it was a flavoprotein that produced hydrogen peroxide. Similar conclusions were obtained about the same time by Kenten and Mann (1952). The important role of this reaction in the respiration of photosynthetic tissue was soon suggested (Zelitch, 1953), because it was already known from the work of Melvin Calvin's group (Schou et al., 1950) that glycolate was an early product of photosynthesis. N. Edward Tolbert, who began this work as a graduate student with Burris, continues to make important research contributions to photorespiration, and has had a leading role in elucidating the biochemistry of the glycolate pathway (Rabson et al., 1962; Tolbert, 1980) and the role of peroxisomes (Tolbert and Yamazaki, 1969).

The plant biochemistry laboratory under the leadership of Burris was one of the pioneers in the use of ^{14}C for metabolic studies with plants. Experiments to determine whether photosynthesis affects the rate of dark, or mitochondrial, respiration in leaves were described by Stutz and Burris in 1951. They supplied $^{14}CO_2$ to tobacco, barley, and rhubarb leaves in the light and found isotopic equilibrium among succinic, malic, and citric acids suggesting that dark respiration occurs in the light. Although this question is still not completely resolved, evidence since then supports the view that dark respiration continues unabated during photosynthesis in leaf tissue. When the detailed functioning of the citric acid cycle in leaves was still not established, Zbinovsky and Burris (1952) added ^{14}C-labeled organic acids to leaves and found that their rapid interconversion was compatible with the functioning of the cycle. Later an important contribution to the elucidation of the glycolate pathway resulted from the report of Wang and Burris (1963) that the conversion of glycine to serine occurred rapidly in the light in wheat leaves. This reaction, the glycine hydroxymethyltransferase reaction, during which glycine is also decarboxylated (Figure 13.7) occurs in the mitochondria of C_3 leaves and accounts for a large part of the photorespiratory CO_2 released in the glycolate pathway (Zelitch, 1971; Tolbert, 1980).

In 1955 John Decker observed that when CO_2 uptake by an illuminated tobacco leaf in a closed system was terminated by suddenly darkening the chamber, there was initially a high rate of CO_2 evolution that decelerated to the slower rate of dark respiration after several minutes. He estimated the initial burst was 4-fold greater than dark respiration. In a classical paper (Decker and Tió, 1959) published in 1959 when he was at the Agricultural Experiment Station of the University of Puerto Rico; Decker used the term "photorespiration" to explain the post-illumination CO_2 burst. Since this rapid CO_2 evolution required a prior exposure to light, he realized that he was observing the aftermath of respiration in the light of a substrate produced during photosynthesis. Photorespiration under normal conditions at high light intensity was estimated to be about 50% of net photosynthesis. Without knowing of Decker's work, I found in 1958 that sulfonic acid analogs of glycolate inhibited the oxidation of glycolate when supplied to tobacco leaves, and from the quantity of glycolate that accumulated during photosynthesis

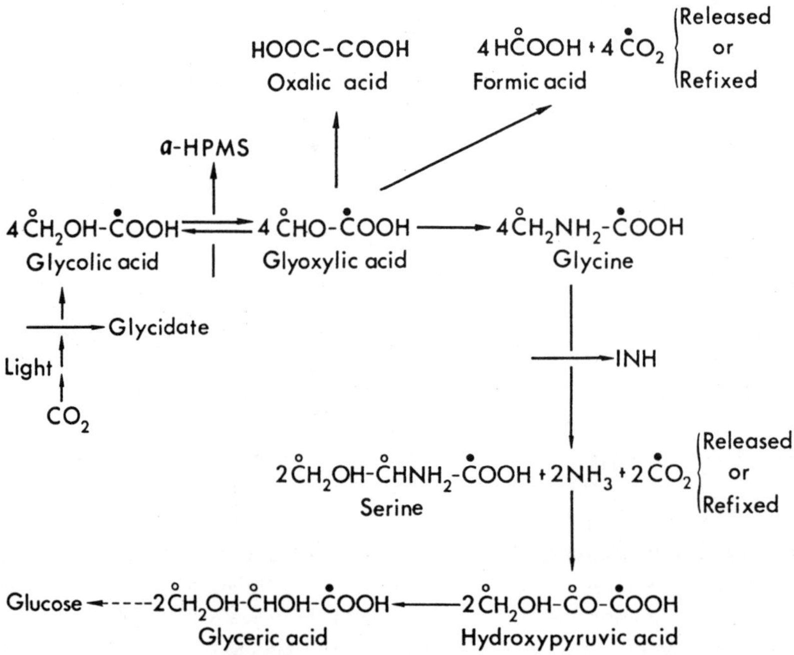

Figure 13.7 Simplified scheme of the glycolate pathway of photorespiration showing the sources of photorespiratory CO_2 and the site of action of several inhibitors.

in $^{14}CO_2$ I was able to conclude that at least 50% of the net CO_2 fixed during photosynthesis was photorespired by the oxidation of glycolate (Zelitch, 1958). It was not until 1963 (Zelitch, 1971) that I realized the post-illumination CO_2 burst and the rapid turnover of glycolate during photosynthesis were different aspects of the same process.

In plants with C_4-type photosynthesis (the first product of CO_2 assimilation is oxaloacetate), there is no detectable photorespiration or oxygen inhibition of photosynthesis, and the rate of glycolate synthesis is slow. The biochemistry of C_4 photosynthesis is a complex system that results in the regulation of enzymes capable of producing glycolate (Zelitch, 1971, 1979). Species with this photosynthetic system (maize, sorghum, sugarcane) have faster photosynthetic rates and generally much greater yields per unit of land area. The work initiated in Burris' laboratory in the early 1950's has spread to the discovery of new enzymes, new organelles, and new biochemical inhibitors, and provided entirely unexpected and unpredicted opportunities for the regulation of metabolic pathways in plants that should increase net CO_2 assimilation in C_3 plants. It has thus provided a major challenge for future research to reverse the oxygen stress on photosynthesis caused by photorespiration, perhaps by genetic means, and thereby, with skillful plant breeding, to greatly increase photosynthesis and crop yields (Zelitch, 1982).

The Nitrogen Problem and Biological Nitrogen Fixation in the Nineteenth Century

In the last century chemists such as Liebig in Germany, Mulder in Holland, and Boussingault in France appreciated that a source of nitrogen was essential for plant growth. The problem could best be stated in the form of two questions (Vickery, 1969): Where does the nitrogen come from? In what form is it used by the plant? Liebig believed that nitrogen came mostly in the form of ammonia brought down in rain, but Boussingault (1856) showed that nitrate added to essentially sterilized artificial soil was sufficient to bring sunflower to maturity if the ash of clover plants was added. Lawes and Gilbert in England analyzed rain water for ammonia and showed the total nitrogen per acre was only about 20 percent as much as the nitrogen taken up by a wheat crop per acre, hence Liebig's view was incorrect. It was known that there were other sources of nitrogen in soil from stubble and plant roots, and under some conditions nitrate was produced from these, but it was unknown how this might be accomplished. Although the term fermentation was applied, this word did not imply the activity of living microorganisms. It merely meant that the reaction was analogous to the conversion of nitrogenous material to ammonia in a manure pile, a fact familiar to every farmer. There was little knowledge or understanding about the action of bacteria by chemists such as Samuel Johnson because Pasteur's work was slowly accepted by his contemporaries, and there was thus no guiding principle or theory that could lead to a correct interpretation (Vickery, 1969).

It is on this basis that *How Crops Feed* discusses the nitrogen problem and concludes that nitrogen gas is not a food to the plant. Contrary results previously described were attributed to contamination of the air by ammonia. The book describes the experiments of Lawes, Gilbert, and Pugh (1857, 1858) in great detail, including a description of their apparatus consisting of a closed system of large glass bells dipping in mercury in which leguminous and nonleguminous plants were grown while supplied with pure air and watered with pure water. Neither kind of plant grew indicating that the plants could not utilize atmospheric nitrogen. These workers later (1861) commented that these were abnormal experimental conditions because in the field leguminous plants obtained more nitrogen than could be accounted for in the combined nitrogen of the soil, rain, water, and air (Fred et al., 1932). This was among the first observations that leguminous plants differed from other plants in the nitrogen source they could assimilate, but the explanation for this apparent discrepancy between laboratory and field conditions did not come quickly.

Besides the lack of appreciation of the role of microorganisms in nitrogen metabolism, part of the delay in solving many of these problems may be attributed to the long, hot, and cumbersome method then used to carry out nitrogen determinations with the Dumas or the Will and Varentrap method then available. It required the heating of a sample in a long glass tube charged with reagents in a furnace, and a skilled chemist with two furnaces might at best do four samples in a day (Vickery, 1969).

The Kjeldahl method was first described in 1883, and chemists like Samuel Johnson adopted it immediately. The Report of the Connecticut Station of 1889 describes his design of a digestion stand and distillation rack that appears obviously the forerunner of modern equipment of the same type. In 1886 Hellriegel reported his great discovery of the use of atmospheric nitrogen by leguminous plants at a meeting in Berlin. His report showed in pot experiments that the weight of oat plants depended on soil nitrate while the growth of pea plants was independent of nitrate. He concluded that leguminous plants are different from other families of plants because they can also obtain nitrogen from the air by means of a symbiotic relationship between living soil microorganisms and the plant that results in the formation of root nodules and causes the assimilation of free nitrogen. Wilfarth was the discoverer of the use of mercury as a catalyst in the Kjeldahl method (Vickery, 1969), and it was used in the extensive report of these experiments by Hellriegel and Wilfarth in 1888 and undoubtedly facilitated this work. Gilbert was chairman of the meeting at which Hellriegel gave his preliminary paper in 1886, and the conclusions were quickly confirmed by Lawes and Gilbert and others (Fred et al., 1932).

In 1888 Beijerinck isolated the root nodule bacterium and named it *Rhizobium*. In 1893 nitrogen fixation by a free-living bacterium was first discovered by Winogradsky in the anaerobic *Clostridium pasteurianum*, and a few years later Beijerinck found the free-living aerobe *Azotobacter*. However, as late as 1948 only *Rhizobium*, *Clostridium*, *Azotobacter*, and *Nostoc* (a blue-green alga) were generally accepted as agents of biological nitrogen fixation (Burns and Hardy, 1975).

This difficulty in detecting biological nitrogen-fixing organisms occurred because, in practice, the direct approach of obtaining growth in nitrogen-free medium seldom gave unequivocal results since ideal conditions were rarely attained. Sometimes this was because of impure reagents including ammonia in the atmosphere, the then unknown repressive effects of fixed nitrogen on biosynthesis of the enzyme responsible for nitrogen fixation (nitrogenase), and the need for anaerobic conditions by facultative organisms. The Kjeldahl method, although convenient and reliable, was not sensitive enough for detecting small differences in total nitrogen before and after growth in nitrogen, especially when sampling errors were large (Burns and Hardy, 1975). Research at Wisconsin has been at the forefront of the study of symbiotic nitrogen fixation, discovery of various nitrogen-fixing systems, detecting biochemical intermediates, devising sensitive methods of assay, analyzing the biochemistry of the nitrogenases of different organisms, and investigating ecological and agricultural aspects of biological nitrogen fixation.

Research in Biological Nitrogen Fixation at Wisconsin

As indicated above, many agents of biological nitrogen fixation were discovered after 1948. I take this opportunity to recall what it was like to accept a graduate assistantship to do research with Robert H. Burris on nitrogen fixation. I arrived in Madison in June, 1948, to begin these studies in a place already well known in this field. Edwin Broun Fred (1887–1981) (Johnson, 1974) was then President of the University. He had joined the Department of Agricultural Bacteriology, later "Bacteriology," from his native Virginia in 1913, and besides being a professor at Wisconsin he had at various times been dean of the graduate school and dean and director of the Agricultural Experiment Station. His interest in research on the legume bacteria began at the Virginia Agricultural Experiment Station and it continued at Wisconsin. Among some of his best known students and collaborators were I. L. Baldwin, who also became a distinguished dean at Wisconsin, P. W. Wilson, and E. McCoy. In 1932 Fred, Baldwin, and McCoy published their classic monograph *Root Nodule Bacteria and Leguminous Plants* (Fred et al., 1932), which is discussed below.

Perry W. Wilson, Professor of Bacteriology, was a student of Fred's who had done extensive work on the carbohydrate-nitrogen relation in symbiotic nitrogen fixation and had shown that hydrogen was a specific inhibitor of nitrogen fixation by red clover plants. Robert H. Burris was a student of Wilson's and had joined the Department of Biochemistry in 1944. In 1948 Burris and Wilson still collaborated a great deal, and it was my good fortune to have research conferences with both Burris and Wilson every few months, usually in Wilson's office, where my progress and future work were discussed in detail. At these meetings Wilson invariably outlined plans for work that would take several years to complete. In the beginning, this was somewhat distressing until one appreciated that this was Wilson's fertile mind at work and one could learn by listening. Toward the end of my career as a graduate student I screwed up my courage enough to tell Wilson that the experiments suggested for the next few months would actually take five years. My remark was simply ignored and he continued suggesting still more experiments.

For many years (1906–1944) Edwin Bret Hart (1874–1953) (Elvehjem, 1954) was Chemist "In Charge" of the Department. He apparently appointed members of the staff without the need of search committees as are common today. He was obviously a shrewd picker of talent, as evidenced by the honors received by his appointees to the faculty. Students were told that his simple dictum was, "Get good people and leave them alone." In 1948 he was an Emeritus Professor with distinguished bearing who came by on occasion, and to keep his hand in things turned off electric lights that were unnecessarily lit—a true conservationist. Although Hart had approached Burris earlier about joining the Department of Biochemistry, the sudden death in 1944 of William Edward Tottingham in mid-semester forced the decision for Burris to take over the graduate course in plant biochemistry and to join the department.

Burris and his graduate students occupied a corner laboratory closest to the railroad tracks, and whenever a steam locomotive came by in the summer when the windows were open, the ammonia concentration greatly increased in the laboratory reagents used to determine it. The only mass spectrometer on the campus available for determining ^{15}N was in the Chemistry Department, where it had been constructed by a graduate student and modified along the way (without written plans). Later this instrument was relocated to a room in the sub-basement of Biochemistry, a room where Babcock had once carried out experiments on magnetism. The walls of the room were painted by the students ("The Green Room"), and because of the skill and diligence of Burris in finding leaks in the glass vacuum system and repairing them, it was possible to do occasional ^{15}N determinations.

Research in Burris' laboratory was done in organic acid metabolism in plants, respiration, and photosynthesis, as discussed previously, and in nitrogen fixation. Several of Mark A. Stahmann's students also shared the laboratory, working on synthetic polypeptides and the biochemistry of plant diseases. A large number of graduate students worked on animal nutrition with Conrad A. Elvehjem, and it was said he had one student for each of the amino acids. Other students did research on enzymes related to bioenergetics, microbial biochemistry, and steroid metabolism, and a student in Genetics told me about the new method then evolving using penicillin combined with replica plating to obtain microbial mutants. There was a sense of the excitement of rapid discovery all around.

For a graduate student arriving in the department from a slower-paced university, it was something like being a baseball rookie who has just made it to the big leagues, thrilled to be there but somewhat apprehensive that he might strike out the first time at bat. Most of the graduate students had been in the military service during the Second World War and were somewhat older and perhaps more mature than usual. Many of us were married and some had children as well. We were in a hurry to make up for lost time, and this eagerness meshed with that of the professors who were in a hurry to answer important scientific questions. Thus the two needs combined to produce an unusually productive period in the Department.

The earliest work at Wisconsin focused on the challenging symbiotic nitrogen-fixing system in leguminous plants and the metabolism of the *Rhizobia* that infect their roots. This work is summarized in the scholarly 1932 book by Fred, Baldwin, and McCoy. Besides a fascinating historical account of the subject, detailed information is given about methods of isolating the root nodule bacteria, their morphology in root nodules, and their biochemical characteristics in laboratory culture. The problem that has "baffled the minds of many investigators" of getting *Rhizobia* to fix nitrogen in a suitable medium apart from the plant has still not completely been resolved. They discuss the specificity of *Rhizobia* to infect certain groups of plants, the so-called cross-inoculation groups, although these groups are not as mutually exclusive as was then believed. They state that the greatest benefit is obtained in treating leguminous seeds with *Rhizobia* cultures when the same crop has not recently been grown in the soil, but the important question of why certain strains (not necessarily the ones to produce the best nitrogen fixation) infect more readily than others is not discussed. This question is still not answered. The section of the book dealing with the chemistry of nitrogen fixation is mainly speculative since little data were then available in the days before the general use of isotope tracers, chromatography, specific biochemical inhibitors, easy methods of enzyme isolation and purification, and the availability of biochemical mutants.

In 1935, Wilson summarized his experiments on the effect of growing leguminous plants in controlled environments in which he changed the light, oxygen, and CO_2 concentrations, in an attempt to account for agronomic observations that affected nitrogen fixation. Methods for quantifying photosynthesis were not then readily available although there was a clear positive effect of photosynthetic activity on increasing nitrogen fixation. More recent studies have shown that increased photosynthesis induced by CO_2 enrichment enhanced nitrogen fixation by five-fold during the complete growth cycle of leguminous plants in the field (Hardy and Havelka, 1975).

Among Burris' first investigations (Burris and Wilson, 1939) was an attempt to determine whether the respiratory system of the root nodule bacteria was unique and whether there was any biochemical basis for differences in strains that might correlate with observed differences in their ability to fix nitrogen in root nodules. These results, unfortunately, were negative in spite of a great deal of hard work. In 1941, Burris took advantage of Harold Urey's concentrating of the stable isotope ^{15}N to do postdoctoral work at Columbia University, and after he showed that actively fixing *Azotobacter* did not incorporate $^{15}N_2$ by exchange he reported that with a ^{15}N assay bacteria-free barley and clover did not fix nitrogen, while inoculated clover did (Burris, 1941). The ^{15}N assay immediately opened up new opportunities in research in nitrogen fixation because sampling errors were now relatively unimportant and there was about a thousand-fold greater sensitivity than was available by the Kjeldahl method. A number of questionable claims in the literature about nitrogen fixation could now be answered unequivocally, and the identification of the "key" (first stable) intermediate produced from molecular nitrogen could be undertaken.

Beginning about 1938 A. I. Virtanen had produced a large number of papers claiming that hydroxylamine was the first stable intermediate because he found large concentrations of oxaloacetic acid in leguminous plants and claimed to have isolated the oxime. Burris and Wilson in 1945 and 1947 (Burris and Wilson, 1945; Wilson and Burris, 1947) summarized a decade of disagreement with Virtanen's hypothesis, and in experiments with $^{15}N_2$ supplied to *Azotobacter* showed that glutamic and aspartic acids were most enriched in ^{15}N and therefore ammonia was the key intermediate. Direct evidence that ammonia was the key intermediate was obtained later with *Clostridium pasteurianum* cells supplied $^{15}N_2$ for a short time, where the free ammonia had as much as ten times the isotopic enrichment of the average nitrogen in the cell supernatant (Zelitch et al., 1951).

Another and still more sensitive method of assaying nitrogen fixation arose from the discovery by Schöllhorn and Burris in 1966 (Schöllhorn and Burris, 1967) and by Dilworth (1966) that nitrogenase preparations from *Clostridium* also catalyzed the reduction of acetylene to ethylene. After publication of these results (Evans, 1969), Hardy and Knight (1969) as well as the Wisconsin and other laboratories used the hydrogen-flame method for detection of the ethylene that was separated by gas chromatography as an assay for nitrogenase activity that was about one thousand times more sensitive than the $^{15}N_2$ method. Because of its ease of use and sensitivity, the acetylene-reduction method has had wide application in laboratory as well as in field and aquatic environments and has permitted the making of accurate estimates of nitrogen fixation under these diverse conditions, including a study of the daily fluctuations in nitrogen fixation and photosynthesis in the blue-green algae of Lake Mendota in Madison (Peterson et al., 1977).

Investigations of the enzymology of nitrogen fixation did not come easily, and many fruitless attempts were made to obtain active cell-free preparations from a variety of organisms. Carnahan et al. in 1960 (Carnahan et al., 1960) described active preparations from *Clostridium* cells dried anaerobically at elevated temperatures (nitrogenase was cold-labile) using high levels of pyruvate to produce a reductant (reduced ferrodoxin) as well as to generate ATP. Burris and his students quickly confirmed this work and obtained active cell-free preparations from *Rhodospirillum* and blue-green algae. This laboratory and others have since obtained active preparations from virtually every system known to fix nitrogen actively, including root nodules.

While purifying nitrogenase, Mortenson (Mortenson et al., 1967) found later it consisted of two proteins. Neither was active alone but the two were effective with proper addition of a strong reductant such as reduced ferredoxin or sodium dithionite as the electron donor and ATP and magnesium ions. Highly pure nitrogenase preparations have been obtained and studied from a number of systems at Wisconsin and elsewhere, and the Wisconsin group has been at the forefront in elucidating the path of electron transfer in the nitrogenase system.

The component known as the dinitrogenase protein has a molecular weight of about 245,000 and contains Fe and Mo, while the component known as the dinitrogenase reductase has a molecular weight of about 64,000 and

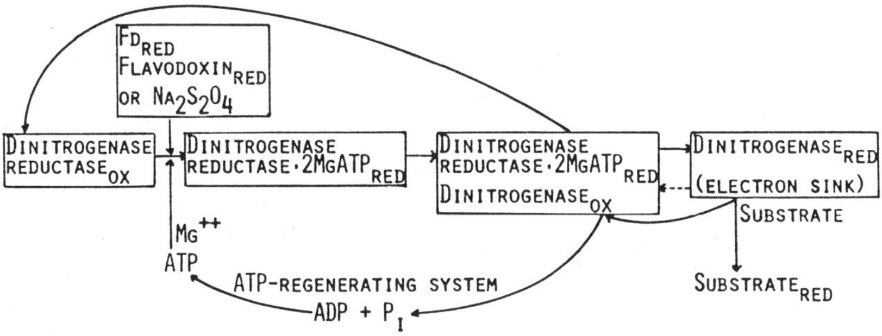

Figure 13.8 Path of electron transfer in the nitrogenase system (from Burris et al., 1981). Dinitrogenase reductase transfers a single electron, and hydrolysis of MgATP accompanies the transfer of electrons from dinitrogenase reductase to dinitrogenase.

is a 4Fe-4S protein. Nitrogenase probably emerged only once during evolution, since about 85% of the tested crosses between dinitrogenase and dinitrogenase reductase from various sources have produced catalytically active nitrogenase systems (Burris et al., 1981). The electron transfer pathway was established mainly from EPR measurements of the nitrogenase system in several laboratories about the same time (Burris et al., 1981). They show that reduced ferredoxin donates electrons to dinitrogenase reductase, which binds MgATP and acquires a low potential and then transfers a single electron at a time to dinitrogenase; the dinitrogenase then binds and reduces nitrogen (Figure 13.8). ATP hydrolysis accompanies the electron transfer, while ATP regeneration follows the electron transfer. It is apparent that a great deal of progress has been made since the remarkable discovery by Hellriegel and Wilfarth almost 100 years ago, and much of the new knowledge can be attributed to research at Wisconsin.

Acknowledgments

Much of the information relating to the Agricultural Experiment Station movement in America was obtained from M. W. Rossiter's fascinating account of this subject (Rossiter, 1975), the work of a professional historian and a lucid writer. The historical information about the nitrogen problem was gleaned largely from the writings of Hubert Bradford Vickery (1969), a biochemist who did considerable research and writing on this subject. His respect for history helped preserve the archival material of Samuel W. Johnson and Thomas B. Osborne at the Connecticut Agricultural Experiment Station which has become a treasure for present-day historians of science. The portraits shown were taken from the slide and picture collection of T. B. Osborne and H. B. Vickery stored in my Department. I am grateful to Joseph S. Fruton,

Yale University, for his helpful comments. Lastly, I would especially like to express my gratitude to the many fine teachers I encountered during my time as a graduate student at the University of Wisconsin, whose example and high standards in research provided the basis for future work. It has been possible to name only a few of these teachers in this essay.

References

Benedict, F. G. (1938) Henry Prentiss Armsby. Biographical Memoirs. *National Acad. Sci.* 19: 271–284.

Burns, R. C. and Hardy, R. W. F. (1975) *Nitrogen Fixation in Bacteria and Higher Plants*, Springer-Verlag, New York.

Burris, R. H. (1941) Failure of barley to fix molecular N^{15}. *Science* 94: 238–239.

Burris, R. H., Arp, D. J., Hageman, R. V., Houchins, J. P., Sweet, W. J., and Tso, M.-Y. (1981) Mechanism of nitrogenase action, in *Current Perspectives in Nitrogen Fixation* (Gibson, A. H. and Newton, W. E., eds.), pp. 56–66, Australian Acad. Sci.

Burris, R. H. and Wilson, P. W. (1939) Respiratory enzyme systems in symbiotic nitrogen fixation. *Cold Spring Harbor Symp. Quant. Biol.* 7: 349–361.

Burris, R. H. and Wilson, P. W. (1945) Biological nitrogen fixation. *Annu. Rev. Biochem.* 14: 685–708.

Carnahan, J. E., Mortenson, L. E., Mower, H. F., and Castle, J. E. (1960) Nitrogen fixation in cell-free extracts of *Clostridium pasteurianum*. *Biochim. Biophys. Acta* 38: 188–189.

Clagett, C. O., Tolbert, N. E., and Burris, R. H. (1949) Oxidation of α-hydroxy acids by enzymes from plants. *J. Biol. Chem.* 1678: 977–987.

Decker, J. P. (1955) A rapid, postillumination deceleration of respiration in green leaves. *Plant Physiol.* 30: 82–84.

Decker, J. P., and Tió, M. A. (1959) Photosynthetic surges in coffee seedlings. *J. Agr. Univ. Puerto Rico* 43: 50–55.

Dilworth, M. J. (1966) Acetylene reduction by nitrogen-fixing preparations from *Clostridium pasteurianum*. *Biochim. Biophys. Acta* 127: 285–294.

Elvehjem, C. A. (1954) Edwin Bret Hart. Biographical Memoirs. *National Acad. Sci.* 28: 117–161.

Evans, H. J. (1969) How legumes fix nitrogen, in *How Crops Grow. A Century Later* (Day, P. R., ed.), *Conn. Agr. Expt. Station Bull.* 708: 110–127, New Haven, Connecticut.

Fred, E. B., Baldwin, I. L., and McCoy, E. (1932) *Root Nodule Bacteria and Leguminous Plants*. University of Wisconsin Press, Madison, Wisconsin. Also see supplemental bibliography, 1939.

Gillispie, C. C., ed. (1981) John Pitkin Norton, in *Dictionary of Scientific Biography* 10: 150–151. Scribners Sons, New York.

Hardy, R. W. F. and Havelka, U. D. (1975) Photosynthate as a major factor limiting nitrogen fixation by field-grown legumes with emphasis on soybeans, in *Symbiotic Nitrogen Fixation in Plants* (Nutman, P. S., ed.), pp. 421–439. Cambridge University Press, London.

Hardy, R. W. F. and Knight, E., Jr. (1967) ATP-dependent reduction of azide and HCN by N_2-fixing enzymes of *Azotobacter vinelandii* and *Clostridium pasteurianum*. *Biochim. Biophys. Acta* 139: 69–90.

Horsfall, J. G. (1969) How Johnson related science to society, in *How Cops Grow. A*

Century Later (Day, P. R., ed.), *Conn. Agr. Exp. Station Bull.* 708: 1–6, New Haven, Connecticut.

Ivins, L. S. and Winship, A. E. (1925) Henry Prentiss Armsby, in *Fifty Famous Farmers*, pp. 328–342. Macmillan, New York.

Jenkins, E. H. (1925) A history of Connecticut agriculture, in *History of Connecticut in Monographic Form* (Osborn, N. G., ed.), pp. 289–425. States History Co., New York.

Johnson, D. O. (1974) *Edwin Broun Fred: Scientist, Administrator, Gentleman*. University of Wisconsin Press, Madison, Wisconsin.

Johnson, S. W. (1868) *How Crops Grow. A Treatise on the Chemical Composition, Structure, and the Life of the Plant*. Orange Judd, New York.

Johnson, S. W. (1870) *How Crops Feed. A Treatise on the Atmosphere and the Soil*. Orange Judd, New York.

Kenten, R. H. and Mann, P. J. G. (1952) Hydrogen peroxide formation in oxidations catalyzed by plant α-hydroxyacid oxidase. *Biochem. J.* 52: 130–134.

Mortenson, L. E., Morris, J. A., and Jeng, D. Y. (1967) Purification, metal composition, and the properties of molybdoferredoxin and azoferredoxin, two of the components of the nitrogen-fixing system of *Clostridium pasteurianum*. *Biochim. Biophys. Acta* 141: 516–522.

Osborne, E. A., ed. (1913) *From the Letter-Files of S. W. Johnson*. Yale University Press, New Haven, Connecticut.

Osborne, T. B. (1911) Samual William Johnson, Biographical Memoirs. *National Acad. Sci.* 7: 205–222.

Peterson, R. B., Friberg, E. E., and Burris, R. H. (1977) Diurnal variation in N_2 fixation and photosynthesis by aquatic blue-green algae. *Plant Physiol.* 59: 74–80.

Rabinowitch, E. I. (1945) *Photosynthesis and Related Processes*, Vol. I., Wiley (Interscience), New York.

Rabson, R., Tolbert, N. E., and Kearney, P. C. (1962) Formation of serine and glyceric acid by the glycolate pathway. *Arch. Biochem. Biophys.* 98: 154–163.

Rossiter, M. W. (1975) *The Emergence of Agricultural Science. Justus Liebig and the Americans, 1840–1880*. Yale University Press, New Haven, Connecticut.

Schöllhorn, R. and Burris, R. H. (1967) Acetylene as a competitive inhibitor of N_2 fixation. *Proc. Natl. Acad. Sci. USA* 58: 213–216.

Schou, L., Beson, A. A., Bassham, J. A., and Calvin, M. (1950) The path of carbon in photosynthesis. XI. The role of glycolic acid. *Plant Physiol.* 3: 487–495.

Stutz, R. E. and Burris, R. H. (1951) Photosynthesis and metabolism of organic acids in higher plants. *Plant Physiol.* 26: 226–243.

Tolbert, N. E. (1980) Photorespiration, in *The Biochemistry of Plants* (Stumpf, P. K. and Conn, E. E., eds.), Vol. 2, pp. 487–523. Academic Press, New York.

Tolbert, N. E., Clagett, C. O., and Burris, R. H. (1949) Products of the oxidation of glycolic acid and *l*-lactic acid by enzymes from tobacco leaves. *J. Biol. Chem.* 181: 905–914.

Tolbert, N. E. and Yamazaki, R. K. (1969) Leaf peroxisomes and their relation to photorespiration and photosynthesis. *Ann. N. Y. Acad. Sci.* 168: 325–341.

Tottingham, W. E., Nagy, R., Ross, A. F., Marek, J. W., and Clagett, C. O. (1943) A primary cause of darkening in boiled potatoes. *J. Agr. Res.* 67: 177–193.

Vickery, H. B. (1969) Samuel W. Johnson and *How Crops Grow*, in *How Crops Grow. A Century Later* (Day, P. R., ed.). *Conn. Agr. Expt. Station Bull.* 708: 7–24, New Haven, Connecticut.

Wang, D. and Burris, R. H. (1963) Carbon metabolism of C^{14}-labeled amino acids in wheat leaves. II. Serine and its role in glycine metabolism. *Plant Physiol.* 38: 430–439.

Warburg, O. (1920) Über die Geschwindigkeit der photochemischen Kohlensaürezersetzung in lebenden Zellen. *Biochem. Z.* 103: 188–217.

Wilson, P. W. (1935) The carbohydrate-nitrogen relation in symbiotic nitrogen fixation. *Univ. of Wisc. Agr. Expt. Station Research Bull.* 129.

Wilson, P. W. and Burris, R. H. (1947) The mechanism of biological nitrogen fixation. *Bact. Rev.* 11: 41–73.

Zbinovsky, V. and Burris, R. H. (1952) Metabolism of infiltrated organic acids by tobacco leaves. *Plant Physiol.* 27: 240–250.

Zelitch, I. (1953) Oxidation and reduction of glycolic and glyoxylic acids in plants. II. Glyoxylic acid reductase. *J. Biol. Chem.* 201: 719–726.

Zelitch, I. (1958) The role of glycolic acid oxidase in the respiration of leaves. *J. Biol. Chem.* 233: 1299–1303.

Zelitch, I. (1971) *Photosynthesis, Photorespiration, and Plant Productivity.* Academic Press, New York.

Zelitch, I. (1979) Photosynthesis and plant productivity. *Chem. Eng. News* 57: 28–48.

Zelitch, I. (1982) The close relationship between net photosynthesis and crop yield. *BioScience* 32: 796–802.

Zelitch, I. and Ochoa, S. (1953) Oxidation and reduction of glycolic and glyoxylic acids in plants. I. Glycolic acid oxidase. *J. Biol. Chem.* 201: 707–718.

Zelitch, I., Rosenblum, E. D., Burris, R. H., and Wilson, P. W. (1951) Isolation of the key intermediate in biological nitrogen fixation by *Clostridium. J. Biol. Chem.* 191: 295–298.

14

Enzymes, Respiration, and Phosphorylation

Paul D. Boyer

Department of Chemistry and Biochemistry
University of California
Los Angeles, California

Quality usually becomes evident in retrospect. The formation and maintenance of an academically excellent department of biochemistry is strikingly difficult. It was not good planning but good fortune that brought me as a student 44 years ago to an unusually fine biochemistry department in a great university. I became immersed in an environment of inquiry and question about how cells work and in a system of dispersing and examining past and, in particular, the present forefronts of biochemical knowledge. In this brief and perhaps too personal account I recall some of my experiences and the events that may help show why Wisconsin biochemistry has had such a dramatic impact on our present understanding of enzymology.

Two individuals with whom I became closely associated in 1939 illustrate well what makes a department great. One was Paul Phillips (Figure 14.1) as my advisor and teacher, to whom I shall always be indebted for his encouragement of my scientific inquiry. The other was Henry Lardy, a fellow graduate student, who then as now (Figure 14.2) was a handsome man, and who then as now was an investigator with an unusual ability to identify the significant fact or approach and to recognize and suggest what others were yet to discern. Henry and I were surrounded by a rich environment of seminars, discussion and evaluation and an expectancy and encouragement of creativity.

My tendency to discard past letters and materials quickly has kept my files minimal, but has lost some personal treasures. Thus in consideration of what I might say of possible interest here, I was pleased to find I still had copies of my notes from classes at Wisconsin. They illustrate well the flavor of the instruction I received here. Some excerpts from Elvehjem's 1940 course on enzymes are given in Figure 14.3. Here we were learning how Krebs obtained the evidence that led to recognition of the citric acid cycle. And here

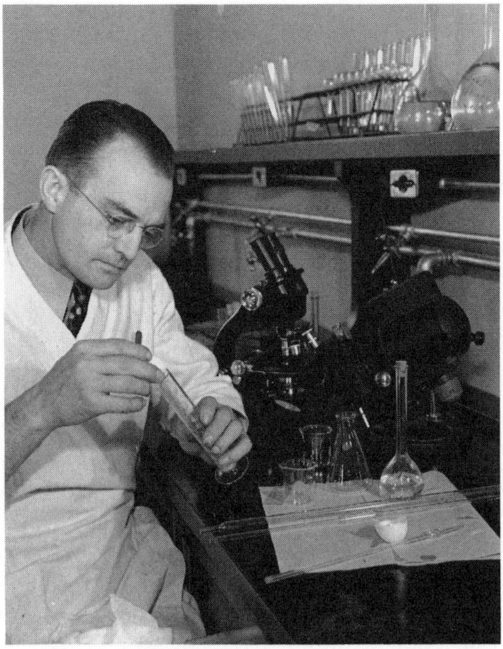

Figure 14.1 Professor Paul Phillips—about 1941.

Figure 14.2 Professor Henry Lardy, 1943.

C.A. Elvehjem lecturing on pyruvate metabolism:

"Krebs and coworkers believe that the citric acid cycle accounts for the major part of pyruvic acid metabolism -- see *Biochem J.* 34: 1442 (1940) for a recent paper."

On enzyme changes in cancer tissue:

"As yet there is not much to talk about."

Figure 14.3 Excerpts from notes for the 1940 course on enzymes.

Marvin Johnson lecturing on oxidation-reduction potentials:

These considerations show why a postulate of a metabolically important reaction
succinate + pyruvate ⟶ lactate + fumarate
is nonsense."

Carl Bauman lecturing on changes involving phosphorus:

"Adenylic acid is the most active phosphate carrier. It forms a triphosphate, A.T.P."

Figure 14.4 Excerpts from notes on 1940 course on enzymes and 1941 course on vitamins and metabolism.

we find the wisdom that for some areas, such as cancer, the available information at that time did not allow the design of meaningful experiments. From other lectures (Figure 14.4) students encountered the penetrating and analytical intellect of Marvin Johnson as he pointed out the limitations that thermodynamics places on postulated metabolic sequences. And it was in these lectures that students learned, as from Carl Bauman, of the remarkable role played by "A.T.P." The periods that appear in my notes were soon dropped.

About this time, Henry and I were isolating ATP from muscle to use in some experiments asking why vitamin E-deficient rats were dystrophic. We found the phosphorylations apparently normal. But the experiments, stimulated by the supportive environment, led to a recognition of the potassium activation of pyruvate kinase. At one seminar where I presented some results, Henry counted the interruptions for questions by Van Potter, Marv Johnson, and others—there were over 30. This illustrates well the splendid research training that Wisconsin's Biochemistry Department has fostered and maintained in a most unusual way.

Not only was the local environment rich, but the value of the special

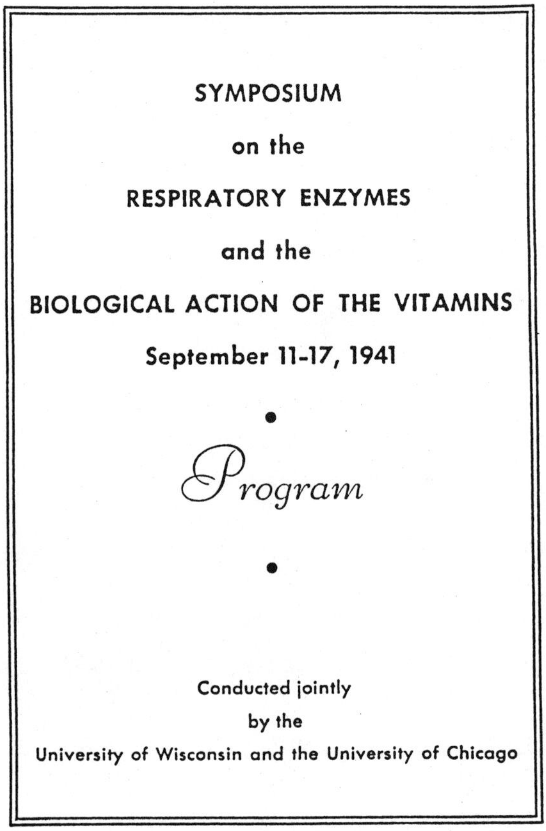

SYMPOSIUM

on the

RESPIRATORY ENZYMES

and the

BIOLOGICAL ACTION OF THE VITAMINS

September 11-17, 1941

•

Program

•

Conducted jointly

by the

University of Wisconsin and the University of Chicago

Figure 14.5 The cover of the program for a "Symposium on Respiratory Enzymes."

symposia, now exemplified by the fine series of Steenbock Symposia, was already evident. Such symposia only occur with a nurtured environment and with good planning. But I took it as a normal event that just a few steps down the hall from the laboratory where I worked I could hear world leaders discuss recent results in enzymology and metabolism. I refer to a symposium on *Respiratory Enzymes* that was held in this lecture hall. In the front of my copy of the published version of this symposium, I found the program, the cover of which is shown in Figure 14.5. The list of participants as given in the frontispiece of the volume is shown in Figure 14.6. This list includes biochemical giants in the area, who provided much of our early understanding of enzymology and phosphorylation. At this symposium I heard Otto Meyerhof, whose penetrating investigations established the framework of glycolysis. I had a hard time understanding Fritz Lipmann's mumblings, but I have come to recognize the value of learning to understand what he said.

A Symposium on

RESPIRATORY ENZYMES

Contributors:

OTTO MEYERHOF
ERIC G. BALL
FRITZ LIPMANN
KURT G. STERN
FRITZ SCHLENK
T. R. HOGNESS
ELMER STOTZ
CARL F. CORI
E. A. EVANS, JR.
PHILIP P. COHEN
K. A. C. ELLIOTT
DEAN BURK
C. J. KENSLER
ERWIN HAAS
H. M. KALCKAR
M. J. JOHNSON
VAN R. POTTER
H. G. WOOD
R. H. BURRIS
C. H. WERKMAN
P. W. WILSON
F. F. NORD
EPHRAIM SHORR
A. E. AXELROD
FREDERICK BERNHEIM
E. S. G. BARRON
FREDRICK J. STARE

The University of Wisconsin Press

MADISON

Figure 14.6 The frontispiece from the volume *A Symposium on Respiratory Enzymes*, University of Wisconsin Press, 1942.

Among the speakers was Carl Cori, who gave us a school of superb biochemists and a tradition of excellence at Washington University. Other speakers were Herman Kalckar, who as soon as anyone recognized the occurrence of oxidative phosphorylation, and Harland Wood, who was embarking on his outstanding career of gaining an understanding of enzymology and metabolism by isotopic tracer studies. Among the Wisconsin participants were Philip Cohen, known to all of us for his later achievements and human warmth; Fred Stare, who became a leader in the nutrition field; Van Potter, who as much as anyone has exemplified the spirit and power of Wisconsin biochemistry; and Marvin Johnson, whose inquiring intellect and insight placed him far above most of us. The volume on *A Symposium on Respiratory Enzymes*

TABLE OF CONTENTS

Figure 14.7 Table of Contents for *Respiratory Enzymes,* Edited by H. A. Lardy, Revised Edition, 1949, Burgess Publishing Co., Minneapolis.

also records an event that illustrates the value of presentation and discussion as fostered by such symposia. Johnson's incisive contribution to the understanding of the adenylate kinase reaction, in which he correctly pointed out the reaction that this enzyme catalyzed, is noted in a quote from page 191 as follows:

> Recently Colowick and I found that phosphorylation of glucose by adenylpyrophosphate when adenosinediphosphate is the phosphate donor needed, besides hexokinase, another protein which occurred only in skeletal muscle and was therefore called myokinase—the exact function of myokinase is not known. Dr. Johnson has suggested that myokinase catalyzes a transfer of phosphate from one molecule of adenosinediphosphate to another molecule, thus forming from two moles of

adenosinediphosphate one mole of adenosinetriphosphate and one mole of adenosinemonophosphate.

Other publications originating from Wisconsin biochemistry around this time had a profound influence on the development of cellular enzymology. The volumes somewhat mistitled *Manometric Techniques* by Umbreit et al. presented the methodology for measuring phosphorylated intermediates and for enzyme assays—the fundamental "know-how" so essential for the execution of experiments. Another volume that collated the knowledge in a rapidly advancing field was *Respiratory Enzymes*, edited by Henry Lardy. The Preface of a revised 1949 edition notes, "The First Edition of *Respiratory Enzymes* was a collection of seminar reports in a field which was comparatively new to biochemists in this country. . . . In the present edition the authors have been assigned subjects with which they are familiar by virtue of research experience." The Table of Contents, given in Figure 14.7, reveals the extent to which Wisconsin faculty had become deeply involved in this important and pioneering area.

Now I would like to change from my early experience that reflects the quality of graduate training available over the years from Wisconsin and take note of the contributions of some Wisconsin faculty in the area I am to review today. As a means of doing this I made use of the 1928 *Science Citation Index* to find the most-cited papers of those faculty. This approach has obvious limitations—sometimes a paper is highly cited because it is wrong. Fortunately this proved not to be the case for the present quest. Another limitation is that the index records only first authors, and thus many prime papers where the faculty members' name often appears last are not included. However, the search does reveal earlier publications that have stood the test of time.

I'll start with a classic paper of Potter and Elvehjem, "A Modified Method for the Study of Tissue Oxidations," from the 1936 *Journal of Biological Chemistry* (Vol. 114: 495). Here we have the description of the Potter-Elvehjem homogenizer that has been and is still used by thousands of researchers. The paper also establishes the use of tissue homogenates for enzymic studies. Potter not only developed innovative methods, but he also applied them incisively and successfully.

For Marvin Johnson the most-cited publication in 1982 was a 1941 *Journal of Biological Chemistry* paper (Vol. 137: 575) on the purification of a yeast peptidase. He was a recognized pioneer in this area. His ability and instruction in microbial fermentations and preparations, an area so ably fostered by W. H. Peterson, added much to the breadth of Wisconsin biochemistry.

I have mentioned earlier my initial encounter with another Wisconsin enzymologist of note, Henry Lardy. Because he has contributed so much to enzymology, in particular to areas of my interest, I mention two of his publications still widely cited in 1982. The first is a 1953 paper (*J. Biol. Chem.* 201: 357) on the catalytic effect of 2,4-dinitrophenol on ATP hydrolysis by mitochondria particles. In his earlier classical review with Elvehjem he pointed out the first clear recognition of the probable "uncoupling" action of this agent.

Lardy recognized the potential usefulness of nature's magic bullets, the antibiotics, as probes for the poorly understood process of oxidative phosphorylation. His widely cited 1958 paper (Lardy et al., 1958) gives a survey of his early studies. Last week I was at a Gordon Conference on Energy Coupling Mechanisms in which the antibiotics oligomycin, aurovertin, bongkreckic acid and efrapeptin, all introduced to the field by Lardy and associates, were used for reported experiments.

Other additions to the Wisconsin faculty in the last quarter century have continued and broadened the research competence in enzymology. One of the most cited of all papers in enzymology (Cleland, 1963) is by William Wallace Cleland who has extended and honed kinetic evaluations of enzyme reactions to establish a leading position for himself and Wisconsin. Many papers from Wisconsin include thanks to Cleland for his help in kinetic analysis.

It is fitting to record here a recognition of another biochemist prominent in the establishment of the Wisconsin tradition for pioneering research on respiratory enzymes, namely David Green. His most-cited 1982 paper (*J. Biol. Chem.* 217: 551) on isolation and properties of the succinic dehydrogenase complex illustrates his interest well. His laboratory established the occurrence of and procedures for obtaining the four recognized complexes of the mitochondrial respiratory chain. A large number of fine and currently active bioenergeticists passed through Green's laboratory and the training impact will long be felt.

Another of the Wisconsin leaders in enzymology is Helmut Beinert; his 1982 most-cited paper is a review from a recent symposium paper entitled "The Structure and Function of Copper Proteins," published in 1980 in *Coordination Chemistry Reviews* 33: 55. Beinert's pioneering and incisive application of electron spin resonance techniques established the role of the iron-sulfur proteins in the respiratory chain and gave us much of our knowledge about the participation of riboflavin and copper in respiratory enzymes. His accomplishments have brought much credit to himself and to Wisconsin for fostering such research.

Among more recent faculty, the 1982 most-cited paper of Perry Frey is entitled "*E. coli* Pyruvate Dehydrogenase Complex Site Coupling in Electron and Acetyl Group Transfer Pathways," published in 1978 in *J. Biol. Chem.* 253: 724. Here he demonstrates the presence and function of two classes of lipoic acid sulfhydryl groups in the functioning complex. Perry brings to Wisconsin his unusual skills in both organic chemistry and enzymology, as demonstrated by his outstanding studies on the stereochemistry of phosphate transfer reactions.

For the productive and promising and, from my perspective, a much younger enzymologist, JoAnne Stubbe, her 1982 most-cited paper is a 1971 study entitled "Analogs of Phosphoenolpyruvate. On the Specificity of Pyruvate Kinase from Rabbit Muscle," published in *Biochemistry* 10: 2669. It is a pleasure to record a publication on pyruvate kinase, as my Ph.D. thesis was, for a large part, concerned with this enzyme, and you never forget your old loves. Stubbe's present interests, as most of you know, center on vitamin B_{12} enzymology. I recall well her crisp presentation at a recent Gordon Conference of penetrating studies on the function of the B_{12} coenzyme.

Another prominent area of enzymology is well represented by a chemically-oriented enzymologist, Marion O'Leary. His 1982 most-cited paper dealt with a recent review on a fascinating topic, namely "Carbon Isotope Fractionation in Plants," in *Phytochemistry* 20: 553 (1981). It is both instructive and valuable to learn how measurements of isotope ratios can provide clues to metabolic pathways of organisms long dead.

I'll close this section with mention of one more present staff member who studies one of the most prominent and important of all chemical reactions in biology, namely photophosphorylation. Bruce Selman's 1982 most-cited publication is entitled "Evidence for a Catalytic Function of the CF_1 Protein Reconstituted with Chloroplast Thylakoid Membranes," published in 1978 in *Biochim. Biophys. Acta* 502: 29. Among the more recent contributions from his laboratory in this area is an important paper demonstrating that the ATP synthase complex functioning in photophosphorylation has a subunit stoichiometry of $\alpha_3\beta_3\gamma\delta\epsilon$. This adds considerably to the evidence that the enzyme complex as found in microorganisms, plants and animals has three catalytic sites.

Lastly, I'll mention briefly some of my own recent research that intertwines closely with my Wisconsin training and with the contributions of Wisconsin biochemistry. Much of my research has concerned oxidative phosphorylation—this remarkable process that was discovered shortly before I became a graduate student. For the first decade after my degree I wisely ignored the area, but could not long resist the lure. Along the way I have learned something about the pitfalls of "wanting to discover" that can handicap critical evaluation of one's own data, but have had the pleasure of finally gaining some useful insights not previously perceived.

For the last decade of my 40 years since graduation, my research group, with others, has been deeply involved in what appears to give a broad outline of how the energy derived from sunlight and from oxidations is used by the ATP synthase complex to make ATP. It has been my good fortune to participate in two previous Steenbock Symposia. One of these was on photosynthesis mechanisms and the second honored Henry Lardy. In the spirit of such symposia I presented some of the evidence for the binding-change mechanism of ATP synthesis, including a figure illustrating how 3 identical catalytic sites may participate in coordinated alternation. Each site is regarded as going through identical conformation, but at any one time each is at a different stage of the catalysis (see Figure 4, p. 471, Steenbock-Lilly Symposium on Biochemistry of Metabolic Processes). In this volume, I included another figure (Figure 6, p. 475) which is clearly speculative but is guiding our present experimental design. This figure illustrates the postulates that during catalysis by the remarkable ATP synthase complex there is a positional interchange of subunits. More specifically, our research suggests that the subunits present in single copies (γ, δ and ϵ subunits) may form an asymmetric core where rotational movement relative to the catalytic β subunits causes the sequential conformational changes required for catalysis. By crude mechanical analogy, the inner core serves as a camshaft to drive the catalytic mechanism.

In these researches it has been of keen interest to us to discern concepts that are new, not only for the ATP synthase, but for the field of enzymology. One is the occurrence of energy-linked binding changes. This appears to be a general phenomenon in bioenergetics. A second concept is that of alternating site participation. The evidence for site-site coupling and catalytic cooperativity demonstrated for ATP synthase and F_1-ATPase catalysis is the most convincing for any enzyme system. Lastly, the concept of required subunit positional interchange during catalysis is novel and exciting to us, as it may give additional insight into a remarkable molecular machine that makes ATP.

The postdoctoral fellows and graduate students who have comprised the research groups that generated these concepts are of the quality and type that I feel meet the standards of the Wisconsin traditions. On an occasion such as this it may not be amiss to mention some of the abilities that Wisconsin biochemists have shown more than most over the years. These are as follows:

To inquire. The driving force of research is the quest to know how living cells and organisms work. Without an insatiable curiosity and the reward of discovery few could continue in research careers.

To examine. The time-honored approach is the hypothesis and the experiment. An unreachable goal for most is to be able to formulate at the outset the hypothesis and to design the experimental approaches that were later to be successful. Accomplishment of most of our researches is the coal we mine while looking for diamonds.

To evaluate. A deep understanding is required to perceive the significant. A sound training, and often some humbling experiences, are required to discover flaws in one's own data and concepts. Progress depends on being able to perceive possibilities and relationships not yet apparent.

To persist. Living things were not designed to be easily understood, and the most frequent answer of an experiment to test a novel question is "no." To recover, regroup and seek other approaches—to persist in inquiring—is a major characteristic of most successful researchers.

To share. Science flourishes from the shared examination of facts, ideas, and experimental designs. Science progresses from the shared effort of many, and an individual scientist must gain much of his own satisfaction from the advancements that are dependent mostly on research in a field as a whole.

Our contribution, combined with those of many others, allows us to glory in the biochemical wonder of living things.

To communicate. A truth, or progress toward a truth, is of limited value until it is ably recorded and presented. Society needs the knowledge of science to continue an upward progression; yet the privilege of being supported by society while we explore the biochemistry of living things is rare and fragile. Communication must be not only with fellow scientists, but with an enlightened citizenry.

In conclusion, those of us who have been part of Wisconsin biochemistry are proud of its accomplishments, tradition and present vigor. I am both honored and pleased to have this opportunity to help commemorate 100 years of Wisconsin's Biochemistry Department.

References

Beinert, H. (1980) *Coord. Chem. Rev.* 33: 55.

Cleland, W. W. (1963) *Biochim. Biophys. Acta* 67: 104.

Frey, P., Ikeda, B. H., Gavino, G. R., Speckhar, D. C., and Wong, S. S. (1978) *J. Biol. Chem.* 253: 7234.

Green, D. E. (1955) *J. Biol. Chem.* 217: 551.

Johnson, M. (1941) *J. Biol. Chem.* 137: 575.

Lardy, H. A., ed. (1949) *Respiratory Enzymes*, revised edition. Burgess Publishing Co., Minneapolis.

Lardy, H. A., Johnson, D., and McMurray, W. C. (1958) *Arch. Biochem. Biophys.* 78: 587.

Lardy, H. A. and Wellman, H. (1953) *J. Biol. Chem.* 210: 357.

O'Leary, M. H. (1981) *Phytochem.* 20: 553.

Potter, V. R. and Elvehjem, C. A. (1936) J. Biol. Chem. 114: 495.

Selman, B. and Durbin, R. D. (1978) *Biochim. Biophys. Acta* 502: 29.

Stubbe, J. and Kenyon, G. L. (1971) *Biochemistry* 10: 2669.

Umbreit, W. W., Burris, R. H., and Stauffer, J. F. (1945) *Manometric Techniques and Related Methods for the Study of Tissue Metabolism*, First Edition. Burgess Publishing Co., Minneapolis.

15

Vitamin C in the Prevention and Cure of Neurolathyrism

Kamaluddin Ahmad

Department of Biochemistry
University of Dhaka
Bangladesh

This occasion to celebrate the centennial of Wisconsin's Biochemistry and to commemorate a great teacher, researcher and patron of biochemistry, Dr. Harry Steenbock, is one of great solemnity. Dr. Steenbock taught me and many of this audience and I join you in paying homage to the memory of this great man. I also respectfully remember other great Wisconsin biochemists who taught me but who are no longer with us. Amongst them are Dr. Conrad Elvehjem, Dr. Karl Paul Link and Dr. William H. Peterson. I consider Karl Paul as one of the greatest teachers I ever had. He used to teasingly tell me, "Your country has lots of rats. I will give you something when you return, to get rid of all those rats." I felt very attached to his affectionate and jovial personality. These are some of the people who brought distinction to Wisconsin Biochemistry. And today, as we mark this important milestone, we pay respect to their immortal souls.

The word lathyrism was introduced by Cantani in 1873 (Cantani, 1873) to describe the disease seen in the Italian populations as a result of ingesting *lathyrus* seeds. It is a crippling condition of the lower limbs, recognized and understood to be irreversible since the days of Hippocrates.

Two kinds of lathyrism are known, osteolathyrism and neurolathyrism. Osteolathyrism is a disease of the animals characterized by skeletal deformity and osteoporosis. It was demonstrated in 1933 here in Dr. Steenbock's laboratory by Beatrice Geiger (Geiger et al., 1933) that rats, when fed seeds of *L. odoratus*, would develop osteolathyrism. Dr. Strong (Schilling and Strong, 1954; Strong, 1956) isolated the active principle, γ-N-glutamyl aminopropionitrile, and later (1954–55) demonstrated that the active portion of the molecule is β-aminopropionitrile. A few synthetic osteolathyrogens are also known. It is also known that osteolathyrogens (Page and Benditt, 1967) block formation of crosslinks in collagen molecules. Osteolathyrism may be affecting some horses and cattle. It is not a public health problem. But neurolathyrism is. It affects hundreds of thousands of poverty stricken people in the Indian subcontinent where a large number of people live on *Lathyrus sativus (khesari)* as the staple dietary constituent, particularly during famine. But for poor people, it is famine most of the year and in most years. In Bangladesh, the situation was particularly serious during the mid-seventies when the price of cereals skyrocketed while the price of *L. sativus* remained low. As the very poor shifted the consumption even further toward *Lathyrus sativus*, a large number of them became crippled for life. As you drive along the roads in the northern districts of Bangladesh, notably Rajshahi and Kushtia, you cannot miss seeing them, limping or crawling their way.

There are various stages of the disease. Those with mild cases just walk by dragging their toes. There is shivering in their thighs. Their two legs touch and cross each other and they walk on their toes. There is much weakness in the waist. Some patients with all the above symptoms cannot walk without a cane; there are others who need two canes; and there are still others who can barely crawl. There are various clinical terms to describe the disease: spastic paralysis, ataxia, scissor gait, etc. Some patients have exaggerated knee and ankle jerks and ankle clonus. Most patients demonstrate appreciable bending at knee joints and the raising of heels while walking.

I became interested in the problem in early 1980 and decided to explore what was known so far. But on reviewing the available literature, I soon found that the real difficulty was the absence of an experimental animal model for neurolathyrism (Sarma and Padmanabam, 1969). None could produce neurolathyrism in any adult animal by feeding them a diet of *Lathyrus sativus*. I fancied that this failure to produce the symptoms in animals might lie in a biochemical difference between man and the experimental animals that were used. One such difference, as we all know, is that human beings cannot synthesize vitamin C in their bodies, whereas most animals, with the exception of guinea pigs, primates and Indian bats, can synthesize this vitamin. I suspected that the experimental diets used to produce this condition in adult monkeys and guinea pigs included vitamin C. So I took four batches of adult guinea pigs—one group had *L. sativus* without vitamin C; another group had *L. sativus* with vitamin C; a third group had *phaseolus radiatus* (which is another kind of legume population in the subcontinent) in place of *L. sativus*; the final group was given *P. radiatus* with vitamin C.

We soon discovered that the symptoms of neurolathyrism developed only in the group of guinea pigs which was fed a diet of *L. sativus* supplemented by all vitamins except vitamin C. The deficiency of vitamin C and the ingestion of *L. sativus* seeds as the principal item of the diet (to supply no less than two-thirds of the caloric content of the diet) for over two months or so are the essential concomitant conditions for the onset of the disease. At the time of seizure of sickness, the serum vitamin C of the affected guinea pigs ranged between 0.2 and 0.4 mg per dl. There were no symptoms of any disability amongst the animals which were fed seeds of *P. radiatus* instead of *L. sativus* and not given any vitamin C. Some of the animals which had neurolathyrism were examined for clinical signs of scurvy. There were none.

This is an appropriate moment for me to return to the history of the researches on neurolathyrism. The late Professor P. S. Sarma of Bangalore in India, another Wisconsin biochemist who worked with Professors Elvehjem and Strong, did a great deal of research. It was found that generalized neurological symptoms (wry necks, head retraction and convulsion) developed in very young chicks within minutes after the administration of an alcoholic extract of *L. sativus*. Using this observation as a test for biological activity, Sarma isolated a neuroactive amino acid, β-N-oxalyl-α,β-diaminopropionic acid (ODAP) from the seeds of *L. sativus* (Rao et al., 1964). But adult birds did not respond to the toxic amino acid even when it was given intraperitoneally in excess of toxic dosage. Curtis and Watkins (1965) found the toxin from *L. sativus* to be a very potent neuroexcitant of the spinal interneurons and Betz cells in the cat spinal cord when administered into the immediate extracellular environment of single neurons. They suggested that the inefficacy of this compound in adult animals to produce neurological symptoms may be due to a blood brain barrier to this toxin. Rao et al. (1967) have shown that very young rats, guinea pigs and pups also respond to this neurotoxin, and this lends support to the concept of a blood brain barrier of this compound in adult animals. It has been further reported that if the ODAP is given to adult monkeys through the lumbar route, thus bypassing the blood brain barrier, it may cause paralysis of the hind legs. According to them, there appeared to be a destruction of the nerve cells of the gray matter of the spinal cord, accompanied by proliferation of the microglial cells. Adult animals (rats, monkeys, etc.) were reported to respond to intraperitoneal injection of the toxin when they were previously fed certain drugs like salicylic acid which could induce acidotic conditions in the animals (Weisberg, 1962).

In spite of all these interesting experiments in animals, observed Dr. Sarma, the role of β-N-oxalyl-α,β-diamino propionic acid in human lathyrism remained in the realm of speculation (Sarma and Padmanaban, 1969).

We come back to the discussion of our own experiments. After having demonstrated neurolathyrism in guinea pigs which were fed *L. sativus* with no vitamin C for four to five weeks, we wanted to find out what would happen if *L. sativus* prepared according to the Indian scientists (Roy et al., 1963) were administered intraperitoneally. We found that vitamin-deficient animals developed neurological symptoms within 30–60 minutes of administering the toxin. No symptoms appeared in those which were not made deficient in this vitamin. Nothing happened to the vitamin C-deficient guinea pigs if extract of *P. raditus* was injected instead of *L. sativus* (Ahmad and Jahan, 1980; Ahmad, 1982; Ahmad and Jahan, 1982). We repeated the same experiments with monkeys and found that adult monkeys, when fed a diet of *L. sativus* seeds deficient only in vitamin C, developed typical symptoms of neurolathyrism—spastic paralysis of the hind legs, ataxia, etc. The same condition also arose in monkeys which were first made deficient in vitamin C (0.3 mg per dl of serum) and then injected intraperitoneally with extracts of *L. sativus* seeds. We have also determined that extracts of leaves, which are often consumed as vegetables, were as toxic in vitamin C-deficient animals as extracts of seeds.

At this stage, we reverted to the experiments with chicks which our Indian colleagues were doing, i.e., the intraperitoneal administration of the toxin preparation to very young chicks. We repeatedly found that the administration of 5 mg of ascorbic acid about 10 minutes before the administration of the toxic extract from *L. sativus* would protect the animals from the expected toxicity. Vitamin C not only protected the animals from the neural symptoms, but also cured the animals of the symptoms if the vitamin was administered soon after the onset of symptoms. The recovery, however, did not take place when the injury appeared to be very severe.

Finally, we were emboldened to look for a human patient. We found Mr. Khabiruddin, a young man of 22 years. He had eaten unleavened *L. sativus (khesari)* bread every day for about two months. He had consumed no vegetables except a paste of boiled red chilis. One day in the first week of May, 1982, Khabir went out to the field with his father to prepare their land for the next crop. Suddenly, he felt paralyzed in both the legs and could not walk. His father and brother who accompanied him to work carried him home in a bullock cart. He was brought to us three months after he became ill. We gave him 500 mg of ascorbic acid intravenously every day for two weeks and continued to give him orally 500 mg of vitamin C for a few weeks more. After two weeks, all the symptoms of neurolathyrism were gone. We kept him for about two months more to watch him. The fellow who could not walk before was playing football before he left. After him, we had seven other patients. Four of them became completely well in 4–6 weeks' time. Others made considerable improvement. I communicated with Dr. M. P. Dwivedi, Professor of Medicine in the Rewa Medical College, Madhya Pradesh, India, and suggested that he might consider vitamin C therapy for his patients. The largest incidence of lathyrism had been recorded in Rewa. Some time later, he wrote back to tell me that he had a patient of lathyrism who could not walk when he was brought to him. After one week's treatment of 1000 mg of vitamin C intravenously every alternate day, the man completely recovered and walked back home on his own. He reported on another case who also completely recovered after similar treatment with vitamin C. But then there were cases in which recovery was not as satisfactory. His experience was that if the illness was more than a year old, then the prospect of complete recovery was slim. We went to the rural areas where lathyrism was seen in large numbers of people, and randomly selected families which had one or more members suffering from lathyrism. As a control group, we randomly selected from other families which did not have anyone suffering from lathyrism but were of equivalent socio-economic status. We found significant differences in serum vitamin C levels.

As we were working on neurolathyrism, we came across a paper of Olney and Ho (1970) who found that in very young mice, monosodium glutamate (MSG) caused neural damage similar to that caused by the *khesari* toxin (ODAP) (Lakshmanan and Padmanaban, 1974). We wondered if the neural damage caused by MSG could also be prevented by vitamin C. We actually found that it could. We experimented to see if MSG would have an effect on vitamin C-deficient adult guinea pigs and monkeys. We found that we could reproduce the symptoms of *L. sativa* toxicity by intraperitoneal administration of MSG in vitamin C-deficient adult guinea pigs and monkeys, and very young chicks.

We wondered how all these could happen. At this stage, we reverted to chick experiments. We found that not only vitamin C, but also D-gulono-γ-lactone and isoascorbic acid could prevent toxicity of both *khesari* toxin (ODAP) as well as MSG, though these structural analogs seemed to be needed in greater amounts (Jahan, 1983).

We speculated on the mechanism of action by ascorbic acid and its analogs. We were familiar with the concept of a blood brain barrier. Could it be that at the point of the blood brain barrier, there are specific sites or holes which could allow or disallow toxin/glutamate molecules from going into synaptosomes? It could be that an external supply of ascorbic acid in young animals fortifies these sites in the yet immature blood brain barrier and does not allow enough toxin or glutamate to enter. Similarly, in the case of comparative deficiency of ascorbic acid in adult animals, the blood brain barrier may become vulnerable or permeable to the entry of these molecules. In those animals in which both glutamate and vitamin C are made in the body, could it be that they balance each other in some kind of homeostatic mechanism?

Isoascorbic acid and D-gulonolactone could do the same thing as ascorbic acid does for the protection against toxicity. Glucono-δ-lactone or glucuronolactone do not do so. Isoascorbic acid is not vitamin C, and D-gulonolactone is not even a reducing agent in the same manner as is ascorbic acid. The role of ascorbic acid in fortifying the blood barrier appears to be different from its role as an antiascorbutic vitamin or a biological reducing agent. While this perhaps could give a lead to further thoughts on the mechanism of the preventive role of L-ascorbic acid in the phenomenon of neurolathyrism, the curative effect of vitamin C in neurolathyrism would need further speculation and experimentation for its understanding. It is yet to be ascertained whether ascorbic acid protects the blood brain barrier from specific entries of ODAP and glutamate into the cerebrospinal fluid or exerts a generalized protective effect on all unwanted entries.

References

Ahmad, K. (1982) in *Adverse Effects of Foods* (Jeliffe, F. F. P. and Jeliffe, D. B., eds.), pp. 71–75. Plenum Press, New York.

Ahmad, K. and Jahan, K. (1982) *Food and Nutrition Bulletin* 4: 65; also see (1980) *Nutrition News* 1(6), Dhaka University: Institute of Nutrition and Food Science.

Ahmad, K. and Jahan, K. (1980) *Abstracts and Programmes*, Third Asian Congress of Nutrition, p. 190, Jakarta.

Cantani, A. (1873) *Morgagni* 15: 745.

Curtis, D. R. and Watkins, J. C. (1965) *Pharm. Rev.* 17: 347.

Geiger, B. J., Steenbock, H., and Parson, H. T. (1933) *J. Nutr.* 6: 427.

Jahan, K. (1983) Ph.D. thesis, University of Dhaka.

Lakshmanan, J. and Padmanaban, G. (1974) *Nature* 249: 459.

Olney, J. W. and Ho, O.-L. (1970) *Nature* 227: 609.

Page, R. C. and Benditt, E. P. (1967) *Biochem.* 6: 1142.

Rao, S. L. N., Adiga, P. R., and Sarma, P. S. (1964) *Biochem.* 3: 432.

Rao, S. L. N., Sarma, P. S., Mani, K. S., Rao, T. R. R., and Sriramchari, S. (1967) *Nature* 214: 610.

Roy, D. N., Nagarajan, V., and Gopalan, C. (1963) *Current Science* 32: 116.

Sarma, P. S. and Padmanaban, G. (1969) in *Toxic Constituents of Plant Foodstuffs* (Liener, I. E., ed.). Academic Press, New York.

Schilling, E. D. and Strong, F. M. (1954) *J. Am. Chem. Soc.* 76: 2843; also see (1955) *J. Am. Chem. Soc.* 77: 2843.

Strong, F. M. (1956) *Nutrition Rev.* 14: 65.

Weisberg, H. F. (1962) *Water, Electrolyte and Acid Base Balance*, pp. 208–259. Williams and Wilkins, Baltimore.

16

Biochemical Contributions to the Understanding of Carcinogenesis from Wisconsin

Elizabeth C. Miller† and James A. Miller

McArdle Laboratory for Cancer Research
University of Wisconsin Medical School
Madison, Wisconsin

Foreword

The centennial anniversary of the Department of Biochemistry of the University of Wisconsin provides a unique opportunity for some of its former students to pay tribute to the unusual intellectual stimulus and strong training for research which they received from the members of its staff. We join wholeheartedly in that tribute, for the background in biochemistry and the understanding of and love of research, which we acquired as graduate students, have been very important factors throughout our careers. It is particularly fitting that this anniversary is the subject of a Steenbock Symposium, since the Department of Biochemistry and its students owe much to Professor Harry Steenbock. His research and his foresight in the founding and direction of the Wisconsin Alumni Research Foundation continue to benefit graduate research at Madison. While the Biochemistry staff of the early 1940's as a whole had a strong role in our research training through the classroom, seminars, and discussions, we are particularly indebted to Professor Carl Baumann, who, as the major professor for each of us, carefully nurtured our interest and ability in research and introduced us to the then emerging field of chemical carcinogenesis.

As noted below, the efforts of several young investigators, including Baumann, in cancer research at the University of Wisconsin were important factors in the establishment in 1940 of the McArdle Laboratory for Cancer Research. Because the University of Wisconsin, primarily through work at the McArdle Laboratory, has played an important role in enlarging the knowledge

†Deceased

of the fundamental nature of cancer, we have chosen carcinogenesis (i.e., the sequence of events that leads to cancer) as the focus of this review. Furthermore, in view of the focus of this symposium, the emphasis is on biochemical research on carcinogenesis carried out at the University of Wisconsin, and no attempt is made to give a broad review of this field.

Introduction

Fossil records indicate that cancer has existed as a human and animal disease from prehistoric times. However, cancer became a much more important disease to society with the introduction of means for the control of infectious disease during the first half of this century. Thus, although cancer occurs at all ages, the total cancer incidence increases markedly with age; 50 percent of the cases in the U. S. are currently diagnosed at age 51 or greater (Young and Pollack, 1982). In the United States, cancers other than basal and squamous cell carcinomas of the skin now affect about one individual in three (Young and Pollack, 1982).

Medical observations relevant to carcinogenesis date from the second half of the 18th century. In 1761 Hill (Redmond, 1970) in London deduced that nasal polyps and cancer were much more common in men who had used excessive amounts of tobacco snuff, and in 1775 Pott (Potter, 1963) in London observed that scrotal skin cancer occurred with much greater frequency among men who had been chimney sweeps in boyhood and were thus exposed to much higher levels of coal tar than other men. In the latter part of the 19th century and early 20th century the incidences of skin cancer among workmen heavily exposed to certain coal tars and high-boiling oils, and of urinary bladder cancer among those exposed to certain aromatic amines in the dyestuffs industry, were recognized to be higher than for unexposed men (Clayson, 1962; Henry, 1947; Rehn, 1895). Starting with the report of Thiersch in 1875, data on the higher incidences of skin cancer among sailors and farmers, as compared to the general population, and among light-skinned persons, as compared to more pigmented individuals, implicated solar radiation (ultraviolet light) as a human carcinogen (Scotto et al., 1982). Early in the twentieth century Frieben and Jagie reported associations between the development of leukemia or skin cancer and exposures to ionizing radiations, and subsequent studies have implicated ionizing radiations as causes of other human cancers (Upton, 1982).

These early clinical observations prompted investigations on tumor induction in experimental animals. In 1915 Yamagiwa and Ichikawa (Yamagiwa and Ichikawa, 1915) in Japan were the first to demonstrate the induction of skin tumors in rabbits by repetitive applications of coal tar. This seminal finding led to the monumental work of Kennaway and his colleagues in London. Their work culminated in the early 1930's with the discovery of dibenz(a,h)anthracene as the first pure chemical carcinogen and the isolation of the carcinogen benzo(a)pyrene from coal tar (Haddow, 1974). At about the same time, Sasaki and Yoshida (1935) and Kinosita (1937) in Japan demonstrated the hepatocarcinogenicities of 2′,3-dimethyl-4-aminoazobenzene (o-aminoazotoluene) and N,N-dimethyl-4-aminoazobenzene for rodents. Soon thereafter Hueper and his colleagues (1938) succeeded in the induction of cancer of the urinary bladder in dogs by long-term administration of the suspected human urinary bladder carcinogen 2-naphthylamine. As with chemicals, the early observations of carcinogenic effects of X-rays and ultraviolet light in humans were subsequently reproduced in experimental animals. In 1910 Marie et al. reported the induction of tumors in rats by X-rays, and in 1928 Findlay demonstrated that repetitive exposures to ultraviolet light caused skin cancer in mice. Likewise, the early observations of viral carcinogenesis date back to 1908, when Ellerman and Bang demonstrated that naturally occurring chicken leukemia could be transferred by cell-free preparations, and 1911, when Rous achieved cell-free transmission of spontaneous chicken sarcomas. Kingery showed cell-free transmission of human warts in 1921. In 1932 Shope demonstrated that papillomas of the rabbit skin were transmissible by virus preparations, and in 1933 the staff of the Roscoe B. Jackson Memorial Laboratory reported a nonchromosomal influence (later identified as a virus) for susceptibility to mouse mammary cancer. However, the transmission of tumors by cell-free preparations appeared to be the exceptional situation; most experimental tumors were not transferrable by these means.

Coupled with these developments toward an understanding of the etiology and nature of cancer were the rapid developments of the biological and chemical sciences during the 20th century. These latter advances made it possible to explore in meaningful ways the biochemical differences between normal and cancer tissues, the metabolism of chemical carcinogens in relation to their carcinogenic activities, the effects of carcinogens on cellular components, and the details involved in the replication and action of the carcinogenic viruses.

Early Cancer Research Investigations at the University of Wisconsin

Certain cancer-related investigations were being pursued on the University of Wisconsin campus as early as the 1920's and 1930's (Rusch, 1983). M. F. Guyer (Department of Zoology) investigated genetic and hormonal factors involved in cancer formation in rats, A. J. Riker (Department of Plant Pathology) was one of the pioneers in the study of crown gall tumors in tobacco plants, and W. E. Stovall (Department of Pathology) was interested in public health studies on cancer. In 1934 the Regents of the University accepted the

Jonathan Bowman Memorial Fund for the support of medical and surgical research on cancer. This bequest led to the appointment in 1935–37 of the first five Bowman Fellows—Harold Rusch (Department of Physiology), Carl Baumann (Department of Physiology), Fred Stare (Department of Physiological Chemistry), Meade Burke (Department of Pathology), and Fred Mohs (Department of Surgery) (Rusch, 1983). The enthusiasm of these investigators, the promise of cancer research in leading the way toward both the prevention and cure of cancers, and the availability of the Bowman and McArdle Funds led in 1940 to the construction of the McArdle Laboratory for Cancer Research. Harold P. Rusch, who was the first director and retained this position until 1973, played the major role in guiding the growth and development of the McArdle Laboratory for 30 years. Van R. Potter, who obtained his Ph.D. with C. A. Elvehjem in the Department of Biochemistry, joined Rusch as the second staff member. Baumann became a member of the staff of the Department of Biochemistry, but he maintained an affiliation with the McArdle Laboratory until the late 1940's. The early interests of Rusch and Baumann were in the induction of cancer by the then newly discovered chemical carcinogens and ultraviolet light and in the modification of cancer induction through nutritional means (see, for example, Baumann et al., 1939; Baumann and Rusch, 1939; Miller et al., 1941; Rusch et al., 1941; Rusch et al., 1942; Bain and Rusch, 1943; Rusch, 1944).

In the mid-1940's three graduates of the U.W. Department of Biochemistry, Roswell K. Boutwell and both of us, joined the staff of the McArdle Laboratory to work in the area of chemical carcinogenesis. Initially Boutwell collaborated with Rusch on the effects of dietary modifications on carcinogenesis of the skin of mice, while we expanded studies on the aminoazo dyes that we had begun with Carl Baumann. In 1948 Charles Heidelberger joined the staff. At the University of California in Berkeley, Heidelberger had synthesized the first ^{14}C-labeled carcinogen, dibenz(a,h)anthracene, and studied its metabolism (Heidelberger and Jones, 1948). His then novel expertise in the use of radioactive isotopes and the synthesis of labeled compounds greatly benefitted both his research and that of investigators at McArdle and in other departments on the Madison campus.

Chemical Carcinogenesis

Metabolism of chemical carcinogens The discovery of the pure chemical carcinogens led to questions with regard to their metabolism by animals susceptible and resistant to their activity. Since methodology suitable for investigation of xenobiotic metabolism was just becoming available, we were among the early investigators in this general area through our interest in the metabolism of the rat liver carcinogen N,N-dimethyl-4-aminoazobenzene (Miller and Miller, 1953). Our early studies thus contributed to the development of knowledge both in the field of chemical carcinogenesis and in the area of xenobiotic metabolism in general. For instance, when Gerald Mueller undertook graduate work at the McArdle Laboratory, he used tissue homogenates, first developed by Potter and Elvehjem (1936) for the quantitative investigation of intermediary metabolism, for the metabolism of the aminoazo dyes. These studies showed that the aerobic oxidative metabolism of N,N-dimethyl-4-aminoazobenzene in rat liver was carried out primarily in the microsomes by an NADPH-dependent enzymatic activity. Furthermore, the microsomal reduction of the azo linkage in this carcinogen was dependent on a flavin-adenine-dinucleotide-containing enzyme that required NADPH for activity (Mueller and Miller, 1948; Mueller and Miller, 1949; Mueller and Miller, 1953). These studies have been credited by Trager (1980) as signalling "the birth of drug metabolism as an independent discipline. . . ."

Our continuing investigations of carcinogen metabolism by rodent liver homogenates led a few years later to the finding with Raymond Brown and Allan Conney that the synthesis of hepatic microsomal oxygenases *in vivo* could be induced rapidly by the administration of certain xenobiotic chemicals (Brown et al., 1954; Conney et al., 1956, 1957). Thus, certain polycyclic aromatic hydrocarbons, as well as crude diets and peroxidized cholesterol, given either orally or parenterally resulted in several-fold increases in the hepatic activities for the oxidative metabolism of the aminoazo dyes and benzo(a)pyrene (Figure 16.1). These increases in enzyme activity were inhibited by inhibitors of protein synthesis and could not be induced *in vitro*. This observation of the induced synthesis of these enzymes of the endoplasmic

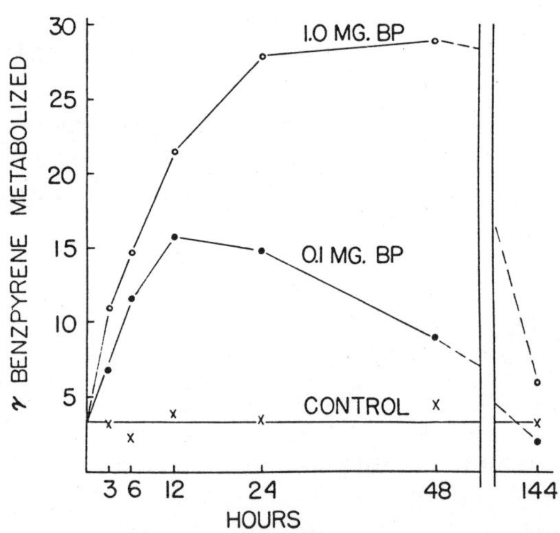

Figure 16.1 The induction of aryl hydrocarbon hydroxylase activity in rat liver by a single intraperitoneal injection of benzo(a)pyrene (BP). From Conney et al., (1957).

reticulum provided a major basis for the extensive development over the past 30 years of the broad literature on the inducibility of cytochrome P-450 monooxygenases by a very wide range of foreign compounds and in many tissues (Conney, 1967; Estabrook and Lindenlaub, 1979; Conney, 1982). In recent years the enzyme systems for the oxidative metabolism of polycyclic aromatic hydrocarbons and their inducibility have been areas of investigation by Charles Kasper of the McArdle Laboratory and of Colin Jefcoate of the Department of Pharmacology at the University of Wisconsin.

With respect to facilitating an understanding of carcinogenesis, the early studies of the overall metabolism of carcinogens, both *in vivo* and *in vitro*, were generally disappointing. Thus, regardless of the carcinogen, the metabolites identified proved to have little or no carcinogenic activity and appeared to be products of metabolic routes by which animals detoxified the carcinogen. However, an observation made by us in 1947 indicated that the carcinogenic aminoazo dyes were also converted *in vivo* to products that bound covalently to liver protein (Miller and Miller, 1947). This interaction between a carcinogen metabolite and a component of the target tissue was of particular interest, since it suggested that the liver might metabolize the dyes to chemically reactive forms. This dye could not be extracted from the liver protein, but alkaline hydrolysis released a polar dye with the same acid-base characteristics. In that day before the availability of ^{14}C- and ^{3}H-labeled carcinogens, we were fortunate to be working with a compound with such a high color potency that the presence of a few micrograms in tissue constituents could be observed by absorbance of visible light. Of further interest was the finding that the proteins of hepatic tumors induced by the dye did not contain detectable levels of the protein-bound form (Miller and Miller, 1947).

Our similar demonstration by fluorometric methods of protein-bound derivatives of benzo(a)pyrene in the skin of mice in 1951 (Miller, 1951) was followed within a few years of findings, through the use of ^{14}C-labeled compounds, of protein-bound derivatives of dibenz(a,h)anthracene by Wiest and Heidelberger (1953) at McArdle, of 2-acetylaminofluorene in our laboratory (Miller and Miller, 1955) and that of Weisburger, Weisburger, and Morris at the National Cancer Institute (1953), and of ethionine by Levine and Tarver at the University of California (1951). Furthermore, in continuation of the studies on the hepatic protein-bound aminoazo dyes, Sorof and Cohen (1951; Sorof et al., 1951) of the Department of Physiological Chemistry noted the high specificity of their binding to a class of soluble rat liver proteins which were essentially lacking in the primary hepatocarcinomas induced by the dyes. Likewise, Abell and Heidelberger (1962) observed specificity in the binding of certain polycyclic aromatic hydrocarbons to a similar class of soluble proteins of mouse skin.

As the information on and interest in DNA as the source of hereditary information in the cell developed in the late 1940's and the 1950's, the possible importance of DNA-bound derivatives of carcinogens became increasingly evident. The first observations on nucleic acid-bound carcinogens came from studies of Wheeler and Skipper (1957) in Birmingham on nitrogen mustard, and these were followed starting in 1960 by reports by Farber and his colleagues in Pittsburgh (Farber and Magee, 1960; Marroquin and Farber, 1962, 1963) of DNA- and RNA-bound derivatives of N,N-dimethyl-4-aminoazobenzene, 2-acetylaminofluorene, ethionine, and dimethylnitrosamine in rat liver. Shortly thereafter Heidelberger and Davenport (1961) published the first report on the binding of a carcinogenic polycyclic hydrocarbon, dibenz(a,h)anthracene, to mouse skin nucleic acids, and comparative studies by Brookes and Lawley (1964) of the Chester Beatty Research Institute on the covalent binding of a number of polycyclic hydrocarbons to mouse skin DNA soon followed. The evidence for the protein- or nucleic acid-bound carcinogens at this time rested on the incorporation of ^{14}C from the carcinogen into the polymer in a nonextractable form. Characterization of the bound carcinogen residues was still at least a decade away.

Metabolic studies with 2-acetylaminofluorene with John Cramer (Cramer et al., 1960) in our laboratory finally led to the first identification of a carcinogen metabolite that appeared to be on the pathway for activation to an "ultimate" carcinogen (i.e., the final active metabolite) (Figure 16.2). This "proximate" carcinogen was N-hydroxy-2-acetylaminofluorene, a previously unrecognized metabolite excreted as a glucuronide in the urine of rats treated with the carcinogen for some weeks. Further studies showed that the N-hydroxy derivative was considerably more carcinogenic than the parent compound (Miller et al., 1961).

However, the lack of nonenzymatic reactivity of N-hydroxy-2-acetylaminofluorene toward proteins and nucleic acids made it evident that further activation of the carcinogen must occur *in vivo*. Having synthesized esters of N-hydroxy-N-methyl-4-aminoazobenzene and N-hydroxy-2-acetylaminofluorene for other reasons, we were impressed by their nonenzymatic electrophilic reactivity with proteins and nucleic acids under physiological conditions, and accordingly we used them as models for reactive metabolites that might be formed *in vivo*. In a series of studies we and our colleagues showed that adducts isolated from the liver nucleic acids or proteins of rats treated with N-hydroxy-2-acetylaminofluorene were identical to those obtained by nonenzymatic reactions of esters of this carcinogen with guanine nucleosides or methionine (Miller, 1970) (Figure 16.3). Correlative studies with DeBaun (DeBaun et al., 1970) pointed to the importance of the sulfuric acid ester of N-hydroxy-2-acetylaminofluorene for hepatic carcinogenesis, and studies with Veronica Maher and Waclaw Szybalski (Maher et al., 1968) demonstrated that this and other reactive electrophilic esters are direct mutagens for bacterial transforming DNA. Subsequent studies in a number of laboratories showed the formation of other electrophiles and of other nucleic acid adducts from N-hydroxy-2-acetylaminofluorene (Miller and Miller, 1981). The relative im-

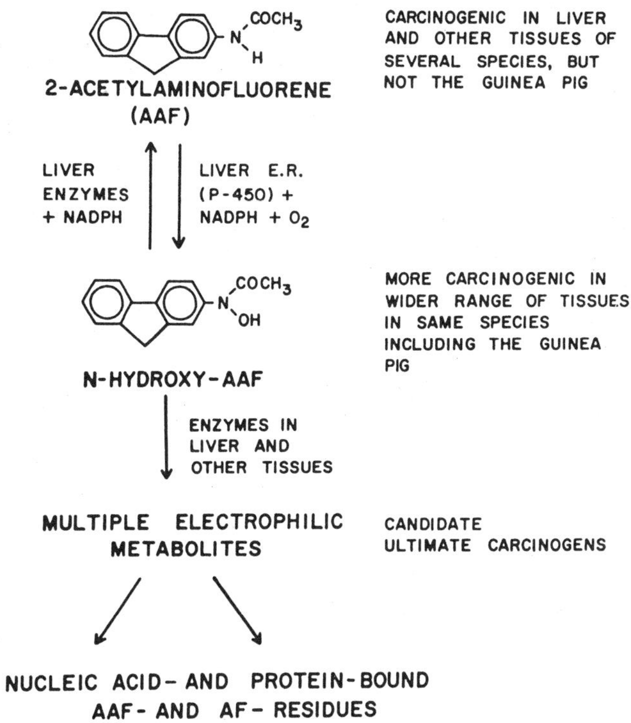

Figure 16.2 The metabolic N-hydroxylation of 2-acetylaminofluorene to a more potent carcinogen.

portance of each electrophile and each adduct in carcinogenesis by this compound is still under study (Miller and Miller, 1981).

From our studies and those of other investigators on carcinogens of a variety of structures, we suggested in 1969 that the common structural feature of chemical carcinogens for the initiation of carcinogenesis is the formation of strong electrophilic reactants (i.e., electron-deficient reactants) that can covalently bind to the informational macromolecules of target tissues (Miller and Miller, 1969) (Figure 16.4). This concept has now been generally accepted as a basic requirement for the initiation of carcinogenesis, and much attention in a number of laboratories has been focused in the past fifteen years on the DNA adducts formed from at least a dozen structurally different families of carcinogens as a consequence of their metabolism to electrophilic reactants (Miller and Miller, 1981). The studies in our laboratory have contributed especially to the knowledge of the activation of aflatoxin B_1, ethyl carbamate, 2-acetylaminofluorene, N-methyl-4-aminoazobenzene, and the alkenyl benzene derivatives safrole and estragole (Miller and Miller, 1981). Recent studies have provided a relatively direct demonstration that in mouse liver 1′-sul-

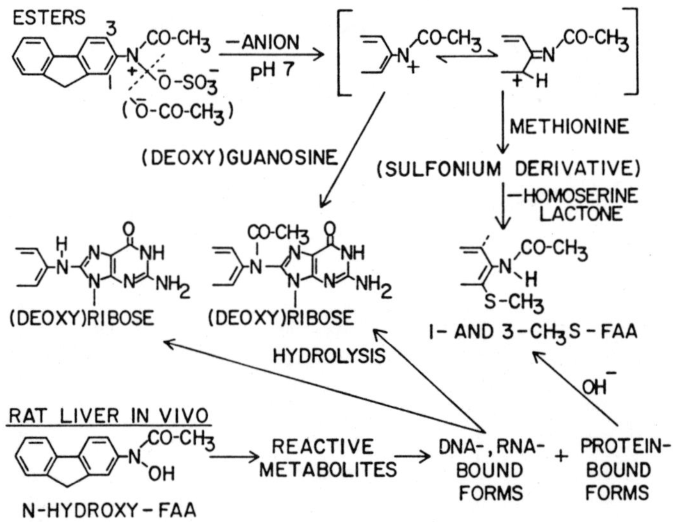

Figure 16.3 The formation of identical adducts from (deoxy)guanosine and methionine on reaction nonenzymatically with esters of N-hydroxy-2-acetylaminofluorene and on degradation of the liver nucleic acids and proteins from rats treated with N-hydroxy-2-acetylaminofluorene. For further details see Miller (1970); Miller and Miller (1981).

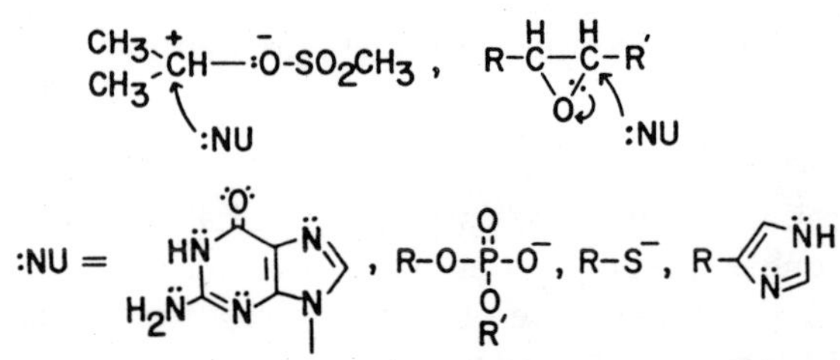

Figure 16.4 Examples of strong electrophilic reactants (positive ions or uncharged molecules with electron-deficient atoms) and their reactions with nucleophiles (:NU) through sharing of electron pairs of electron-rich atoms.

foöxysafrole is both the major ultimate electrophilic reactant and the major carcinogenic metabolite formed from 1′-hydroxysafrole, a proximate carcinogenic metabolite of safrole (Boberg et al., 1983).

Carcinogens of possible clinical interest Members of the Clinical Oncology Division of the Department of Surgery (1952 to 1973) and, since 1973, of the Wisconsin Clinical Cancer Center have also had a continuing interest in carcinogenesis by chemicals, especially those of possible clinical interest.

In the 1950's and early 1960's George Bryan, James Price, and Raymond Brown, simultaneously with Bonser, Boyland, and their colleagues (Bonser et al., 1958; Boyland, 1958) in England, examined the carcinogenicity for the mouse urinary bladder of aromatic amine metabolites of tryptophan and assessed tryptophan metabolism in humans in relation to a possible etiologic role of tryptophan in human urinary bladder cancers (Bryan et al., 1964; Bryan, 1969). Subsequent studies by Bryan, Price, and their colleagues focused on the carcinogenicity and metabolism in rodents of nitrofuran derivatives (Bryan, 1978), which had been introduced as antibacterial and antiprotozoal agents for clinical use. Similarly, the carcinogenic activity of bracken fern, sometimes used as a food for humans and often consumed by grazing animals, has been studied extensively by Pamukcu, Bryan, and Price in experimental animals (Pamukcu and Bryan, 1979).

On the multi-stage nature of carcinogenesis A major observation in the early 1940's was that skin carcinogenesis induced by chemicals can be separated into two qualitatively different stages. The early studies of Rous (Rous and Kidd, 1941) in this country and of Mottram (1944) and Berenblum (1941) in England were refined by Berenblum and Shubik in 1947. The principle of Berenblum and Shubik's experiment, as further developed by Boutwell (1964), is shown in Figure 16.5. Thus, a small "initiating" dose of a polycyclic aromatic hydrocarbon applied to mouse skin was not sufficient by itself to cause the appearance of tumors. However, subsequent repetitive applications of a non-carcinogenic "promoting agent" caused a high incidence of tumors to develop. These two stages of skin carcinogenesis are now widely known as "initiation" and "promotion." In working with this system at McArdle and using purified diets, Boutwell and Rusch (Boutwell, 1964; Boutwell et al., 1949) showed that severe caloric restriction had no effect on the initiation stage but greatly in-

```
                                        TUMOR INCIDENCE
I                                             0
I ←—P—→                                      HIGH
I (UP TO 1-YR        ←—P—→                   HIGH
   INTERVAL)
←—P—→   I                                     0
←—P—→                                         0

I = 50 µG 7,12-DIMETHYLBENZ(A)ANTHRACENE
←—P—→ = 10 µG TETRADECANOYL PHORBOL
   ACETATE, 2X/WK FOR 12 OR MORE WKS
```

Figure 16.5 Diagrammatic representation of the two-stage induction of skin tumors in the mouse. Papillomas develop in 12–20 weeks and carcinomas in about one year. See Boutwell (1964) for further details.

hibited promotion. These findings, with those of Tannenbaum (1959) in Chicago, indicated that the modulating effects of diet on mouse skin tumorigenesis were expressed primarily during the promotion stage. Boutwell, Brush, and Rusch (1948) also showed a role of the pituitary-adrenal axis in tumor inhibition in calorie-restricted mice.

Boutwell (1964) pioneered in the selective breeding of mice that were either especially sensitive to or resistant to the induction of skin tumors by the application of a single dose of initiator and repetitive doses of croton oil. Skin-tumor sensitive mice developed from his original stock are now commercially available.

In 1964 Boutwell showed that the promotion stage could be subdivided into two stages. The first stage, which he termed "conversion," was dependent on the classical promoting agent croton oil. The second stage, which Boutwell termed "propagation," could be induced by croton oil or other agents or processes that stimulated cell division, some of which were unable to elicit "conversion." Boutwell viewed conversion as a specific intracellular change that resulted in a dormant tumor cell and propagation as a process that facilitated replication of these cells. Exploitation of the three-stage process was hampered by the limited number and the lack of purity of the promoting agents then available. However, the concept has been recently confirmed and extended by Slaga, a former postdoctoral student with Boutwell, and his colleagues (Slaga et al., 1980) with the potent pure agents now available.

By the mid-1960's the structure of the major potent tumor-promoting component of croton oil had been identified as 12-O-tetradecanoylphorbol-13-acetate (TPA) by Hecker (1971) and his colleagues in Heidelberg and by Van Duuren and his associates (Van Duuren and Orris, 1965) in New York, and a series of phorbol diesters which differed in their tumor-promoting activities by orders of magnitude were available (Hecker, 1971). Hecker developed an interest in this problem from his contacts with Boutwell while doing postdoctoral work with Mueller. Starting in the mid-1960's and using first croton oil and later TPA, Boutwell and his colleagues (Hennings and Boutwell, 1970; Boutwell, 1974) carried out a long-term program to elucidate the biochemical changes involved in tumor promotion in mouse skin. The responses in the mouse epidermis to the application of the phorbol ester included sequential increases in phospholipid, RNA, protein, and DNA synthesis within the first day (Figure 16.6). These biochemical responses to a single dose of promoter were dissipated within a few days, but further doses amplified the responses. Thus, Hennings and Boutwell (1970) viewed skin tumor promoters as reversible derepressors of gene activity.

In pursuing the elucidation of early biochemical changes, Boutwell, O'Brien and their colleagues (Boutwell, 1977, 1978; Boutwell et al., 1979) observed striking increases in mouse epidermal cells of ornithine decarboxylase activity, which reached peak values in the mouse epidermal cells of 200–300 times the basal level at 5 hr after application of TPA. As shown later by Verma

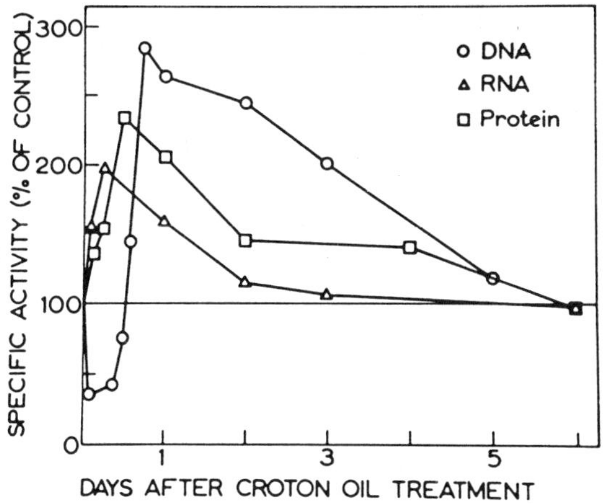

Figure 16.6 The sequential increases in RNA, protein, and DNA in the skin of mice after a single topical application of croton oil. From Hennings and Boutwell (1970).

and Boutwell (1977), application before the phorbol ester of retinoic acid greatly reduced this increase in epidermal ornithine decarboxylase activity (Figure 16.7). Similar applications of retinoic acid before each dose of TPA markedly inhibited skin tumor promotion. These and other studies from Boutwell's laboratory (Boutwell, 1977, 1978; Boutwell et al., 1979) have provided

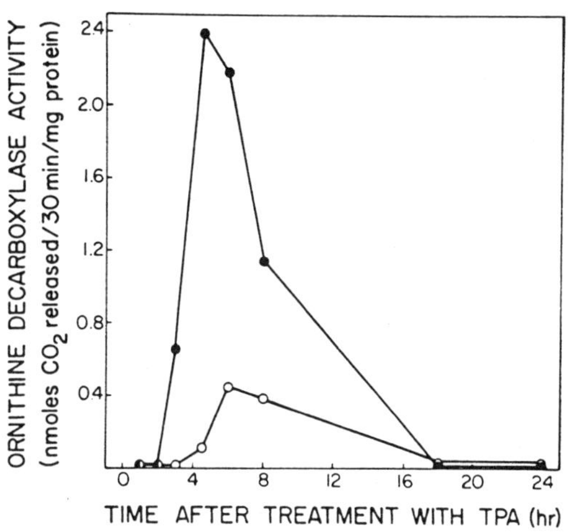

Figure 16.7 The ornithine decarboxylase activity of the skin of mice given a single topical application of 17 nmol of TPA with (open circles) or without (solid circles) a treatment, 1 hour previously, with 1.7 nmol of trans-retinoic acid. From Verma and Boutwell (1977).

a strong correlation between tumor promotion and the early rapid increase in ornithine decarboxylase activity in mouse skin. The mechanisms through which the increased activity of ornithine decarboxylase may be related to tumor promotion presumably include the role of this enzyme as a rate-limiting factor for the synthesis of the polyamines and a role for the polyamines in the regulation of DNA, RNA, and protein synthesis.

Because Hecker had been a postdoctoral fellow with Mueller, Mueller early had a supply of pure TPA. His demonstration with Kajiwara (Mueller and Kajiwara, 1965) in 1965 that very low levels of TPA inhibited replication of HeLa cells in culture was the forerunner for the broad pleiotropic effects of TPA observed by Mueller and his colleagues (Mastro and Mueller, 1974; Mueller and Wertz, 1982) and by many other investigators (Hecker et al., 1982) on the expression of genes and on the differentiation of a variety of cultured cells.

By the early 1960's limited data for experimental cancers in several tissues were consistent with division of the processes leading to these cancers into stages of initiation and promotion (Boutwell, 1964). However, acceptance of the concept of separable initiation and promotion stages as a general property of chemical- and radiation-induced carcinogenesis, regardless of the cell type or tissue or the initiating carcinogen, has occurred only in the past ten years, as better-defined initiation-promotion systems were developed for several tissues (Hecker et al., 1982). Except for mouse skin, the most studied tissue has been the rodent liver, for which a large body of data on the pathological and biochemical changes which accompany the neoplastic process has developed over the past 30 years (Emmelot and Scherer, 1980; Farber, 1980; Pitot and Sirica, 1980). One of the prominent investigators of the initiation-promotion sequence in liver carcinogenesis has been Henry Pitot of the McArdle laboratory.

One of Pitot's contributions in this area was the development of a protocol in which a single treatment of rats which had been partially hepatectomized 24 hours previously with diethylnitrosamine or 7,12-dimethylbenz(a)anthracene initiated the development of hepatic carcinomas that became evident on long-term treatment (promotion) with phenobarbital (Pitot et al., 1982). Based on earlier data by others, especially from Emmelot and Scherer (1980) in Amsterdam, Pitot and his colleagues (Campbell et al., 1982) have developed histochemical and stereological methods for the quantification of the number and size of foci of altered hepatic parenchymal cells as measures of initiation and promotion, respectively. Although the exact relationship of these relatively numerous clones of histochemically detectable cells is not yet clear, there is a strong quantitative correlation between the number of the foci and the probability of later development of hepatic carcinomas. Of recent interest is the finding of Pitot, Poland, and their colleagues (Pitot et al., 1980) that the environmental toxin 2,3,7,8-tetrachloro-*p*-dioxin is a very potent promoter of hepatic carcinogenesis in the rat.

Malignant transformation of cultured cells The reports of Ilse Lasnitzki (1963) in England during the late 1950's on the massive hyperplasia and squamous metaplasia induced in organ cultures of rodent prostates by small doses of carcinogenic hydrocarbons resulted in Heidelberger's decision to spend a year in her laboratory. Although Heidelberger's early goal of establishing a system for malignant transformation of mouse prostate organ cultures was not successful, work in his laboratory with these cultures led to the development of clones of mouse prostate cells which could be maintained indefinitely in culture, which had a low spontaneous transformation rate, and which could be malignantly transformed on exposure to chemical carcinogens (Heidelberger, 1973) A few years later C. Reznikoff, Heidelberger and their colleagues established the now famous C3H/10T1/2 cells from C3H mouse embryos for quantitative transformation of fibroblasts in culture (Reznikoff et al., 1973). Both the lines established in Heidelberger's laboratory and the hamster embryo fibroblast assay developed by Berwald and Sachs in Israel in 1963 have been studied intensively in many laboratories in attempts to relate the metabolism of chemical carcinogens and the biochemical and biological changes in the cells to malignant transformation (Heidelberger, 1973, 1981).

Viral Carcinogenesis

As noted above, the study of viral carcinogenesis developed in parallel to the study of chemical carcinogenesis during the first few decades of this century. A major interest in viral carcinogenesis in the early part of the century was in the identification of tumors that could be propagated by cell-free preparations and the conditions under which the transmission occurred. With the development of this information and of cellular and molecular biology, the properties of the oncogenic viruses (i.e., those that can cause malignant transformation) and the mechanisms by which they elicit the development of tumors became major focuses of research.

About 40 years after Rous' discovery of the virus, interest in the Rous sarcoma virus was revived. Among the investigators attracted to this field was Howard Temin, then a graduate student with Renato Dulbecco at the California Institute of Technology. Temin, with Harry Rubin, undertook quantitative analyses of the infection of cultured chicken cells by the virus. Out of these studies came Rubin and Temin's conclusion in 1959 that "... it appears that the genetic information required for virus production has been incorporated into the cell's genome during the first division ..." (Rubin and Temin, 1959). They further inferred that "... since the genetic information of the virus is probably incorporated into the host genome, the latter is modified in a manner which cannot be distinguished from a mutation."

These ideas about the mechanism of action of Rous sarcoma virus provided the basis on which Temin continued his research after moving to the McArdle Laboratory in 1960. By using a low ratio of virus to cells for infection Temin (1962) demonstrated the formation of morphologically transformed, nonvirus-producing cells which yielded sarcomas on injection into chickens. Thus, the process of transformation was separated from virus production.

The central role of the incorporation into infected cells of one or two regularly inherited copies of genes from the virus and the concept that the expression of these viral-derived genes was a key factor in the transformation of normal cells to malignant cells were expressed as shown in Figure 16.8 from a review by Temin in 1966. Data on a possible role of DNA synthesis in the infection of cells by Rous sarcoma virus and in the induction of malignancy were presented by Temin starting in 1964. One approach used actinomycin D, 5-fluorodeoxyuridine, amethopterin, and other inhibitors to show that preventing DNA synthesis in cells soon after they were exposed to the virus prevented virus infection and the initiation of virus production (Temin, 1963, 1964b, 1964c). Similarly, exposure to visible light of cells which had incorporated 5-bromodeoxyuridine immediately after infection with limited doses of Rous sarcoma virus in the G_1 stage of the cell cycle readily inactivated the virus and prevented synthesis of new virus (Boettiger and Temin, 1970). However, after this early phase of infection was completed, inhibition of DNA synthesis had no effect on virus production. More direct evidence for an integrated DNA-containing provirus in the cellular DNA was obtained through early DNA-RNA hybridization studies (Temin, 1964a). Some similar results were reported by Bader from the National Cancer Institute (1964, 1965). However, since, as shown by Crawford and Crawford in 1961, RNA appeared to be the only nucleic acid in the virus, the concept of an integrated DNA provirus met with considerable resistance.

In 1970 Temin and Mizutani reported the occurrence in Rous sarcoma virus of an RNA-directed DNA polymerase. As shown in Table 16.1 the

$\text{CELL}_{\text{NORMAL}} + \text{RSV}_{\text{RNA}}$

↓ FORMATION OF NEW GENES

$\text{CELL}(\text{RSV}_{\text{DNA}})_{\text{NORMAL}}$

↓ ACTIVATION OF NEW GENES

$\text{CELL}(\text{RSV}_{\text{DNA}} + \text{GENE PRODUCTS})_{\text{NORMAL}}$

↓ EXPRESSION OF NEW GENES

$\text{CELL}(\text{RSV}_{\text{DNA}} + \text{GENE PRODUCTS})_{\text{TUMOR}}$

Figure 16.8 The postulated stages in the change from a normal cell to a malignant cell after infection with Rous sarcoma virus, as proposed by Temin in 1966.

Table 16.1 RNA-dependent DNA polymerase activity in Rous sarcoma virions*

Incubation conditions	[^{3}H]TTP incorporation (% of control)
Control (complete system with disrupted virions)	(100)
Minus virions	2
Disrupted virions pretreated with RNase	0
Minus dATP, dCTP, or dGTP	15–35
Minus Mg^{++}	3

*Adapted from Temin and Mizutani (1970).

activity of the polymerase was destroyed by RNase and required all of the deoxynucleoside triphosphates. An entirely analogous finding was reported simultaneously for the Rauscher murine leukemia virus by Baltimore (1970) from the Massachusetts Institute of Technology. Thus, although the final proof for the DNA provirus depended upon later work involving nucleic acid hybridization and infectious DNA experiments (Cooper and Temin, 1975), the demonstration of the RNA-directed DNA polymerase resulted in general acceptance of the integrated DNA provirus as an intermediate in the replication and induction of malignancy by this class of carcinogenic RNA viruses, which are now known as retroviruses.

The demonstration of RNA-directed DNA synthesis and the idea that it may be a relatively common biological phenomenon have had major effects on molecular biological methodology and biological concepts. Thus, this background led Temin (1971) to formulate his "protovirus" hypothesis, even before his demonstration with Kang (Kang and Temin, 1973) of RNA-directed DNA synthesis in normal cells. The protovirus hypothesis suggested that specific DNA sequences in normal nuclear DNA may in some cases be copied as RNA. At some frequency this RNA may then be transcribed and integrated into nuclear DNA, either in the same cell or in another cell (Figure 16.9). This new DNA transcript could be either an exact copy of the original DNA sequence or a modified sequence, and it was conceived of as integrating either next to the original DNA sequence or at some other chromosomal locus. In any case, to the extent that such DNA-to-RNA-to-DNA transcriptions and integration of the DNA took place, the model provided opportunities for formation of multiple copies of genes, modified copies of genes, and copies of genes that were under different control than was the original sequence in the DNA. In the last few years the ideas from Temin's protovirus hypothesis and the oncogene hypothesis of Huebner and Todaro (1969) of the National Cancer Institute have been fused with much new data from the laboratories of a number of investigators to yield new insights into possible modes of origin of spontaneous tumors and those caused by radiation or chemicals (see below).

PROTOVIRUS REGION OF NORMAL CELLULAR DNA

↓ DNA-DIRECTED RNA POLYMERASE

RNA COPY

↓ RNA-DIRECTED DNA POLYMERASE

DNA COPY

↓

NEW PROTOVIRUS COPY INTEGRATED INTO
DNA OF SAME OR DIFFERENT CELL

Figure 16.9 The steps in protovirus evolution, as proposed by Temin in 1971.

In the past decade the study of viral carcinogenesis at McArdle has been enlarged with the addition of three younger staff members—Bill Sugden, who is studying the Epstein-Barr virus; Rex Risser, who is interested in the Moloney and Abelson viruses; and Janet Mertz, who is examining the expression of the SV40 genome.

Studies on the Control of Growth and Differentiation

Normal cells have characteristic functions and grow under homeostatic control, while cancer cells frequently have a reduced capacity for expression of the normal functions of the homologous cells and are under less strict growth control. The biochemical features underlying these differences between normal and neoplastic cells have been investigated in many laboratories through a variety of approaches. The resulting knowledge of the biochemical differences between normal and neoplastic cells has had important impacts in several areas, including cancer treatment, but this comparative biochemistry is considered here only in relation to its contributions to our understanding of carcinogenesis.

Intermediary metabolism and its control in normal and tumor cells The principal contributor in this area from the University of Wisconsin was Potter, who from his training with Elvehjem, von Euler, and Krebs approached the problem of carcinogenesis through comparative analyses of intermediary metabolism. When Potter joined the McArdle Laboratory in 1940, the data on the biochemical differences between normal and tumor tissues, especially between tumor tissues and their cells of origin, were very limited. Potter's early studies emphasized quantitative comparisons, especially of the succinoxidase system, the Kreb's citric acid cycle, and oxidative phosphorylation in normal and malignant tissues (Potter, 1944, 1951). Beginning in the early 1950's Potter and his colleagues also directed much attention to the enzymatic synthesis of nucleic acid precursors in normal and neoplastic tissues. Among other findings, the latter studies resulted in his observation with Schmitz and Hurlburt (Schmitz et al., 1954) in 1954 of the mono-, di-, and triphosphates of cytidine, guanosine, and uridine as normal tissue metabolites.

Potter's approach to research was to develop generalizations based on the current forefront of knowledge as standards against which to design and interpret his experiments. Potter's early ideas encompassed "protein deletion" and "catabolic deletion" hypotheses (Potter, 1958). These hypotheses were based on the many enzymatic differences between tumor tissues and their tissues of origin, on the inherent need of tumor cells for the enzymatic machinery for DNA replication and cell division, and on the presumed lesser need of tumor cells for more differentiated functions. As the data accumulated, studies in Potter's laboratory, as well as in a number of other laboratories, made it evident that there were exceptions with one tumor or another to each enzyme suggested as a key to the maintenance of the normal, as contrasted to the malignant, state. As Potter wrote with De Verdier in 1960 (De Verdier and Potter, 1960):

> The considerations mentioned by Dr. Greenstein and the immense amount of data that he compiled compel us to realize that tumors must have a combination of uniformity and diversity in their enzyme pattern. To be realistic we must admit that the definitive enzymatic difference between a cancer cell and the normal cell from which it arose is yet to be established *in a single instance.* If this is true, we may as well seek for a cancer cell that differs from a normal prototype as little as possible and try to narrow the spectrum of enzymatic differences that exist in order to decide which differences are definitive.

Accordingly, when Harold Morris of the National Cancer Institute found that rats fed low doses of hepatocarcinogens developed well-differentiated hepatomas which could be transplanted and which retained differentiated characteristics (Morris and Wagner, 1968), Potter (1961) was the one who recognized the potential of these tumors for giving new insights into the *critical* changes that occur during the development of malignancy. Potter coined the term "minimal-deviation tumor," and he convinced Morris to expand his collection and to make the well-differentiated hepatic tumors available to other investigators. After Henry Pitot joined Potter as a postdoctoral fellow in 1959 and one year later became an independent investigator at the McArdle Laboratory, they worked both collaboratively and independently to demonstrate the broad scope of enzymatic capacities of the minimal-deviation hepatomas (Figure 16.10). For any given enzyme, some of the tumors had activities similar to those of normal adult liver while others deviated widely from normal. The studies of Potter and Pitot, with both tissues from the animal and from cell cultures, also demonstrated the broad spectrum of regulation of enzyme activities that can accompany malignancy, as well as the wide

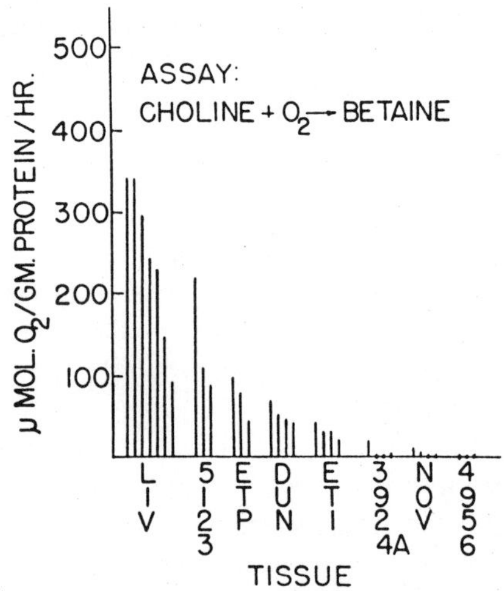

Figure 16.10 Comparative activities of normal adult rat liver and a variety of hepatic tumors for oxidation of choline to betaine. LIV refers to normal adult rat liver and each of the other abbreviations to a strain of transplantable hepatocarcinoma. From Pitot (1960).

spectrum of enzyme activities and enzyme regulation which are in the "normal range" of fetal, non-growing adult, or normal regenerating liver (Pitot, 1960; Pitot et al., 1964; Potter, 1964; Potter and Watanabe, 1968; Watanabe et al., 1969; Potter, 1973; Pitot, 1975). These studies led Potter (1978) recently to the concept that tumors are clones of cells which are arrested in various stages of development from the stem cell—i.e., that "oncogeny is blocked ontogeny." Potter (1978) recognized that the mechanisms by which such blocked ontogeny and altered feedback could theoretically develop covered a wide range of mechanisms, which might include both alteration of the genome by virus insertion or chemical attack and alteration of cell membranes and their receptors. Potter also emphasized that "the point is not that cancer tissue resembles foetal tissue in some respects, it is that in important ways cancer tissue *does not* resemble foetal tissue. The difference lies in the word *blocked.*" He suggested that "the concept must include 'partially blocked ontogeny' in which a substantial *fraction* of the tumor cell *population* undergoes extensive differentiation but does not progress to the normal organized terminal state" (Potter, 1978).

Factors controlling cellular replication and differentiation Three other investigators at McArdle have also examined aspects of the regulation of cellular replication and differentiation. In the course of his early studies on the culture of normal and virus-transformed chicken fibroblasts, Temin (1967) observed that a major role of serum in facilitating the growth of the normal cells in culture was to supply multiplication-stimulating factor(s) which were not needed or were needed in lesser amounts by the virus-transformed cells. Dulak and Temin (Dulak and Temin, 1973; Temin et al., 1974) further observed that a line of cultured malignant rat liver cells secreted large amounts of material with multiplication-stimulating activity similar to that of serum and demonstrated that the activity resided in a family of small polypeptides. Subsequent to this study, a series of multiplication-stimulating factors for cultured cells has been identified by other investigators as products of various cultured normal and malignant cells.

Of similar importance for an understanding of carcinogenesis have been the extensive studies of Mueller and his associates (Mueller, 1971, 1975) on the elucidation of factors that control the cell cycle and the elucidation of cellular mechanisms that determine whether a cell in G_0 will stay in that quiescent state or will reenter the cycle and prepare for cell division. The studies from Mueller's laboratory have emphasized that the cell cycle is a tightly coupled sequence of events with control points which determine specific cellular activities. Both Mueller and Jeffrey Ross and their associates have used cultured Friend erythroleukemic cells as models for the control of erythroid differentiation. Ross and Sautner (1976) demonstrated the requirement of heme for this erythroid differentiation, and subsequent studies by Mueller and his colleagues (Lo et al., 1981) have examined the interplay of other factors in this process.

Perspectives

Although a few observations suggested some years ago that chemical carcinogens or X-radiation and specific viruses might in some cases complement each other in the induction of malignancy, until quite recently the data on the induction of malignancy by chemicals and radiations and by viruses have not overlapped in identifiable ways. Each process has been quite a separate area of study. Fortunately, new knowledge obtained over the past few years now suggests a more unified view of carcinogenesis, regardless of the inciting agent, although the details of the interactions are not yet clear (Figure 16.11).

An important new principle is the finding, starting with the work of Stehelin and his colleagues (Stehelin et al., 1976) in San Francisco and of Frankel and Fischinger (1976) at the National Institutes of Health, that apparently normal cells contain DNA sequences in their nuclear DNA that are homologous to major portions of the transforming sequences in the nucleic acids of strongly-transforming retroviruses. Nearly twenty of these DNA sequences have now been identified in some normal or cancer cells, and some have been identified in an altered form in a series of cancer cells (Bishop, 1983). Since the transforming sequences from the viral nucleic acids were

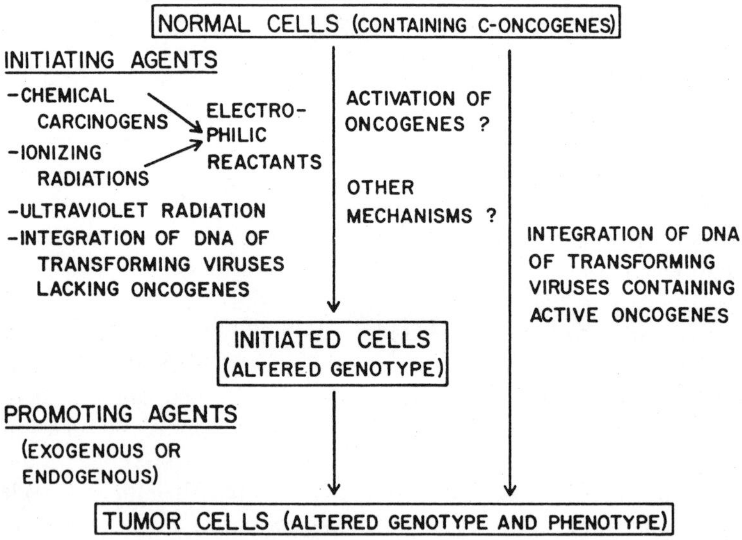

Figure 16.11 A current view of the malignant transformation of cells by transforming viruses, chemical carcinogens, and carcinogenic radiations.

earlier termed oncogenes, the viral sequences are now known as *v*- (for viral-) oncogenes. Their cellular homologs are called *c*- (for cellular-) oncogenes or proto-oncogenes. In an evolutionary sense the cellular oncogenes apparently came first, and these sequences were apparently incorporated, as an infrequent event, into infecting retroviruses, which then became strongly transforming viruses (Linial and Blair, 1982). The term proto-oncogene for each of these sequences in normal cellular DNA may be a misnomer if, as seems likely, these sequences are eventually found to code for proteins essential for normal cellular activities.

As noted above, it has been evident for some years that normal cells can be converted to tumor cells by the integration into the DNA of a transforming virus containing an active oncogene. The Rous sarcoma virus is a prototype of such viruses, and its oncogene, called *v-src*, was the first to be recognized (Linial and Blair, 1982). Integration of a DNA copy of a variant Rous virus which contains a temperature-sensitive mutation in the *src* gene results in cells that have the tumor phenotype at permissive temperatures and the normal, nonmalignant phenotype at higher, nonpermissive temperatures (Linial and Blair, 1982). Thus, the protein product of *src* is required, not only for the conversion of normal to malignant phenotype, but it is also required for maintenance of the malignant phenotype. Similarly, for other oncogenic retroviruses and for certain of the oncogenic DNA viruses, including polyoma and SV40, DNA sequences have been identified that are essential for the malignant transformations caused by these viruses (Topp et al., 1980; Linial and Blair, 1982).

The molecular mechanisms of transformation are less well defined for the tumors induced by chemicals and radiations. As noted above, there is strong biological evidence for at least a two- or three-step process for carcinogenesis induced by chemicals, ionizing radiations, and ultraviolet light, and this situation can be demonstrated both in the whole animal and in cell cultures. The initiation stage of carcinogenesis induced by each of these agents appears to involve modification of cellular DNA, although the specific types of DNA modification that result in initiation are not yet identified. Thus, while much emphasis has been placed on single base-pair changes, additions, or deletions, alterations that may place a gene (*c*-oncogene?) under different control could also result in initiation. By analogy, the integration of DNA of a transforming virus can also be considered as an initiating event. The integration of viral information into cellular DNA has been termed insertional mutagenesis (Bishop, 1983) in recognition of the new relationships between the DNA sequences. According to current views, the insertion of viral sequences may alter the nucleotide sequence of the *c*-oncogene of the normal cell and/or promote its transcription at greater than normal levels (Bishop, 1983). Of obvious relevance to this idea are the new cytogenetic findings on some human and mouse tumors (Rowley and Testa, 1982; Yunis, 1983). Thus, new chromosome-banding procedures are revealing increasing numbers of specific chromosomal abnormalities that are uniquely associated with specific malignancies. The fact that these chromosomal abnormalities appear frequently at or near the locus of a *c*-oncogene provides further support for the role of modifications of a *c*-oncogene or the activation of a *c*-oncogene as a factor in carcinogenesis.

The natures of the protein products coded for by oncogenes and thus apparently responsible for the transformation and for the maintenance of the malignant state are important questions. The first of these products to be identified were protein kinases that phosphorylate specific tyrosine residues in proteins (Bishop, 1983). Two closely related oncogenes code for products that phosphorylate one of the threonine residues in the oncogene product and also bind guanosine triphosphate, while a third oncogene product is a DNA-binding protein localized in the cell nucleus (Bishop, 1983). Another oncogene codes for a protein with a strong homology to the platelet-derived growth factor (Waterfield et al., 1983). In relation to the protein kinase activities of some of the oncogene products, it is of considerable interest that investigators, including Ashendel, Boutwell, and their colleagues (Ashendel et al., 1983), have provided evidence that the membrane receptor for the tumor-promoting phorbol ester TPA has protein kinase activity.

Until recently most studies on the transformation of cultured cells by the integration of oncogenes utilized the highly aneuploid NIH 3T3 cells, which can also assume the malignant phenotype as a response to alterations in their method of culture (Weinberg, 1982). Most diploid cell lines appeared not to be susceptible. However, recent studies show that malignant transformation of primary cultures of diploid fibroblasts can be induced by the introduction of two oncogenes; the order of transfection with the two oncogenes appears not to be important (Land et al., 1983; Ruley, 1983). These data, together with the apparent relationship between a phorbol ester receptor and a protein kinase activity, may indicate a role for some oncogenes in the promotion phase of carcinogenesis.

Research during the next few years should greatly clarify the roles of the oncogenes in malignant transformations and provide new perspectives on the nature of the overall process. However, it is important to recognize, as emphasized in a recent review by Temin (1983), that cancer, the disease, is still not understood. Only further research, possibly in currently uncharted areas, is likely to provide the knowledge to permit effective prevention and/or interruption of the process of malignant transformation on a broad scale.

Acknowledgements

The authors are indebted to the National Cancer Institute, U.S. Public Health Service (CA-07175 and CA-22484) for support of their research. We also wish to express our deep thanks to Professors Harold P. Rusch, Van R. Potter, and Henry C. Pitot for their efforts in establishing and maintaining the stimulating research environment at McArdle that we and many others have enjoyed during the past four decades.

References

Abell, C. W. and Heidelberger, C. (1962) Interaction of carcinogenic hydrocarbons with tissues. VIII. Binding of tritium-labeled hydrocarbons to the soluble proteins of mouse skin. *Cancer Res.* 22: 931–946.

Ashendel, C. L., Staller, J. M., and Boutwell, R. K. (1983) Protein kinase activity associated with a phorbol ester receptor purified from mouse brain. *Cancer Res.* 43: 4333–4337.

Bader, J. P., (1964) The role of deoxyribonucleic acid in the synthesis of Rous sarcoma virus. *Virol.* 22: 462–468.

Bader, J. P. (1965) The requirement for DNA synthesis in the growth of Rous sarcoma and Rous-associated viruses. *Virol.* 26: 253–261.

Bain, J. A. and Rusch, H. P. (1943) Carcinogenesis with ultraviolet radiation of wavelength 2,800–3,400 Å. *Cancer Res.* 3: 425–430.

Baltimore, D. (1970) RNA-dependent DNA polymerase in virions of RNA tumour viruses. *Nature* 226: 1209–1211.

Baumann, C. A., Jacobi, H. P., and Rusch, H. P. (1939) The effect of diet on experimental tumor production. *Am. J. Hygiene* 30: 1–6.

Baumann, C. A. and Rusch, H. P. (1939) Effect of diet on tumors induced by ultraviolet light. *Am. J. Cancer* 36: 213–221.

Berenblum, I. (1941) The cocarcinogenic action of croton resin. *Cancer Res.* 1: 44–48.

Berenblum, I. and Shubik, P. (1947) A new quantitative approach to the study of the stages of chemical carcinogenesis in the mouse's skin. *Brit. J. Cancer* 1: 383–391.

Berwald, Y. and Sachs, L. (1963) *In vitro* cell transformation with chemical carcinogens. *Nature* 200: 1181–1184.

Bishop, J. M. (1983) Cellular oncogenes and retroviruses. *Ann. Rev. Biochem.* 52: 301–354.

Boberg, E. W., Miller, E. C., Miller, J. A., Poland, A., and Liem, A. (1983) Strong evidence from studies with brachymorphic mice and pentachlorophenol that 1′-sulfoöxysafrole is the major ultimate electrophilic and carcinogenic metabolite of 1′-hydroxysafrole in mouse liver. *Cancer Res.* 43: 5163–5173.

Boettiger, D. and Temin, H. M. (1970) Light inactivation of focus formation by chicken embryo fibroblasts infected with avian sarcoma virus in the presence of 5-bromodeoxyuridine. *Nature* 228: 622–624.

Bonser, G. M., Clayson, D. B., and Jull, J. W. (1958) Some aspects of the experimental induction of tumours of the bladder. *Brit. Med. Bull.* 14: 146–152.

Boutwell, R. K. (1964) Some biological aspects of skin carcinogenesis. *Progr. Exp. Tumor Res.* 4: 207–250.

Boutwell, R. K. (1974) The function and mechanism of promoters of carcinogenesis. *CRC Crit. Rev. Toxicol.* 2: 419–443.

Boutwell, R. K. (1977) The role of the induction of ornithine decarboxylase in tumor promotion, in *Origins of Human Cancer, Book B* (Watson, J. D., Hiatt, H. H., and Winsten, J. A., eds.), pp. 773–783. Cold Spring Harbor Laboratory, Cold Spring Harbor, New York.

Boutwell, R. K. (1978) Biochemical mechanism of tumor promotion, in *Carcinogenesis* (Slaga T. J., Sivak, A., and Boutwell, R. K., eds.), Vol. 2. Mechanisms of Tumor Promotion and Cocarcinogenesis, pp. 49–58. Raven Press, New York.

Boutwell, R. K., Brush, M. K., and Rusch, H. P. (1948) Some physiological effects associated with chronic caloric restriction. *Am. J. Physiol.* 154: 517–524.

Boutwell, R. K., Brush, M. K., and Rusch, H. P. (1949) The stimulating effect of dietary fat on carcinogenesis. *Cancer Res.* 9: 741–749.

Boutwell, R. K., O'Brien, T. G., Verma, A. K., Weekes, R. G., DeYoung, L. M., Ashendel, C. L., and Astrup, E. G. (1979) On the biochemical mechanisms of tumor promotion in mouse skin, in *Naturally Occurring Carcinogens—Mutagens and Modulators of Carcinogenesis* (Miller, E. C., Miller, J. A., Hirono, I., Sugimura, T., and Takayama, S., eds.), pp. 287–300. Japan Scientific Societies Press, Tokyo.

Boyland, E. (1958) The biochemistry of cancer of the bladder. *Brit. Med. Bull.* 14: 153–158.

Brookes, P. and Lawley, P. D. (1964) Evidence for the binding of polynuclear aromatic hydrocarbons to the nucleic acids of mouse skin: Relation between carcinogenic power of hydrocarbons and their binding to deoxyribonucleic acid. *Nature* 202: 781–784.

Brown, R. R., Miller, J. A., and Miller, E. C. (1954) The metabolism of methylated aminoazo dyes. IV. Dietary factors enhancing demethylation *in vitro*. *J. Biol. Chem.* 209: 211–222.

Bryan, G. T. (1969) Role of tryptophan metabolites in urinary bladder cancer. *Am. Ind. Hyg. Assoc. J.* 30: 27–34.

Bryan, G. T., ed. (1978) *Carcinogenesis*, Vol. 4. Nitrofurans: Chemistry, Metabolism, Mutagenesis, and Carcinogenesis. Raven Press, New York.

Bryan, G. T., Brown, R. R., and Price, J. M. (1964) Mouse bladder carcinogenicity of certain tryptophan metabolites and other aromatic nitrogen compounds suspended in cholesterol. *Cancer Res.* 24: 596–602.

Campbell, H. A., Pitot, H. C., Potter, V. R., and Laishes, B. A. (1982) Application of quantitative stereology to the evaluation of enzyme-altered foci in rat liver. *Cancer Res.* 42: 465–472.

Clayson, D. B. (1962) *Chemical Carcinogenesis.* Little, Brown and Co., Boston.

Conney, A. H. (1967) Pharmacological implications of microsomal enzyme induction. *Pharmacol. Rev.* 19: 317–366.

Conney, A. H. (1982) Induction of microsomal enzymes by foreign chemicals and carcinogenesis by polycyclic aromatic hydrocarbons: G. H. A. Clowes Memorial Lecture. *Cancer Res.* 42: 4875–4917.

Conney, A. H., Miller, E. C., and Miller, J. A. (1956) The metabolism of methylated aminoazo dyes. V. Evidence for induction of enzyme synthesis in the rat by 3-methylcholanthrene. *Cancer Res.* 16: 450–459.

Conney, A. H., Miller, E. C., and Miller, J. A. (1957) Substrate-induced synthesis and other properties of benzpyrene hydroxylase in rat liver. *J. Biol. Chem.* 228: 753–766.

Cooper, G. M. and Temin, H. M. (1975) Infectious DNA from cells infected with Rous sarcoma virus, reticuloendotheliosis virus or Rous-associated virus-O. *Cold Spring Harbor Symp Quant. Biol.* 39: 1027–1032.

Cramer, J. W., Miller, J. A., and Miller, E. C. (1960) N-Hydroxylation: A new metabolic reaction observed in the rat with the carcinogen 2-acetylaminofluorene. *J. Biol. Chem.* 235: 885–888.

Crawford, L. V. and Crawford, E. M. (1961) The properties of Rous sarcoma virus purified by density gradient centrifugation. *Virol.* 13: 227–232.

DeBaun, J. R., Miller, E. C., and Miller, J. A. (1970) N-Hydroxy-2-acetylaminofluorene sulfotransferase: Its probable role in carcinogenesis and protein-(methion-S-yl)-binding in rat liver. *Cancer Res.* 30: 577–595.

De Verdier, C.-H. and Potter, V. R. (1960) Alternative pathways of thymine and uracil metabolism in the liver and hepatoma. *J. Natl. Cancer Inst.* 24: 13–29.

Dulak, N. C. and Temin, H. M. (1973) Multiplication-stimulating activity for chicken embryo fibroblasts from rat liver cell conditioned medium: A family of small polypeptides. *J. Cell. Physiol.* 81: 161–170.

Ellerman, V. and Bang, O. (1908) Experimentelle Leukämie bei Hünem. *Centralblatt f. Bakteriologie, I. Abt. Orig.* 45: 595–609.

Emmelot, P. and Scherer, E. (1980) The first cell stage in rat liver carcinogenesis. A quantitative approach. *Biochim. Biophys. Acta* 605: 247–304.

Estabrook, R. W. and Lindenlaub, E., eds. (1979) *The Induction of Drug Metabolism,* Symposia Medica Hoechst 14. F. K. Schattauer Verlag, Stuttgart/New York.

Farber, E. (1980) The sequential analysis of liver cancer induction. *Biochim. Biophys. Acta* 605: 149–166.

Farber, E. and Magee, P. N. (1960) The probable alkylation of liver riboucleic acid by the hepatic carcinogens dimethylnitrosamine and ethionine. *Biochem. J.* 76: 58P.

Findlay, G. M. (1928) Ultraviolet light and skin cancer. *Lancet* 2: 1070–1073.

Frankel, A. E. and Fischinger, P. J. (1976) Nucleotide sequences in mouse DNA and RNA specific for Moloney sarcoma virus. *Proc. Natl. Acad. Sci. USA* 73: 3705–3709.

Haddow, A. (1974) Sir Ernest Laurence Kennaway FRS, 1881–1958: Chemical causation of cancer then and today. *Perspectives Biol. Med.* 17: 543–588.

Hecker, E. (1971) Isolation and characterization of the cocarcinogenic principles from croton oil, in *Methods in Cancer Research* (Busch, H., ed.), Vol. 6, pp. 439–484. Academic Press, New York.

Hecker, E., Fusenig, N. E., Kunz, W., Marks, F., and Thielmann, H. W., eds. (1982) *Carcinogenesis—A Comprehensive Survey,* Vol. 7, Cocarcinogenesis and Biological Effects of Tumor Promoters. Raven Press, New York.

Heidelberger, C. (1973) Chemical oncogenesis in culture. *Adv. Cancer Res.* 18: 317–366.

Heidelberger, C. (1981) Initiation and promotion, mutagenesis and transformation of C3H/10T1/2 mouse embryo fibroblasts. *Gann Monograph Cancer Res.* 27: 207–219.

Heidelberger, C. and Davenport, G. R. (1961) Local functional components of carcinogenesis. *Acta Unio Internat. Contra Cancrum* 17: 55–63.

Heidelberger, C. and Jones, H. B. (1948) The distribution of radioactivity in the mouse following administration of dibenzanthracene labeled in the 9 and 10 positions with carbon 14. *Cancer* 1: 252–260.

Hennings, H. and Boutwell, R. K. (1970) Studies on the mechanism of skin tumor promotion. *Cancer Res.* 30: 312–320.

Henry, S. A. (1947) Occupational cutaneous cancer attributable to certain chemicals in industry. *Brit. Med. Bull.* 4: 389–401.

Huebner, R. J. and Todaro, G. J. (1969) Oncogenes of RNA tumor viruses as determinants of cancer. *Proc. Natl. Acad. Sci., USA* 64: 1087–1094.

Hueper, W. C., Wiley, F. H., and Wolfe, H. D. (1938) Experimental production of bladder tumors in dogs by administration of beta-naphthylamine. *J. Ind. Hygiene Toxicol.* 20: 46–84.

Kang, C.-Y. and Temin, H. M. (1973) Early DNA-RNA complex from the endogenous RNA-directed DNA polymerase activity of uninfected chicken embryos. *Nature New Biol.* 242: 206–208.

Kingery, L. B. (1921) The etiology of common warts. Their production in the second generation. *J. Am. Med. Assoc.* 76: 440–442.

Kinosita, R. (1937) Studies on the cancerogenic chemical substances. *Jap. Path. Soc. Trans.* 27: 665–727.

Land, H., Parado, L. F., and Weinberg, R. A. (1983) Tumorigenic conversion of primary embryo fibroblasts requires at least two cooperative oncogenes. *Nature* 304: 596–602.

Lasnitzki, I. (1963) Growth pattern of the mouse prostate gland in organ culture and its response to sex hormones, vitamin A, and 3-methylcholanthrene. *Natl. Cancer Inst. Monogr.* 12: 381–403.

Levine, M. and Tarver, H. (1951) Studies on ethionine. III. Incorporation of ethionine into rat proteins. *J. Biol. Chem.* 192: 835–850.

Linial, M. and Blair, D. (1982) Genetics of retroviruses, in *Molecular Biology of Tumor Viruses, RNA Tumor Viruses* (Weiss, R., Teich, N., Varmus, H., and Coffin, J., eds.), second edition, pp. 648–783. Cold Spring Harbor Laboratory, Cold Spring Harbor, New York.

Lo, S. C., Aft, R., and Mueller, G. C. (1981) Role of nonhemoglobin heme accumulation in the terminal differentiation of Friend erythroleukemia cells. *Cancer Res.* 41: 864–870.

Maher, V. M., Miller, E. C., Miller, J. A., and Szybalski, W. (1968) Mutations and decreases in density of transforming DNA produced by derivatives of the carcinogen 2-acetylaminofluorene and N-methyl-4-aminoazobenzene. *Mol. Pharmacol.* 4: 411–426.

Marie, P., Clunet, J., and Raulet-Lapointe, G. (1910) *Bull. Ass. franc. Cancer* 3: 404. (Quoted from A. Glücksmann. 1958. Carcinogenesis of skin tumours induced by radiation. *Brit. Med. Bull.* 2: 178–180.)

Marroquin, R. F. and Farber, E. (1962) The apparent binding of radioactive 2-acetylaminofluorene to rat-liver ribonucleic acid *in vivo*. *Biochim. Biophys. Acta* 55: 403–405.

Marroquin, R. F. and Farber, E. (1963) The *in vivo* labelling of liver ribonucleic acid by *p*-dimethylaminoazobenzene-1′-C^{14}. *Proc. Am. Assoc. Cancer Res.* 4: 41.

Mastro, A. M. and Mueller, G. C. (1974) Synergistic action of phorbol esters in mitogen-activated bovine lymphocytes. *Exper. Cell Res.* 88: 40–46.

Miller, E. C. (1951) Studies on the formation of protein-bound derivatives of 3,4-benzpyrene in the epidermal fraction of mouse skin. *Cancer Res.* 11: 100–108.

Miller, E. C. and Miller, J. A. (1947) The presence and significance of bound aminoazo dyes in the livers of rats fed *p*-dimethylaminoazobenzene. *Cancer Res.* 7: 468–480.

Miller, E. C. and Miller, J. A. (1955) Biochemical investigations in hepatic carcinogenesis. *J. Natl. Cancer Inst.* 15: 1571–1590.

Miller, E. C. and Miller, J. A. (1981) Searches for ultimate chemical carcinogens and their reactions with cellular macromolecules. *Cancer* 47: 2327–2345.

Miller, E. C., Miller, J. A., and Hartmann, H. A. (1961) N-Hydroxy-2-acetylaminofluorene: A metabolite of 2-acetylaminofluorene with increased carcinogenic activity in the rat. *Cancer Res.* 21: 815–824.

Miller, J. A. (1970) Carcinogenesis by chemicals—An overview. G. H. A. Clowes Memorial Lecture. *Cancer Res.* 30: 559–576.

Miller, J. A. and Miller, E. C. (1953) The carcinogenic aminoazo dyes. *Adv. Cancer Res.* 1: 339–396.

Miller, J. A. and Miller, E. C. (1969) Metabolic activation of carcinogenic aromatic amines and amides via N-hydroxylation and N-hydroxy esterification and its relationship to ultimate carcinogens as electrophilic reactants, in *The Jerusalem Symposia on Quantum Chemistry and Biochemistry* (Bergmann, E. and Pullman, B., eds.), Vol. 1, Physicochemical Mechanisms of Carcinogenesis, pp. 237–261. The Israel Academy of Sciences and Humanities, Jerusalem.

Miller, J. A., Miner, D. L., Rusch, H. P., and Baumann, C. A. (1941) Diet and hepatic tumor formation. *Cancer Res.* 1: 699–708.

Morris, H. P. and Wagner, B. P. (1968) Induction and transplantation of rat hepatomas with different growth rates (including "minimal deviation" hepatomas), in *Methods in Cancer Res.* (Busch, H., ed.), Vol. 4, pp. 125–152. Academic Press, New York.

Mottram, J. C. (1944) A developing factor in experimental blastogenesis. *J. Pathol. Bacteriol.* 56: 181–187.

Mueller, G. C. (1971) Biochemical perspectives of the G_1 and S intervals in the replication cycle of animal cells: A study in the control of cell growth, in *The Cell Cycle and Cancer* (Baserga, R., ed.), pp. 269–307. Marcel Dekker, Inc., New York.

Mueller, G. C. (1975) Biochemical events in the S phase of replicating animal cells, in *Mammalian Cells: Probes and Problems, Proc. First Los Alamos Life Sciences Symposium* (Richmond, C. R., Peterson, D. F., Mullaney, P. F., and Anderson, E. C., eds.), pp. 128–144. U.S. Energy Research and Development Administration, Washington, D.C.

Mueller, G. C. and Kajiwara, K. (1965) Regulatory steps in the replication of mammalian cell nuclei, in *University of Texas M. D. Anderson Hospital and Tumor Institute, Developmental and Metabolic Control Mechanisms and Neoplasia*, pp. 452–473. Williams and Wilkins Co., Baltimore.

Mueller, G. C. and Miller, J. A. (1948) The metabolism of 4-dimethylaminoazobenzene by rat liver homogenates. *J. Biol. Chem.* 176: 535–544.

Mueller, G. C. and Miller, J. A. (1949) The reductive cleavage of 4-dimethylaminoazobenzene by rat liver: The intracellular distribution of the enzyme system and its requirement for triphosphopyridine nucleotide. *J. Biol. Chem.* 180: 1125–1136.

Mueller, G. C. and Miller, J. A. (1953)The metabolism of methylated aminoazo dyes. II. Oxidative demethylation by rat liver homogenates. *J. Biol. Chem.* 202: 579–587.

Mueller, G. C. and Wertz, P. W. (1982)A possible role of protein alkylation in phorbol ester action, in *Carcinogenesis—A Comprehensive Survey* (Hecker, E., Fusenig, N. E., Kunz, W., Marks, F., and Thielman, H. W., eds.), Vol. 7, Cocarcinogenesis and Biological Effects of Tumor Promoters, pp. 499–511. Raven Press, New York.

Pamukcu, A. M. and Bryan, G. T. (1979) Bracken fern, a natural urinary bladder and intestinal carcinogen, in *Naturally Occurring Carcinogens-Mutagens and Modulators of Carcinogenesis* (Miller, E. C., Miller, J. A., Hirono, I., Sugimura, T., and Takayama, S., eds.), pp. 89–99. Japan Scientific Societies Press, Tokyo.

Pitot, H. C. (1960) The comparative enzymology and cell origin of rat hepatomas. II. Glutamate dehydrogenase, choline oxidase, and glucose 6-phosphatase. *Cancer Res.* 20: 1262–1268.

Pitot, H. C. (1975) Metabolic controls and neoplasia, in *Cancer* (Becker, F. F., ed.), Vol. 3, pp. 121–154. Plenum Publishers, New York.

Pitot, H. C., Goldsworthy, T., Campbell, H. A., and Poland, A. (1980) Quantitative evaluation of the promotion by 2,3,7,8-tetrachlorodibenzo-*p*-dioxin of hepatocarcinogenesis from diethylnitrosamine. *Cancer Res.* 40: 3616–3620.

Pitot, H. C., Goldsworthy, T., Moran, S., Sirica, A. E., and Weeks, J. (1982) Properties of incomplete carcinogens and promoters of hepatocarcinogenesis, in *Carcinogenesis—A Comprehensive Survey* (Hecker, E., Fusenig, N. E., Kunz, W., Marks, F., and Thielmann, H. W., eds.), Vol. 7, Cocarcinogenesis and Biological Effects of Tumor Promoters, pp. 85–98. Raven Press, New York.

Pitot, H. C., Peraino, C., Morse, P. A., Jr., and Potter, V. R. (1964) Hepatomas in tissue culture compared with adapting liver *in vivo. Natl. Cancer Inst. Monogr.* 13: 229–245.

Pitot, H. C. and Sirica, A. E. (1980) The stages of initiation and promotion in hepatocarcinogenesis. *Biochim. Biophys. Acta* 605: 191–215.

Potter, M. (1963) Percivall Pott's contribution to cancer research. (Including a reproduction of Pott's original article). *Natl. Cancer Inst. Monogr.* 10: 1–13.

Potter, V. R. (1944) Biological energy transformation and the cancer problem. *Adv. Enzymol.* 4: 201–256.

Potter, V. R. (1951) Studies on the reactions of the Krebs citric acid cycle in tumor, with homogenates, slices, and in vivo technics. *Cancer Res.* 11: 565–570.

Potter, V. R. (1958) The biochemical approach to the cancer problem. *Fed. Proc.* 17: 691–697.

Potter, V. R. (1961) Transplantable animal cancer, the primary standard. Guest Editorial. *Cancer Res.* 21: 1331–1333.

Potter, V. R. (1964) Biochemical studies on minimal deviation hepatomas, in *Cellular Control Mechanisms and Cancer* (Emmelot, P. and Mühlbock, O., eds.), pp. 189–210. Elsevier Publishing Co., Amsterdam.

Potter, V. R. (1973) Biochemistry of cancer, in *Cancer Medicine* (Holland, J. F. and Frei, E., eds.), pp. 178–192. Lea and Febiger, Philadelphia.

Potter, V. R. (1978) Phenotypic diversity in experimental hepatomas: The concept of partially blocked ontogeny. *Brit. J. Cancer* 38: 1–21.

Potter, V. R. and Elvehjem, C. A. (1936) A modified method for the study of tissue oxidations. *J. Biol. Chem.* 114: 495–504.

Potter, V. R. and Watanabe, M. (1968) Some biochemical essentials of malignancy: The challenge of diversity, in *Proceedings of the International Conference on Leukemia-Lymphoma* (Zarafonetis, ed.), pp. 36–46. Lea and Febiger, Philadelphia.

Redmond, D. E., Jr. (1970) Tobacco and cancer: The first clinical report, 1761. *New Engl. J. Med.* 282: 18–23.

Rehn, L. (1895) Blasengeschwülste bei Fuchsin-Arbeitern. Arch. klin. Chir. 50: 588–600. Translated in Shimkin, M. G. (1980) *Some Classics of Experimental Oncology: 50 Selections, 1775–1965,* NIH Publication No. 80-2150, p. 44–51.

Reznikoff, C. A., Bertram, J. S., Brankow, D. W., and Heidelberger, C. (1973) Quantitative and qualitative studies of chemical transformation of cloned C3H mouse embryo cells sensitive to post-confluence inhibition of cell division. *Cancer Res.* 33: 3239–3249.

Ross, J. and Sautner, D. (1976) Induction of globin mRNA accumulation by hemin in cultured erythroleukemic cells. *Cell* 8: 513–520.

Rous, P. (1911). Transmission of a malignant new growth by means of a cell-free filtrate. *J. Am. Med. Assoc.* 56: 198.

Rous, P. and Kidd, J. G. (1941) Conditional neoplasms and subthreshold neoplastic states: A study of the tar tumors of rabbits. *J. Exper. Med.* 73: 365–390.

Rowley, J. D. and Testa, J. R. (1982) Chromosome abnormalities in malignant hematologic diseases. *Adv. Cancer Res.* 103–148.

Rubin, H. and Temin, H. M. (1959) Radiation studies on lysogeny and tumor virus, in *University of Texas M. D. Anderson Hospital and Tumor Institute, Radiation Biology and Cancer,* pp. 359–381. University of Texas Press, Austin, Texas.

Ruley, H. E. (1983) Adenovirus early region 1A enables viral and cellular transforming genes to transform primary cells in culture. *Nature* 304: 602–606.

Rusch, H. P. (1944) Extrinsic factors that influence carcinogenesis. *Physiol. Rev.* 24: 177–204.

Rusch, H. P. (1983) Cancer Research at the University of Wisconsin, 1934–1979. Private printing.

Rusch, H. P., Kline, B. E., and Baumann, C. A. (1941) Carcinogenesis by ultraviolet rays with reference to wavelength and energy. *Arch. Pathol.* 31: 135–146.

Rusch, H. P., Kline, B. E., and Baumann, C. A. (1942) The nonadditive effect of ultraviolet light and other carcinogenic procedures. *Cancer Res.* 2: 183–188.

Sasaki, T. and Yoshida, T. (1935) Experimentelle Erzeugung des Lebercarcinoms durch Fütterung mit o-Amidoazotoluol. *Virchow's Arch. Path. Anat.* 295: 175–200.

Schmitz, H., Hurlbert, R. B., and Potter, V. R. (1954) Nucleotide metabolism. III. Mono-, di-, and triphosphates of cytidine, guanosine, and uridine. *J. Biol. Chem.* 209: 41–53.

Scotto, J., Fears, T. R., and Fraumeni, J. F., Jr. (1982) Solar Radiation, in *Cancer Epidemiology and Prevention* (Schottenfeld, D. and Fraumeni, J. F., Jr., eds.), pp. 254–276. W. B. Saunders Co., Philadelphia.

Shope, R. E. (1932) Transmissible tumor-like condition in rabbits. *J. Exp. Med.* 56: 793–802.

Slaga, T. J., Fischer, S. M., Nelson, K., and Gleason, G. L. (1980) Studies on the mechanism of skin tumor promotion: Evidence for several stages in promotion. *Proc. Natl. Acad. Sci. USA* 77: 3659–3663.

Sorof, S. and Cohen, P. P. (1951) Electrophoretic and ultracentrifugal studies on the soluble proteins of various tumors and of livers from rats fed 4-dimethylaminoazobenzene. *Cancer Res.* 11: 376–381.

Sorof, S., Cohen, P. P., Miller, E. C., and Miller, J. A. (1951) Electrophoretic studies on the soluble proteins from livers of rats fed aminoazo dyes. *Cancer Res.* 11: 383–387.

Staff of Roscoe B. Jackson Memorial Laboratory. (1933) The existence of nonchromosomal influence in the incidence of mammary tumors in mice. *Science* 78: 465–466.

Stehelin, D., Varmus, H. E., Bishop, J. M., and Vogt, P. K. (1976) DNA related to the transforming gene(s) of avian sarcoma viruses is present in normal avian DNA. *Nature* 260: 170–173.

Tannenbaum, A. (1959) Nutrition and cancer, in *The Physiopathology of Cancer* (Homburger, F., ed.), pp. 517–562. Hoeber-Harper, New York.

Temin, H. M. (1962) Separation of morphological conversion and virus production in Rous sarcoma virus infection. *Cold Spring Harbor Symp. Quant. Biol.* 27: 404–414.

Temin, H. M. (1963) The effects of actinomycin D on growth of Rous sarcoma virus in vitro. *Virology* 20: 577–582.

Temin, H. M. (1964a) Homology between RNA from Rous sarcoma virus and DNA from Rous sarcoma virus-infected cells. *Proc. Natl. Acad. Sci. USA* 52: 323–329.

Temin, H. M. (1964b) The participation of DNA in Rous sarcoma virus production. *Virology* 23: 486–494.

Temin, H. M. (1964c) Nature of the provirus of Rous sarcoma. *Natl. Cancer Inst. Monogr.* 17: 557–570.

Temin, H. M. (1966) Genetic and possible biochemical mechanisms in viral carcinogenesis. *Cancer Res.* 26: 212–216.

Temin, H. M. (1967) Studies on carcinogenesis by avian sarcoma viruses. VI. Differential multiplication of uninfected and of converted cells in response to insulin. *J. Cell. Physiol.* 69: 377–384.

Temin, H. M. (1971) The protovirus hypothesis: Speculations on the significance of RNA-directed DNA synthesis for normal development and for carcinogenesis. *J. Natl. Cancer Inst.* 46: iii–vii.

Temin, H. M. (1983) We still don't understand cancer. *Nature* 302: 656.

Temin, H. M. and Mizutani, S. (1970) RNA-dependent DNA polymerase in virions of Rous Sarcoma virus. *Nature* 226: 1211–1213.

Temin, H. M., Smith, G. L., and Dulak, N. C. (1974) Control of multiplication of normal and Rous sarcoma virus-transformed chicken embryo fibroblasts by purified multiplication-stimulating activity with nonsuppressible insulin-like and sulfation factor activities, in *Control of Proliferation in Animal Cells* (Clarkson, B. and Baserga, R., eds.), pp. 19–26. Cold Spring Harbor Laboratory, Cold Spring Harbor, New York.

Topp, W. C., Lane, D., and Pollack, R. (1980) Transformation by SV40 and polyoma virus, in *Molecular Biology of Tumor Viruses* (Tooze, J., ed.), Part 2, DNA Tumor Viruses, pp. 205–296. Cold Spring Harbor Laboratory, Cold Spring Harbor, New York.

Trager, W. F. (1980) Mechanisms of oxidative functionalization reactions, in *Concepts in Drug Metabolism* (Jenner, P. and Testa, B., eds.), Part A, pp. 177–209. Marcel Dekker, New York.

Upton, A. C. (1982) Physical carcinogenesis: Radiation—history and sources, in *Cancer, A Comprehensive Treatise* (Becker, F. F., ed.), second edition, Vol. 1, pp. 551–567. Plenum Press, New York.

Van Duuren, B. L. and Orris, L. (1965) Tumor-enhancing principles of *Croton tiglium L. Cancer Res.* 25: 1871–1875.

Verma, A. K. and Boutwell, R. K. (1977) Vitamin A acid (retinoic acid) a potent inhibitor of 12-O-tetradecanoyl-phorbol-13-acetate-induced ornithine decarboxylase activity in mouse epidermis. *Cancer Res.* 37: 2196–2201.

Watanabe, M., Potter, V. R., Pitot, H. C., and Morris, H. P. (1969) Systemic oscillations in metabolic activity in rat liver and hepatomas: Effect of hydrocortisone, glucagon, and adrenalectomy on diploid and other hepatoma lines. *Cancer Res.* 29: 2058–2100.

Waterfield, M. D., Scrace, G. T., Whittel, N., Stroobant, P., Johnsson, A., Wasteson, A., Westermark, B., Heldin, C.-H., Huang, J. S., and Deuel, T. F. (1983) Platelet-derived growth factor is structurally related to the putative transforming $p28^{sis}$ of simian sarcoma virus. *Nature* 304: 35–39.

Weinberg, R. A. (1982) Oncogenes of spontaneous and chemically induced tumors. *Adv. Cancer Res.* 36: 149–163.

Weisburger, E. K., Weisburger, J. H., and Morris, H. P. (1953) Studies on the metabolism of 2-acetylaminofluorene-9-C^{14}. *Arch. Biochem. Biophys.* 43: 474–484.

Wheeler, G. P. and Skipper, H. E. (1957) Studies with mustards. III. *In vivo* fixation of C^{14} from nitrogen mustard-$C^{14}H_3$ in nucleic acid fractions of animal tissues. *Arch. Biochem. Biophys.* 72: 465–475.

Wiest, W. G. and Heidelberger, C. (1953) The interaction of carcinogenic hydrocarbons with tissue constituents. II. 1,2,5,6-Dibenzanthracene-9-,10-C^{14} in skin. *Cancer Res.* 13: 250–254.

Yamagiwa, K. and Ichikawa, K. (1915) Experimentelle Studie über die Pathogenese der Epithelialgeschwülste. *Mitt. Med. Facultat Kaiseri. Univ. Tokyo* 15(2): 295–344. Also see Experimental study of the pathogenesis of carcinoma (1918) *J. Cancer Res.* 3: 1–29.

Young, J. L., Jr. and Pollack, E. S. (1982) The incidence of cancer in the United States, in *Cancer Epidemiology and Prevention* (Schottenfeld, D. and Fraumeni, J. F., Jr. eds.), pp. 138–165. W. B. Saunders Co., Philadelphia.

Yunis, J. J. (1983) The chromosomal basis of human neoplasia. *Science* 221: 227–236.

17

Nucleotide Synthesis and the Years at the Enzyme Institute: A Personal Account

H. Gobind Khorana

Departments of Biology and Chemistry
Massachusetts Institute of Technology
Cambridge, Massachusetts

When I arrived in Vancouver, British Columbia, in the fall of 1952, carbodiimides (examples in Figure 17.1) were very prominent in my thinking as potentially useful synthetic reagents. I had come across this class of compounds accidentally in German literature a few years earlier while spending a year with Prelog in Zurich. Carbodiimides were little known then and had been but little studied. During the following three years (1950–1952) in Cambridge, England, I worked on peptides and also became aware of nucleotides, an area in which Todd's laboratory was very active at that time. I began to think increasingly about carbodiimides, and some interesting results were obtained with them in Cambridge. During the years after my move to British Columbia (1952–1960) carbodiimides, especially dicyclohexylcarbodiimide (Figure 17.1), proved very useful in synthetic work on nucleotides. The work done in that period provided the basis of work that was accomplished during my subsequent ten years at the Enzyme Institute and I therefore first review the main lines of work pursued in Vancouver.

Synthesis of ATP and related nucleoside polyphosphates The first encouraging observation was that dicyclohexylcarbodiimide (DCC) (see Figure 17.1) reacted instantaneously with dibenzyl phosphate to form tetrabenzyl pyrophosphate. This formed the basis of an efficient method for pyrophosphate bond formation, a problem of main concern in the synthesis of biologically interesting pyrophosphates such as nucleotide coenzymes.

Soon after arrival in British Columbia, I attempted nucleotide condensations without any protecting groups. I was successful in a one-step synthesis of ATP from adenosine 5′-phosphate (AMP) and inorganic phosphate in the presence of DCC (Figure 17.2).

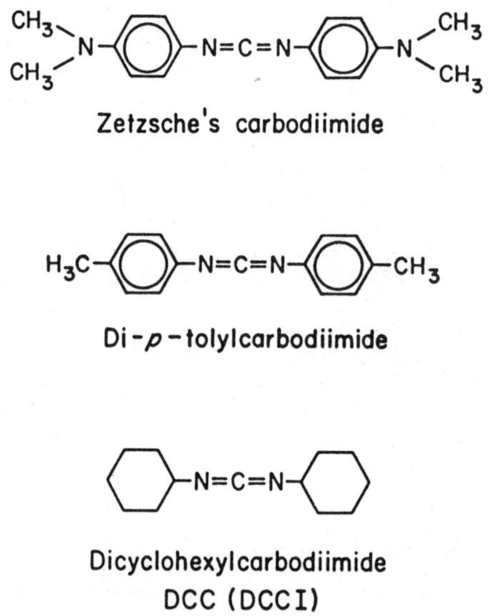

Figure 17.1 The structures of three carbodiimides: Zetzsche carbodiimide, di-*p*-tolylcarbodiimide and dicyclohexylcarbodiimide. The last has found the most applications.

An enzymatic assay of the chemically synthesized ATP was required. In British Columbia at that time I was completely isolated; there were no biochemists around and I had no contact with the biochemical world at large. A new and fruitful contact, however, did emerge following a visit I made to the University of California at Berkeley in the fall of 1953 at the invitation of the Biochemistry Department. I talked about my ATP synthesis and I met a num-

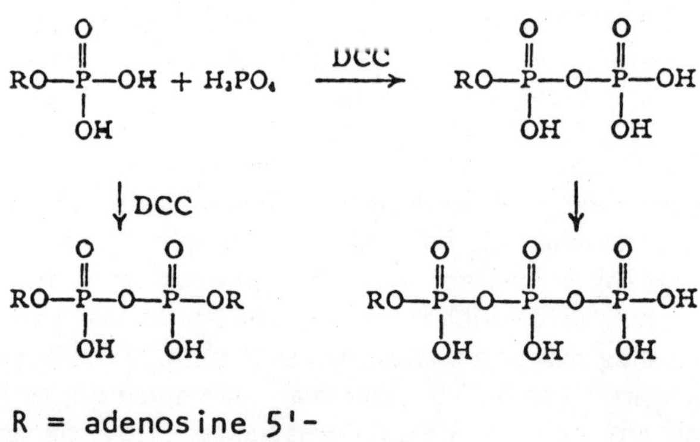

R = adenosine 5'-

Figure 17.2 One-step synthesis of ADP and ATP from completely unprotected AMP using dicyclohexylcarbodiimide and inorganic phosphate.

ber of biochemists that had been trained in Madison. One of these, Joe Nielands, suggested to me that I write to one Van Potter in Madison, who could assay my ATP sample. On return to B.C., I immediately wrote to Van Potter and there ensued a correspondence that was a continuing source of encouragement to me. My synthetic ATP preparation was assayed by Van Potter and his colleagues and the result was very satisfying. When preparing the manuscript describing this work, I sought Van Potter's approval to a simple statement that the synthetic sample was enzymatically active. Van Potter suggested a much more positive statement. This appeared in my paper in *J. Am. Chem. Soc.* (1954). Van Potter's original letter to me on this is in Figure 17.3.

Correspondence with Van Potter continued and this contact with an established and generous biochemist meant a great deal to me. Van Potter

THE UNIVERSITY OF WISCONSIN
THE MEDICAL SCHOOL
MADISON 6

McARDLE MEMORIAL LABORATORY
FOR CANCER RESEARCH

December 15, 1953

Dr. H. G. Khorana
British Columbia Research Council
At the University of British Columbia
Vancouver 8, Canada

Dear Dr. Khorana:

Thank you for your letter of Dec. 11. I am glad to know my letters reached you. The three statements that you propose to incorporate into your paper are entirely agreeable to me. I would however modify the statement on activity because it now reads.."sample showed enzymic activity". The changed part would read."According to the enzymatic method of LePage and Potter ( ), the activity of the enzyme system containing the synthetic sample was equal or better than the activity when the system was supplemented with comparable amounts of the standard ATP employed routinely by these workers." I think the distinction is apparent.

I shall look forward to hearing of your progress in the synthesis of the other di- and triphosphates.

Sincerely yours,

Van R Potter

Van R. Potter
Professor of Oncology

VRP:ms

Figure 17.3 Letter from Van Potter to Gobind Khorana, 1954.

was a prompt and considerate correspondent. Further, there were relatively few workers in the nucleic acid field then, although the momentum was rapidly gaining. Van Potter was engaged with LePage, Schmitz, Herbert and others in pioneering studies of acid-soluble nucleotides in different tissues including many aspects of nucleotide biochemistry and metabolism. The news and information that his letters contained were particularly welcome during the years in British Columbia. Little did I know at that time that Van would have a decisive impact on my future some years later.

Discoveries of new nucleotide coenzymes: development of their syntheses
The '50's were the golden age of discoveries of biosynthetic pathways and these became synonymous with the discoveries of new nucleotide intermediates. Van Potter in his book *Nucleic Acid Outlines* characterized this phase as "nucleotide explosion" (Figure 17.4). At that time, nucleotide coenzymes and related derivatives were highly inaccessible. The chemical methods for their synthesis that were developed in Vancouver began to be used increasingly in many laboratories. To illustrate, I shall only cite Gene Kennedy's work (at that time at the University of Chicago). Gene was studying the synthesis *in vitro* of phosphatidylcholine, and the main question was the nature of the active intermediate carrying phosphorylcholine. ATP samples that he was trying were giving variable results because new purification procedures were being developed at that time in different industrial laboratories. Perhaps a nucleotide of the general structure (Figure 17.5) was involved. Gene Kennedy synthesized, using DCC, the pyrophosphate derivatives of phosphorylcholine formed by linkage to each one of the four 5′-nucleotides. In this way, the cytidine nucleotide derivative CDP-choline was shown to be

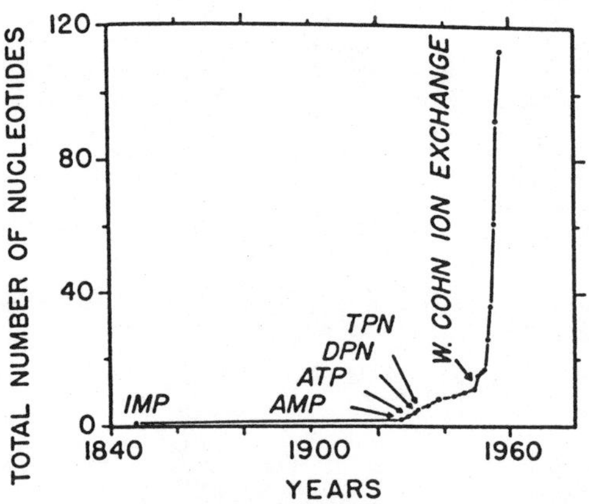

Figure 17.4 A plot of discoveries of new nucleotides as a function of time. Taken from Van R. Potter (1960) in *An Outline of Nucleic Acid*, Vol. I. Burgess Publishing Co., Minneapolis.

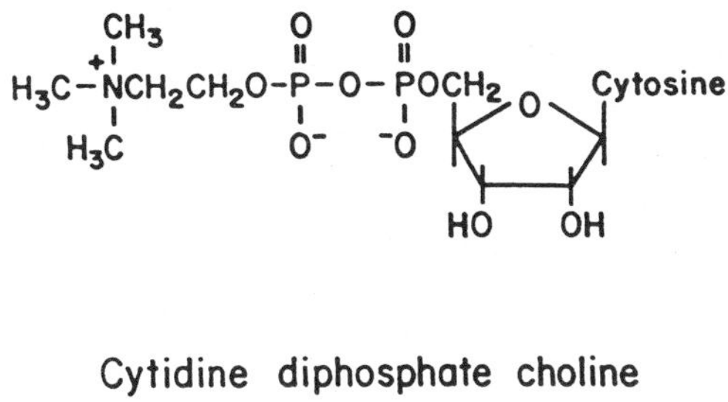

Figure 17.5 The structure of cytidine diphosphate choline.

specifically required in phospholipid synthesis. The work served as a great catalyst in promoting an awareness of our chemical methods among the biochemical community.

Specific synthesis of nucleotide coenzymes Condensations between different components to form nucleotide pyrophosphates as promoted directly by DCC were not always satisfactory, especially where one of the components, such as glucose α-1-phosphate and flavin mononucleotide, could undergo a competing intramolecular reaction to form a cyclic phosphate. So we developed a specific method for the activation of 5′-mononucleotides. DCC again was the activating agent, and several classes of substituted phosphoramidates, such as that shown in Figure 17.6, were prepared. These, on subsequent reaction with the anion of a second phosphate component, formed the desired products in much better yields. Practically all the known coenzymes, including coenzyme A, and nucleotide pyrophosphate sugars were synthesized by John Moffatt in my laboratory in the late '50's.

Synthesis of polynucleotides The chemical structures of RNA and DNA were elucidated in the early '50's. In Vancouver, I began to think about the

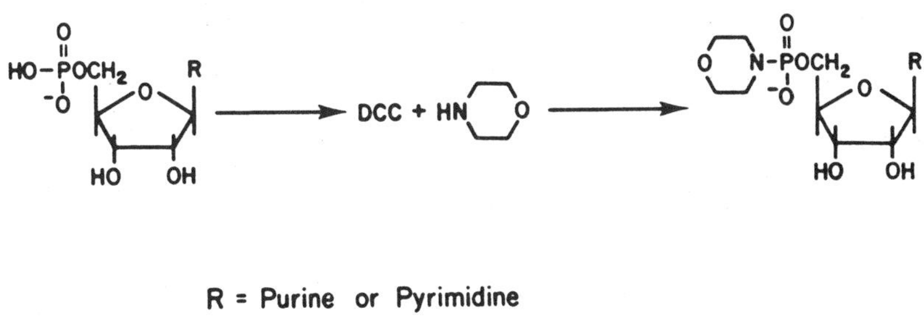

Figure 17.6 Preparation of ribonucleoside-5′-phosphoromorpholidates. The reactive nucleotide intermediates used in the synthesis of nucleotide coenzymes and, in particular, Coenzyme A.

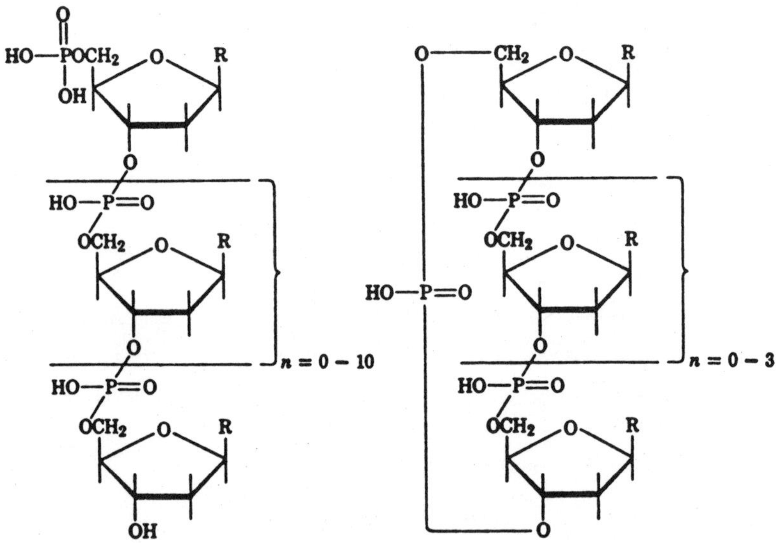

Figure 17.7 General structures of the two classes (linear and cyclic) of thymidine oligonucleotides produced by the reaction of thymidine 5′-phosphate with *p*-toluenesulfonyl chloride.

possibilities for the chemical synthesis of the internucleotide bond. I began some work in 1953 but this did not get very far. However in the fall of 1954, Gordon Tener, a former student of Frank Strong, joined me, and our efforts on the problem increased. By 1956, we had discovered a strikingly efficient new method for making dinucleotides and for polymerizing thymidine nucleotide to thymidine polynucleotides (Figure 17.7). Systematic work on different aspects of the problem of synthesis of oligonucleotides containing specific sequences continued for several years in Vancouver and in Madison after my move to the Enzyme Institute. A large number of co-workers in the laboratory made important contributions. The basic principles of the overall methodology are illustrated in Figures 17.8–17.11 and these deal with: (a) Synthesis of internucleotide bonds from the two mononucleotide components, one bearing a 5′-phosphate group and one a 3′-hydroxyl group; (b) protection of the unwanted functional sites in nucleosides; (c) suitable activation of the phosphate groups; (d) the stepwise synthesis of oligodeoxynucleotides; (e) stepwise synthesis of ribo-oligonucleotides. A large number of oligodeoxynucleotides and all the sixty-four possible ribotrinucleotides were synthesized. These were directly important in our work on the genetic code.

Synthetic polynucleotides and the mode of action of phosphodiesterases

The temptation to combine synthetic work with some enzymatic studies was

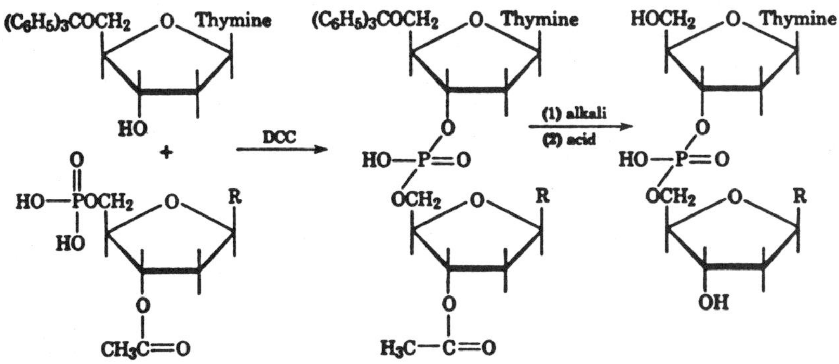

Figure 17.8 Condensation of two protected components to form a dinucleotide. On the top left is the 5′-protected thymidine and at bottom left is the 3′-protected thymidine 5′-phosphate. Condensation to form the internucleotide bond (middle) is effected by activation of the phosphate group. The protecting groups are removed by mild acid and alkali to give the unprotected dinucleotide (right). Mild alkaline treatment alone would free the 3′-hydroxyl end group for further chain elongation by condensation with protected mono- or oligonucleotides carrying 5′-phosphate groups.

very strong in the late '50's, especially because very little was known at that time about the mode of action of enzymes that attack nucleic acids. Further, such enzymes appeared to be extremely promising reagents for characterization of the synthetic oligonucleotides. The classic analytical methods of organic chemistry were simply not suitable for the purpose. With the availability of thymidine oligonucleotides in purified form, we began to study the action of venom phosphodiesterase on them. The mode of action, as depicted in Figure 17.12, was demonstrated. Thus, the stepwise action began from the end of the chain bearing the 3′-hydroxyl group and successively released nucleoside 5′-phosphate units. Soon after, we found that the action of the spleen phosphodiesterase (Figure 17.12) was complementary, in that it was an exonuclease whose action began from the end of the chain bearing the 5′-hydroxyl group and released nucleoside 3′-phosphate units.

The Years at the Enzyme Institute

The move to the Enzyme Institute David Green and Henry Lardy had shared the space in the original small building on University Avenue since the creation of the Enzyme Institute. In 1959, David Green was able to obtain funds for a new building to house his research group. Therefore, his original space was going to be available and the University of Wisconsin decided to make a new appointment at the Enzyme Institute. C. A. Elvehjem was the President of the University at that time, and John Willard was the Dean of

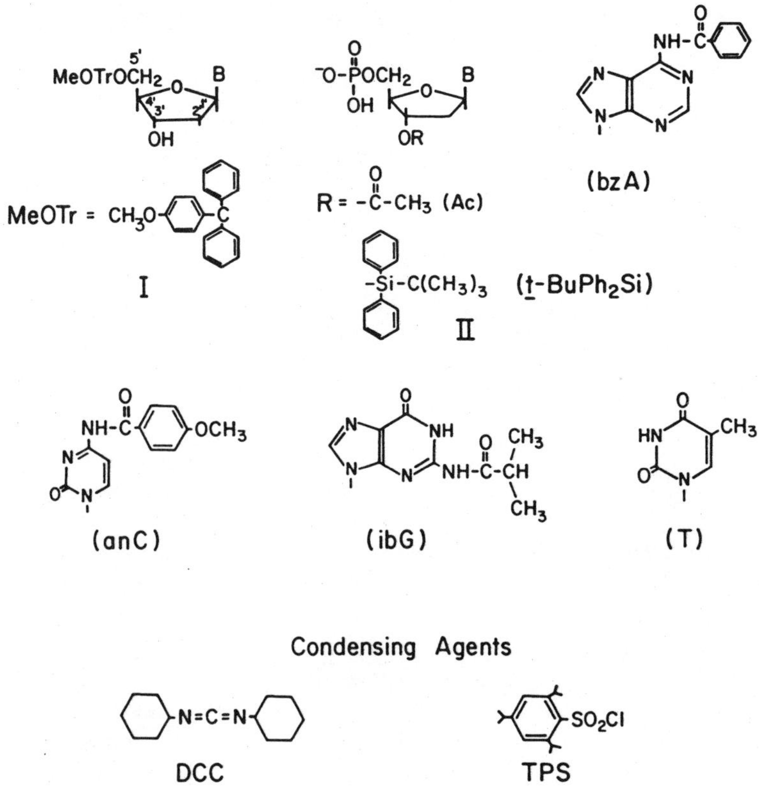

Figure 17.9 Groups used for the protection of deoxyribonucleosides and deoxyribonucleotides and the condensing agents used in the chemical synthesis of polynucleotides. The abbreviations used for the different groups are also shown.

the Graduate School. Van Potter was the chairman of the Enzyme Institute Committee and he was charged with the search for candidates for this appointment.

That spring (1959) I met my friend, the late Charles Heidelberger, then a colleague of Van Potter at the McArdle Memorial Laboratory, at the Federation meetings in Atlantic City. It was he who first asked if I wished to be considered for the position at the Enzyme Institute. Soon the dialogue with Madison began. I remember that during my first visit to the campus in June, 1959, I gave an Enzyme Institute seminar. (In those days, this series of seminars, made possible by David Green's research grants, were held in room 125 of the Biochemistry Department on Wednesday evenings). During that visit to Madison, I sensed that Van Potter was driven by completely selfless motivation. As he said, he had no direct interest or any personal involvement at the Enzyme Institute. Throughout my negotiations, I deeply felt that Van

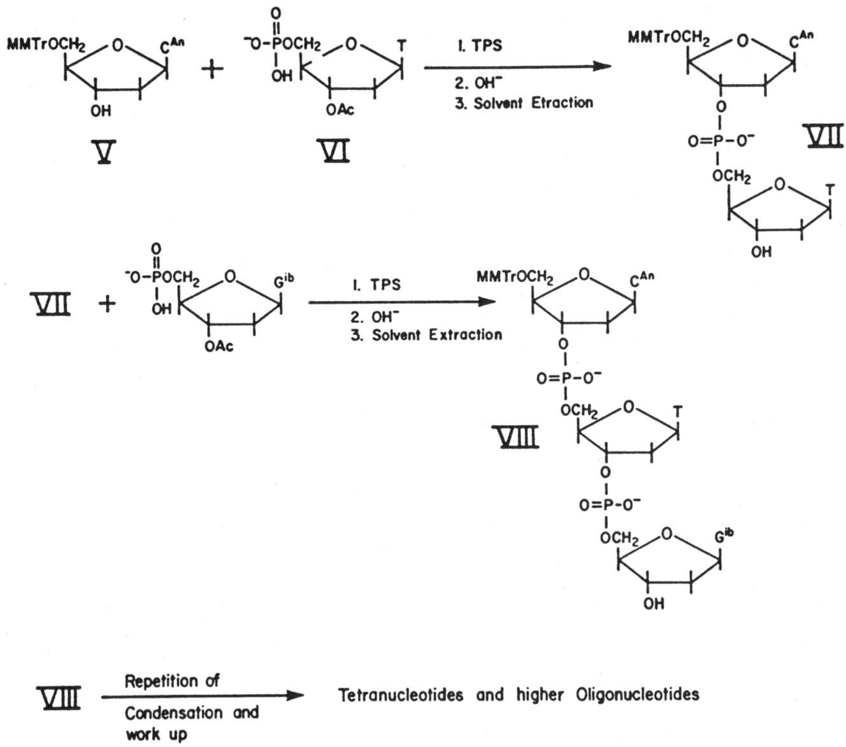

Figure 17.10 Principles in the stepwise synthesis of tri- and higher oligonucleotides.

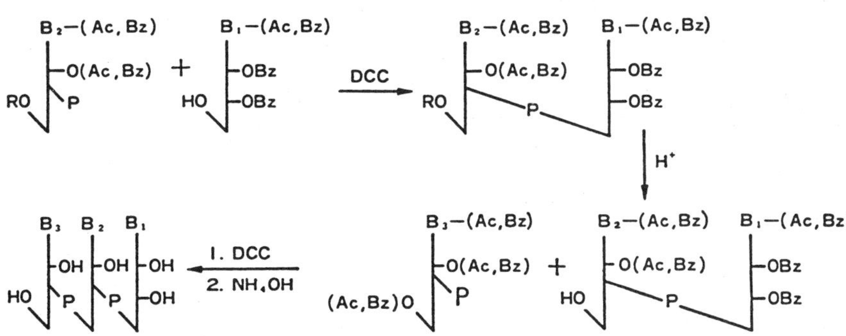

B1, B2, B3 = PURINES OR PYRIMIDINES
R = METHOXYTRITYL, Ac = ACETYL, Bz = BENZOYL
DCC = DICYCLOHEXYLCARBODIIMIDE

Figure 17.11 Generalized method for the synthesis of ribotrinucleotides using protected intermediates.

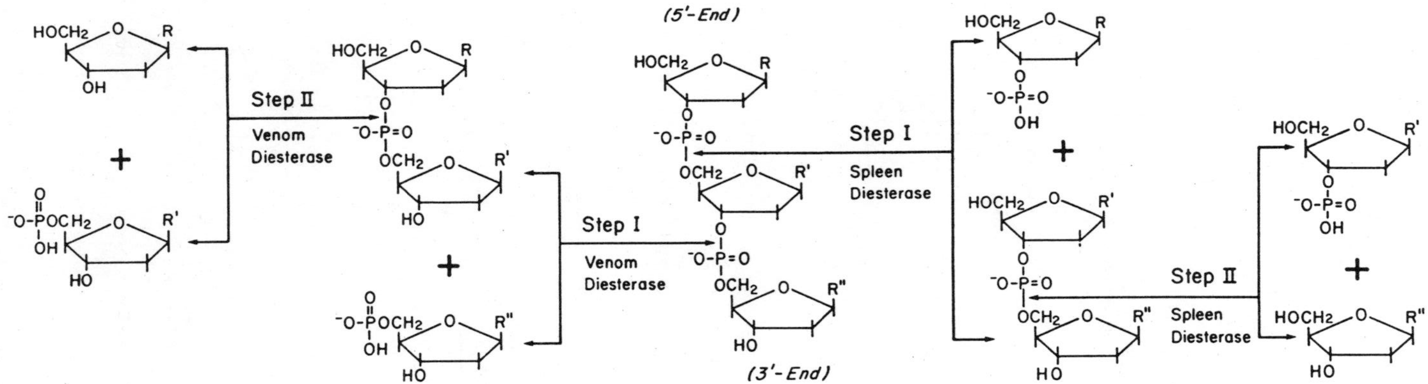

(Deoxyribo– or Ribo–trinucleotide)

Figure 17.12 Illustration of the enzymatic degradation of an oligonucleotide chain (a trinucleotide) possessing the required 3′-hydroxyl and 5′-hydroxyl groups at the termini. Snake venom phosphodiesterase is an exonuclease which attacks a chain at the end bearing a 3′-hydroxyl group and successively removes ribo- or deoxyribonucleoside 5′-phosphate units from that end. Spleen phosphodiesterase is also an exonuclease, but its mode of action is complementary to that described above, in that it begins at the end of the chain bearing the 5′-hydroxyl group and it successively releases ribo- or deoxyribonucleoside 3′-phosphate units.

was my spokesman to the administration and that he would get the administration "to go all out" to make it attractive for me to come to the Enzyme Institute. At some point I expressed to Van my concerns that perhaps I could not fit into a chemistry department or a biochemistry department, and that I would probably not feel at home in either. Van Potter emphasized that at the Enzyme Institute I could develop in whatever way or direction I wanted. I began to feel that the Enzyme Institute indeed offered complete freedom and independence and, thus, promised to be a unique opportunity to develop a research program in organic as well as biochemistry.

Polynucleotide synthesis and the genetic code Julius Adler also had accepted an appointment in the Biochemistry Department at UW, and he and I arrived in Madison within the same week in September, 1960. We soon began to see a great deal of each other and became good friends. I was greatly influenced in my biochemical thinking by Julius' insights into biochemistry and molecular genetics.

A year later (September, 1961) the news of the experiment by Nirenberg and Matthaei on cell-free polyuridylate-directed synthesis of polyphenylalanine reached the Madison campus. In those days Julius and I used to meet frequently for lunch on Saturdays and the new doors opened by the above discovery for a biochemical attack on the problem of the genetic code began to be the subject of excited discussions at these lunches. Both of us were also interested, although for somewhat different reasons, in DNA-dependent RNA polymerase, which had been discovered earlier that year. As our first goal, we decided to study transcription of our well-characterized synthetic thymidine polynucleotides. We were hoping that, by using deoxypolythymidylates of defined chain length, we would obtain ribopolyadenylate messengers of the corresponding chain length. These experiments began when, later that fall, Arturo Falaschi, a postdoctoral fellow, arrived from Italy. As the analysis of the transcription products progressed, we realized that no matter what the size of the oligothymidylate used as the template, the transcripts were invariably large; ribopolyadenylates formed in every case were evidently longer than at least 100 nucleotides. This unexpected result seemed discouraging at first. However, on reflection, it provided the means to amplify the "information" contained in the short synthetic polynucleotides.

Soon after, I showed, with Arthur Kornberg, that our short synthetic deoxypolynucleotides containing alternating A and T units served as primers for the DNA polymerase (now designated polymerase I) and gave extensive synthesis of the high molecular weight $(A\text{-}T)_n$ polymer. The reiteration process, which seemed to operate for both the RNA and DNA polymerases, provided, in essence, an efficient means for the amplification and multiplication of the chemically synthesized templates (Figure 17.13), and led to the general strategy outlined in Figure 17.14 for our proposed work on the genetic code.

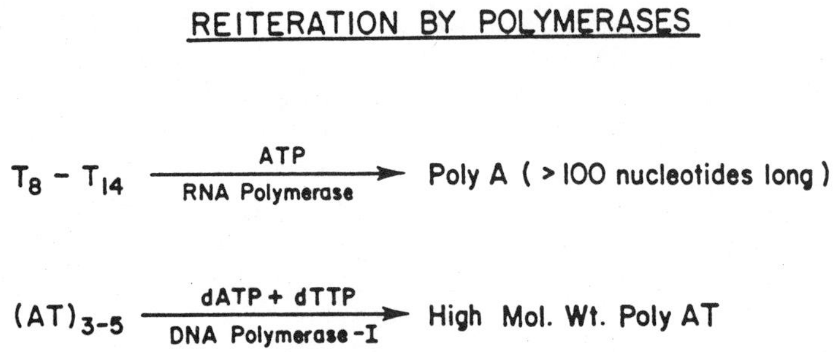

Figure 17.13 Reiteration by DNA-dependent RNA polymerase in transcribing short chains of deoxyoligo-adenylates and reiteration by DNA polymerase I in the synthesis of high molecular weight $d(AT)_n$ by using short oligomers as primers.

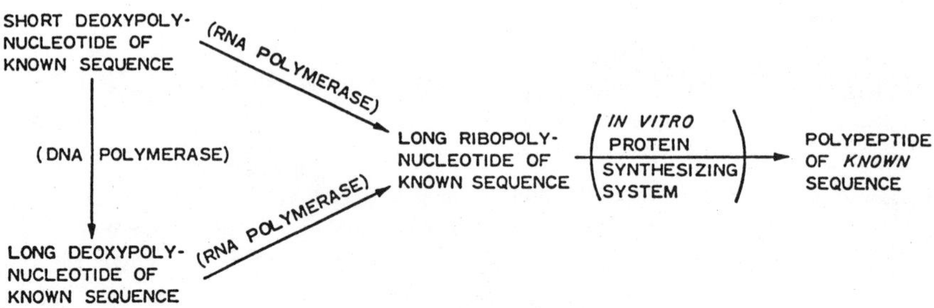

Figure 17.14 Sequence of enzymatic reactions for the synthesis of specific polypeptides using chemically synthesized specific deoxyribopolynucleotides.

A variety of short polymers, containing repeating di-, tri- and tetranucleotide sequences, were chemically prepared (Table 17.1). The arguments and rationale for the selection of the particular sequences in each one of the three classes of polymers have been described previously.

DNA polymerase I used all the three sets of oligonucleotides as primers in polymerization reactions shown in Figure 17.15. Many features of these reactions were truly remarkable: (a) In all these reactions the enzyme showed complete fidelity in reproducing the sequences; (b) the synthesis was extensive, 50-to 200-fold, and the products were of high molecular weight (500,000 to over 1,000,000)—thus, the enzyme amplified and multiplied the sequence information present in the chemically synthesized oligonucleotides. (c) Finally, from the standpoint of an organic chemist, the most satisfying aspect was that the DNA polymers thus made could be used again and again for further synthesis. We never had to go back to the time-consuming chemical

Table 17.1 Synthetic deoxyribopolynucleotides with repeating sequences

Repeating dinucleotides		Repeating trinucleotides		
$\begin{bmatrix}(TC)_6\\(AG)_6\end{bmatrix}$	$\begin{bmatrix}(TG)_6\\(AC)_6\end{bmatrix}$	$\begin{bmatrix}(TTC)_4\\(AAG)_4\end{bmatrix}$	$\begin{bmatrix}(CCT)_{3-5}\\(GGA)_{3-5}\end{bmatrix}$	$\begin{bmatrix}(TAC)_{4-6}\\(TAG)_{4-6}\end{bmatrix}$
Repeating tetranucleotides		$\begin{bmatrix}(TTG)_{4-6}\\(CAA)_{4-6}\end{bmatrix}$	$\begin{bmatrix}(CGA)_{4-6}\\(CGT)_{3-5}\end{bmatrix}$	$\begin{bmatrix}(ATC)_{3-5}\\(ATG)_{3-5}\end{bmatrix}$
$\begin{bmatrix}d(TTAC)_4\\d(GTAA)_2\end{bmatrix}$	$\begin{bmatrix}d(TCTA)_3\\d(TAGA)_2\end{bmatrix}$			$\begin{bmatrix}(CCA)_{3-5}\\(GGT)_{3-5}\end{bmatrix}$

synthesis for obtaining that particular sequence again. DNA polymerase assured the continuity of these sequences.

Figure 17.16 catalogues the different double-stranded DNA polymers that were prepared in this way and characterized. Thus, there were three classes of polymers: two double-stranded polymers with repeating dinucleotide sequences, four polymers with repeating trinucleotide sequences, and two polymers with tetranucleotide sequences.

The availability of a variety of DNA-like polymers with completely defined nucleotide sequences made possible an attack on many aspects of the studies of the biological reaction sequence DNA $\longrightarrow$ RNA $\longrightarrow$ protein. There ensued a highly fruitful and exciting phase in our research. Thus, by relatively straightforward biochemical methods and minor innovations, we were able to use the transcribing enzyme RNA polymerase to prepare a large variety of ribopolynucleotide messengers. Because all the DNA-like polymers contain

$$d[TG]_6 + d[AC]_6 + \begin{bmatrix}dTTP\\dATP\\dCTP\\dGTP\end{bmatrix} \rightarrow \text{poly(dTG:CA)}$$

$$d[TTC]_4 + d[AAG]_3 + \begin{bmatrix}dTTP\\dATP\\dCTP\\dGTP\end{bmatrix} \rightarrow \text{poly(dTTC:GAA)}$$

$$d[TATC]_3 + d[TAGA]_2 + \begin{bmatrix}dTTP\\dATP\\dCTP\\dGTP\end{bmatrix} \rightarrow \text{poly(dTATC:GATA)}$$

Figure 17.15 Types of reactions catalyzed by DNA polymerase. All of the DNA-like polymers are written so that the colon separates the two complementary strands. The complementary sequences in the individual strands are written so that antiparallel base pairing is evident.

DNA-LIKE POLYMERS WITH REPEATING NUCLEOTIDE SEQUENCES

Repeating dinucleotide sequences	Repeating trinucleotide sequences	Repeating tetranucleotide sequences
Poly(d-TC:GA)	Poly(d-TTC:GAA)	Poly(d-TTAC:GTAA)
Poly(d-TG:CA)	Poly(d-TTG:CAA)	Poly(d-TATC:GATA)
	Poly(d-TAC:GTA)	
	Poly(d-ATC:GAT)	

Figure 17.16 Enzymatically synthesized DNA-like polymers. Three groups are shown, those containing repeating di-, tri- and repeating tetra-nucleotide sequences.

two, or a maximum of three, different bases in individual strands, it was possible, by giving the nucleoside triphosphates required for copying only one strand, to restrict the action of the RNA polymerase to that strand. In this way, each one of the DNA-like polymers described yielded two RNA messengers, again with the expected repeating sequences. Thus, by a combination of chemical and enzymatic methods, we had at our disposal a variety of high molecular weight ribopolynucleotides of defined sequences.

Polypeptide synthesis *in vitro* and the genetic code Although far from being proven, the genetic code had been suspected in the early '60's to be non-overlapping with a trinucleotide sequence representing each amino acid. If this were so, polymers with repeating dinucleotide sequences (AB) which contain two triplets, ABA and BAB, in alternating sequence should direct incorporations of two amino acids in strictly alternating sequence. Repeating trinucleotide polymers $(ABC)_n$ contain three repeating triplets, depending upon the starting point; these are ABC, BCA, and CAB. Here one would predict one amino acid to be incorporated at a time to form a homopolypeptide chain, and a maximum of three such polypeptides should result. Similar considerations for polynucleotides with repeating tetranucleotide sequences $(ABCD)_n$ showed that *in vitro* polypeptide synthesis should give products containing repeating tetrapeptide sequences, irrespective of the starting point in the reading of the messages. All these predictions were fully borne out experimentally for all the repeating polymers without a single exception. Only single examples of each type of repeating polymer are given in Figure 17.17 to illustrate the three types of results. Complete surveys of the results have been given elsewhere. The general conclusions from these results were as follows: (a) DNA does specify the sequence of amino acids in proteins, and this information is relayed through RNA (this was the first time that a direct sequence correlation between DNA and a protein had been established); (b) all the results prove the three-letter and non-overlapping properties of the code; (c)

Repeating DNA Polymer		Single-stranded RNA Transcript		*In vitro* Translation Product(s)
d[TC:GA]	RNA Polymerase → UTP+CTP	Poly(rUC)	→	$(Ser\text{-}Leu)_n$
d[TTC:GAA]	RNA Polymerase → UTP+CTP	Poly(rUUC)	→	$(Phe)_n$, $(Ser)_n$, $(Leu)_n$
d[TATC:GATA]	RNA Polymerase → UTP+CTP+ATP	Poly(rUAUC)	→	$(Tyr\text{-}Leu\text{-}Ser\text{-}Ile)_n$

Figure 17.17 Examples of *in vitro* transcription and translation using DNA-like polymers as templates.

finally, information on codon assignments can also be derived from these results.

In further work, by a combination of the above approach and the ribosome-tRNA-trinucleotide binding technique developed by Nirenberg and Leder, as well as extensive mutational analyses, it proved possible to deduce the total structure of the genetic code.

In addition to the studies bearing directly on the genetic code, defined synthetic polynucleotides provided new opportunities for many aspects of work on many related problems of protein synthesis such as codon-anticodon interactions, nature of genetic suppression, misreading of the genetic code as induced by various antibiotics, direct translation of single-stranded DNA, and initiation and termination of protein synthesis. Finally, there was the large complementary problem of the "adapter" molecules, the transfer RNA's (see below), which contained the anticodons corresponding to the codons for each amino acid. The first problem was one of nucleotide sequence determination. This was solved first for the major yeast alanine tRNA by Holley and coworkers. The synthesis of the gene specifying this tRNA soon became the next major interest of our research.

Total Synthesis of Genes

The methodology and application to the synthesis of the DNA for the yeast alanine tRNA The amplification of synthetic oligonucleotides by the polymerases was limited to repeating nucleotide sequences, and the formidable problem of constructing large polynucleotides containing *specific* nucleotide sequences lay ahead, unsolved. It was abundantly clear that, following the elucidation of the genetic code, attention of molecular biologists would be focussed increasingly on problems of controls for transcription and translation, regulation of gene expression, protein and nucleic acid interactions and, of necessity, sequencing of nucleic acids. Therefore, development of methods for the synthesis of biologically specific double-stranded DNA was seen as a central chemical problem in relation to the directions of molecular biology.

There was a practical limit on the length of oligonucleotides that could be made by purely chemical methods. Clearly, chemical synthesis had to be coupled to a different kind of concept which would bring about end-to-end joining of short, chemically synthesized chains. The idea that we investigated very early in this work was the ability of polynucleotides to form hydrogen-bonded bihelical complexes. When aligned in an antiparallel fashion, the terminal phosphate group of one oligonucleotide chain could phosphorylate a properly apposed 3′-hydroxyl group of the next chain. Indeed, as far back as 1961, Peter Gilham, working in my laboratory, demonstrated the end-to-end joining of a hexanucleotide, d-A_6, to form d-A_{12} by using d-T_{12} as the template. The activating agent then used was a water-soluble carbodiimide. The yield of the joined product was low in those experiments, but conceptually it was an encouraging result. A number of workers, especially Leslie Orgel and coworkers, have performed such template-guided polynucleotide synthesis in recent years. This concept formed the basis of successful gene syntheses described below.

The complete sequence of a transfer RNA appeared in the spring of 1965 from Holley's laboratory at Cornell University. This discovery focussed our interest on the total synthesis of the DNA specifying that tRNA sequence. In late 1965, Hans Weber undertook the very demanding task of synthesizing an icosadeoxynucleotide, complementary to a part of the alanine tRNA sequence. Soon after, Henry Buchi began the synthesis of a complementary icosanucleotide. The relationships of the two icosanucleotides to one another and to the alanine tRNA are illustrated in Figure 17.18. To be able to estimate

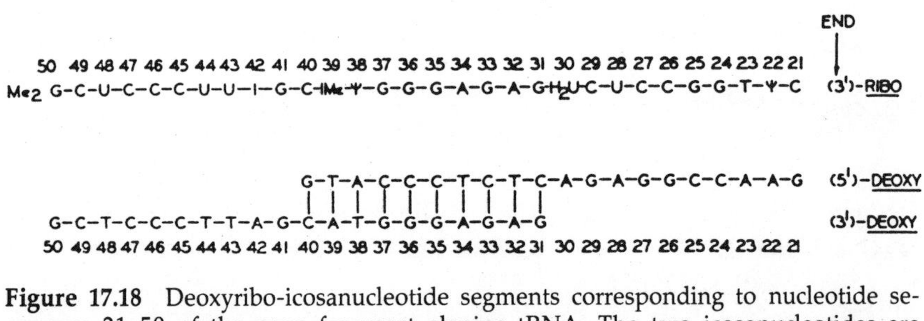

Figure 17.18 Deoxyribo-icosanucleotide segments corresponding to nucleotide sequences 21–50 of the gene for yeast alanine tRNA. The two icosanucleotides are complementary to each other through halves of their length in the required antiparallel manner.

closely the lengths of the overlapping regions, one needed to know a great deal more about the stabilities of the hydrogen-bonded double-helical structures than we did at that time. So we spent considerable effort on determining Tm's of short, hydrogen-bonded synthetic polynucleotides. We decided to synthesize icosanucleotides with decanucleotide overlaps (Figure 17.18) with the belief that the ability to form stable double-stranded structures would allow the end-to-end joining reactions.

In 1965, when the above syntheses were undertaken, the T4-polynucleotide kinase and the DNA ligase, which were to be the standard reagents in our synthetic methodology, had not yet been discovered. We began the syntheses only because of the conviction that there was no alternative to the template concept for the end-to-end joining of the chemically synthesized oligonucleotides. By the time the syntheses of the icosanucleotides had been achieved, the above two enzymes had been discovered. The immediate task then was to delineate the minimal size range of the hydrogen-bonded duplexes that polynucleotide ligase would require for joining reactions. We synthesized short oligonucleotides corresponding to the single-stranded "arms" of the presumed hydrogen-bonded icosanucleotides (Figure 17.19). When the 5′-ends were labelled with polynucleotide kinase, the ligase was shown to catalyze the correct joining of the 5′-[^{32}P] phosphate of the oligonucleotides

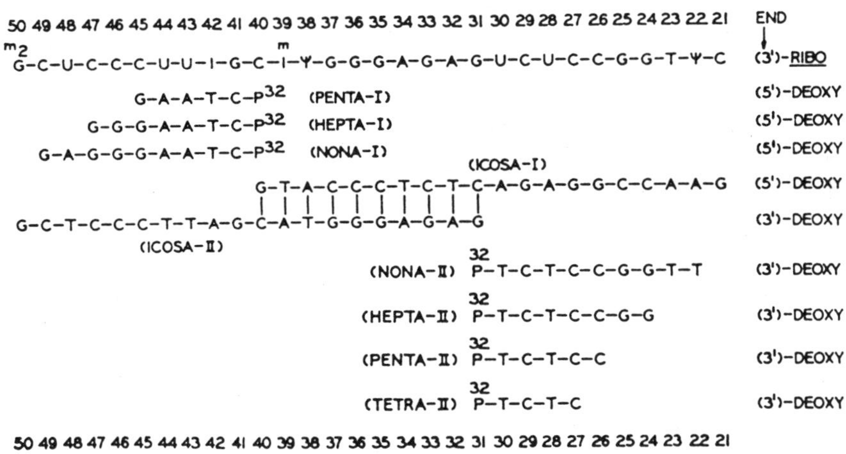

Figure 17.19 Chemically synthesized deoxyribopolynucleotides corresponding to sequences 21–50 of yeast ala-tRNA. The icosanucleotide-I (Icosa-I) represents a sequence complementary to nucleotides 21–40 of the tRNA and has polarity opposite to that of the tRNA; the nona-, hepta, and penta-nucleotides (Nona-I, Hepta-I, etc.) similarly contain sequences complementary to nucleotides 41–49 or less, and again have polarity opposite to that of the tRNA. The deoxyribopolynucleotides (Icosa-II, Nona-II, Hepta-II, Penta-II and Tetra-II) are segments, complements of the complement, and therefore contain the same sequences and polarity as the tRNA itself. ^{32}P represents the 5′-phosphate end group wherever shown.

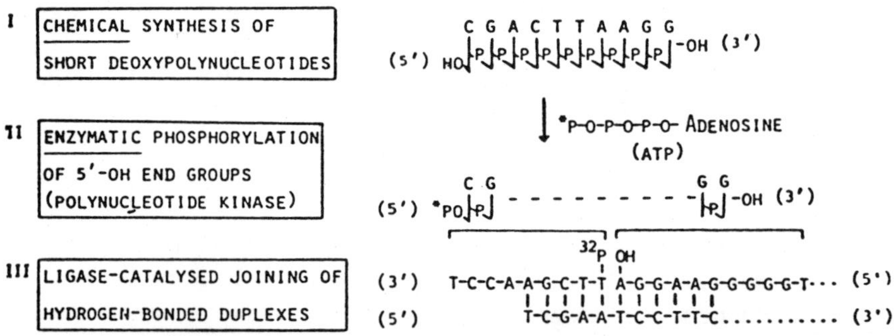

Figure 17.20 Three-step strategy for the total synthesis of double-stranded DNA.

to the 3′-hydroxyl end groups of the adjoining icosanucleotides (Figure 17.19). The results showed that, in general, only short chains, mostly in the range of 10–12 nucleotides, would be needed as the "units" in enzyme-catalyzed joining reactions. One could now clearly formulate a three-phase strategy (Figure 17.20) for the step-by-step synthesis of long DNA duplexes. In principle, the synthesis of genes appeared feasible.

The total synthesis of the 77 base-pair long DNA, the structural gene for the yeast alanine tRNA (Figure 17.21), was completed in the spring of 1970. It was nice to be asked by the Biochemistry Department to give a seminar on this work. This I gave in early June, only a few months before my move to M.I.T.

The *E. coli* tyrosine suppressor tRNA gene Further work with the yeast alanine tRNA gene, namely its transcriptional controls and structure-function studies, was not feasible at that time, because of inadequate progress in the molecular biology of the yeast system. Instead, we undertook the total synthesis of an *E. coli* tyrosine suppressor tRNA gene. Detailed arguments for the choice of this tRNA have been presented elsewhere. With passage of time (1968–1972), additional findings validated this choice. One critical finding was the discovery of a precursor to this tRNA (Figure 17.22). This structure, containing a 5′-triphosphate end group, unambiguously defined the site of initiation of transcription. We therefore could try to identify the promoter (preinitiation region), the initiation and termination of transcription, as well as processing of the precursor. In fact, one could envisage a linear arrangement for the different regions, including control elements of the gene, as shown in Figure 17.23.

The total synthesis of the gene consisted of the following phases: (a) Total synthesis of the 126 base-pair long structural gene corresponding to the sequence in Figure 17.22A; (b) determination of the nucleotide sequence of the promoter region flanking the 5′-end of the precursor region; (c) deter-

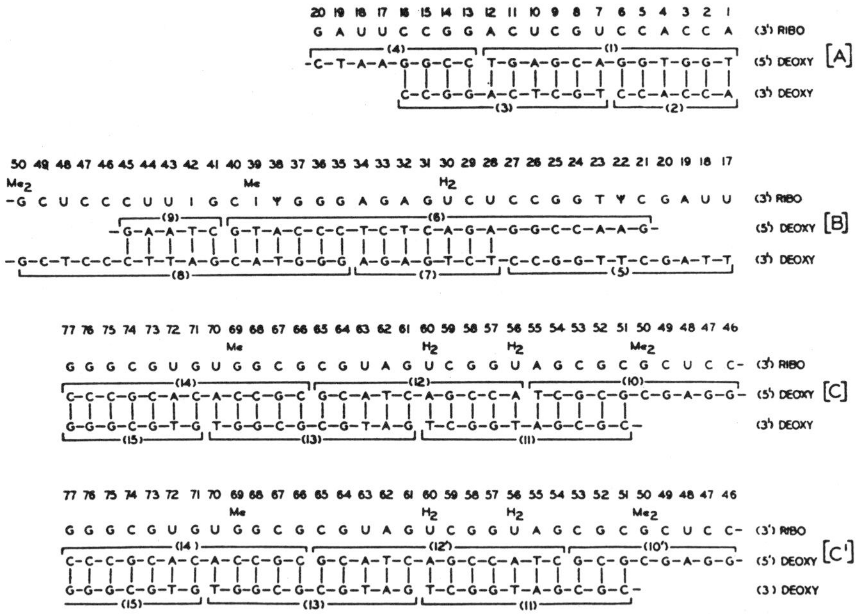

Figure 17.21 Total plan for the synthesis of the yeast alanine tRNA gene. The chemically synthesized segments are shown by horizontal brackets, the serial number of the segment being in parentheses inserted into brackets. A total of 17 segments (including 10′ and 12′) varying in chain length from penta-to icosanucleotides were synthesized.

mination of the nucleotide sequence downstream from the CCA end of the tRNA gene to learn about the processing signal or termination; (d) synthesis of all of the oligonucleotide segments required for different sections of the gene; (e) DNA ligase-catalyzed joining of the oligonucleotides to form the partial duplexes followed by the joining of the fragment duplexes and satisfactory characterization of the gene; (f) its integration into a suitable vector to test for amber suppression *in vivo*. In its final form, the work required the synthesis of 39 oligonucleotides comprising all sections of the gene (Figure 17.24). These were joined to form the gene shown in Figure 17.25. In order to insert the gene into vectors, we had provided *E. coli* cohesive ends at the two termini of the gene (Figure 17.25). The gene was inserted into the λ vector (Figure 17.26) carrying amber mutations in the proteins A and B. Amber suppression did indeed occur, as shown by the production of viable phage plaques only when the phage contained the synthetic gene.

The conception of the above work, detailed plans and early phases of the experimental work were carried out between 1967 and 1970 in Madison. Synthesis of the structural gene was well under way and the sequence work on the control elements had also been started. The work was continued at

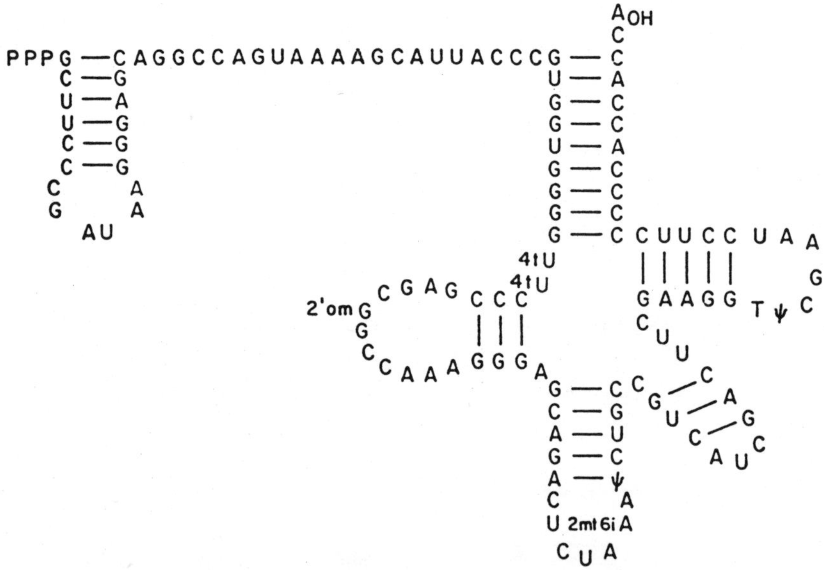

Figure 17.22 The primary nucleotide sequence of an *E. coli* tyrosine tRNA precursor. The sequence is shown using the standard clover-leaf structure for the tRNA portion and the possible hairpin at the 5′-end.

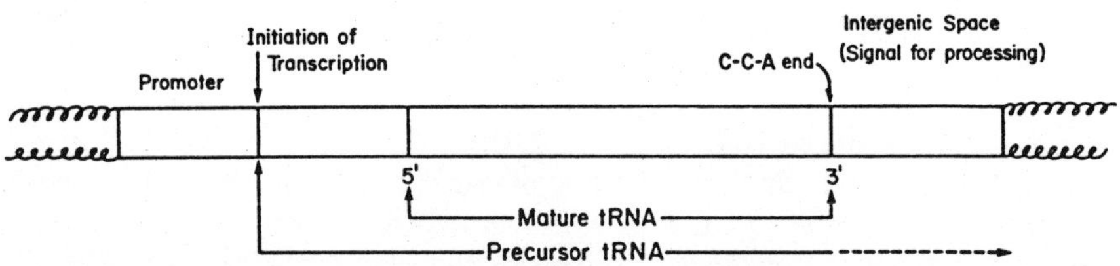

Figure 17.23 Linear arrangement of the different parts (structural gene and control region) of the tyrosine suppressor tRNA gene.

M.I.T., where the discovery of the precursor in 1974 gave it a further stimulus. Development of new DNA sequencing methods also accelerated its pace. The total project was completed in the summer of 1976.

Postscript

Synthetic polynucleotides and molecular biology As a result of the recent efforts of a number of groups led by former members of my group, chemical synthesis of short DNA fragments is now extremely rapid and ef-

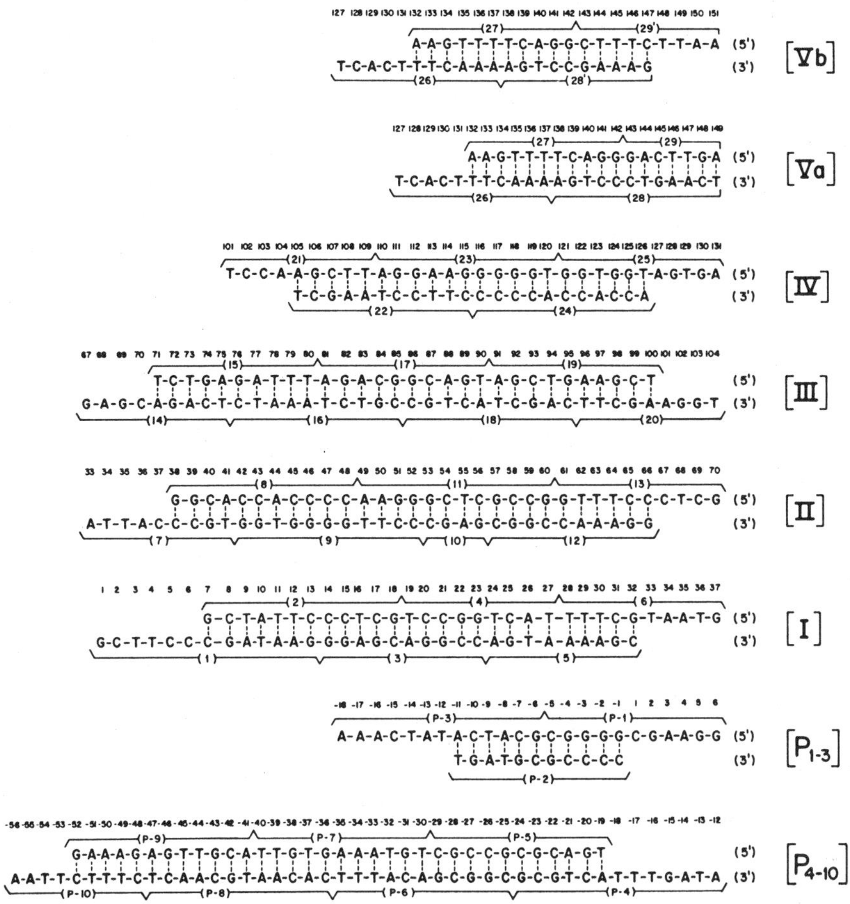

Figure 17.24 Plan for the enzymatic joining of the segments in the total synthesis of the gene for the precursor of tyrosine tRNA. Grouping of the total chemically synthesized segments into eight duplexes also includes the promoter and processing sequence at the 3′ end.

ficient. Chemical synthesis is playing an important and powerful role in molecular biology and in recombinant DNA methodology. Indeed, the scope of work in these general areas is greatly enhanced by synthesis. Some of the possibilities offered by synthesis may be mentioned.

One large field is the total synthesis of genes. Following the above-described synthesis of transfer RNA genes, total or partial synthesis of many genes has been reported. These include the genes for the peptide hormones somatostatin, human growth hormone, epidermal growth hormone, insulins and α-thymosin, interferons and many other genes. It seems evident that

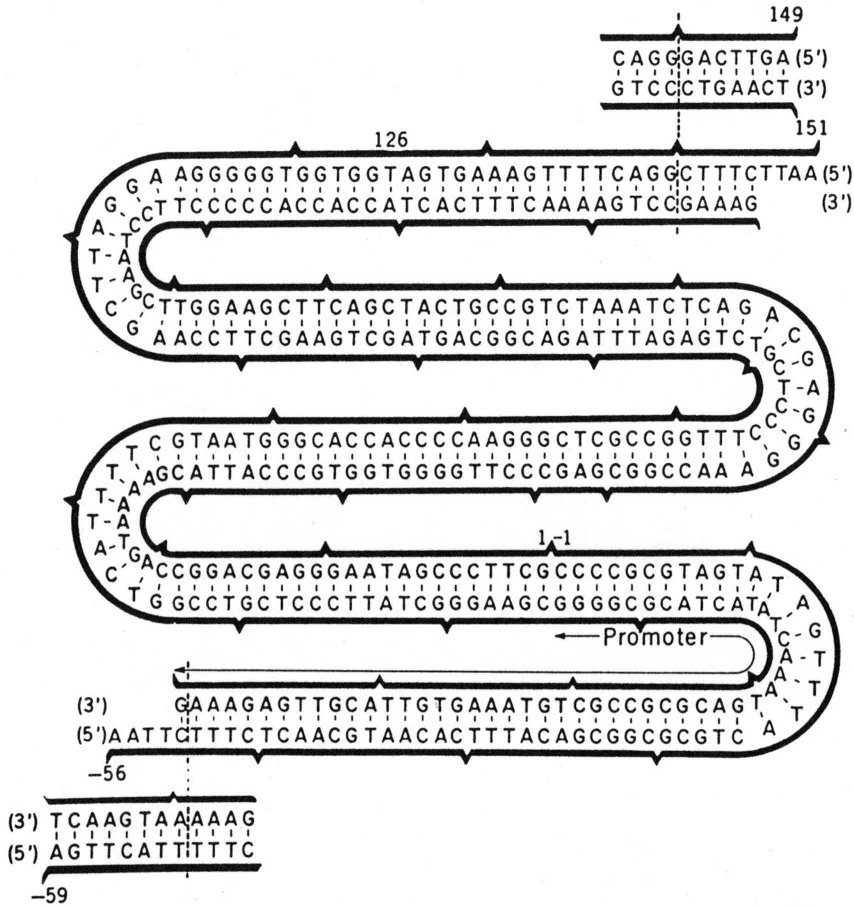

Figure 17.25 Synthetic tyrosine suppressor tRNA gene. At bottom left terminus is the $EcoR_1$ endonuclease-specific sequence. Then follow a 51-nucleotide-long promoter region, a 126-nucleotide-long DNA corresponding to the precursor RNA and, finally, a 25-nucleotide-long region flanking the 3′-end of tRNA. Of the latter, 16 nucleotides belong to the EcoRI natural sequence including the EcoRI endonuclease-specific sequence. Additional duplex segments shown on top right and bottom left are the sequences as they naturally are found in these regions and continue in the two directions.

synthesis will encompass larger and larger genes, as well as their modified counterparts. However, it is important to add that the main prospects offered by synthesis are in gene dissection and manipulation. For these purposes, total synthesis of genes may often not be necessary. Instead, natural genes may be modified, altered, shortened or edited for increased gene expression. This may be accomplished by insertion or modification of only small segments

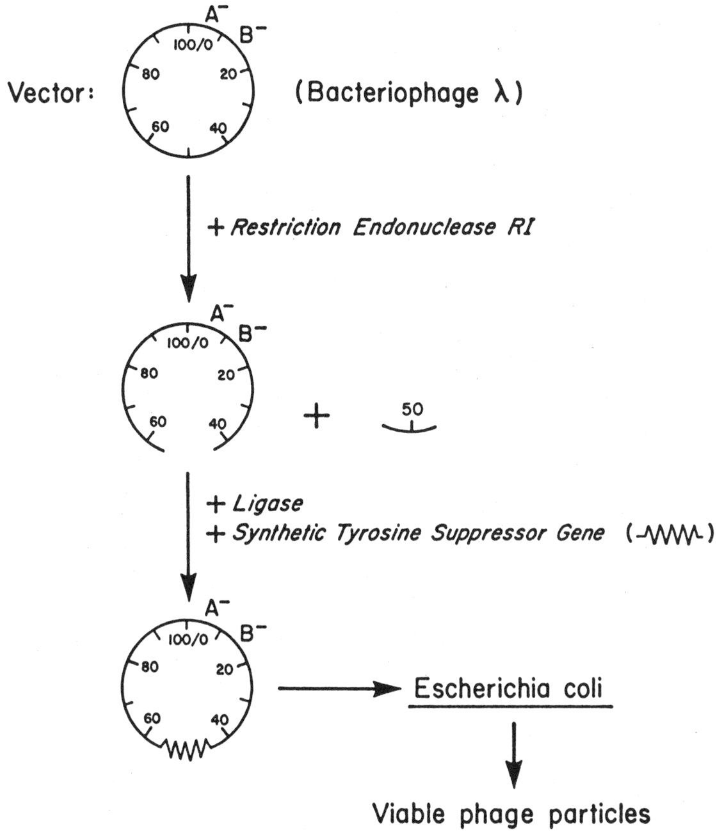

Figure 17.26 Cloning of the synthetic gene for the tyrosine suppression tRNA. The vector used was a derivative of bacteriophage λ with two amber mutations (A^-, B^-). Digestion with restriction endonuclease R_1 excised a segment as shown. Subsequent addition of the synthetic gene and ligase gave the circular phage, which formed plaques by virtue of suppression of the amber mutations.

of the natural genes. Thus a combination of restriction cuts, linkers and synthetic DNA fragments will, in general, be used most often for gene alterations and gene expression.

Related to all these studies are the syntheses of gene linkers, of primers for cDNA synthesis, of probes for hybridization purposes in identification, isolation and cloning of genes. Work in this area has been enormously productive.

Another field of fundamental importance is that of interactions between nucleic acids and proteins. Clearly, this is important for understanding the nature of molecular interactions in gene expression. Synthesis provides a unique and versatile approach for these studies, in that structure-function relationships of the DNA region concerned may be studied using systematic alterations by chemical synthesis. An example is the interaction between *E. coli* RNA polymerase and the promoters recognized by it. An example of such studies in our own laboratory is the tyrosine tRNA gene promoter. Another example of the use of the synthetic approach is the work by Marvin Caruthers and colleagues on the interaction between the lactose operator and its repressor in *E. coli*.

Finally, a highly powerful approach to the study of structure-function relationships in proteins involves controlled mutagenesis at the gene level. A number of approaches, variously called localized, directed and site-specific mutagenesis, have been introduced. Site-specific mutagenesis is carried out in a number of ways, but all depend on precision that requires chemical synthesis. One approach of wide scope, the cassette replacement method, involves the excision of a restriction DNA fragment from the gene of interest and its replacement by a synthetic counterpart containing the required nucleotide changes. For extensive mutagenesis this is the method of choice.

Into different things In the early '70's I began to dabble into biological membranes. The work and interest first involved the *E. coli* envelope, especially lipopolysaccharide and phospholipids, and development of photoaffinity labeling methods for studies of membrane proteins. A number of integral membrane proteins, glycophorin A, cytochrome b_5 and Ca^{2+}/Mg^{2+}-ATPase of sarcoplasmic reticulum were studied. However, a major interest in the past several years has been the light-transducing retinal-based protein, bacteriorhodopsin. A single membrane-embedded protein carries out light-dependent proton translocation from the inside to the outside of the extreme halophile, *Halobacterium halobium*. This results in an electrochemical gradient which is used by the cell for the synthesis of ATP. It is a beautiful system for study of light-transduction and of proton pumping mechanism. Methods of membranology, protein chemistry and recombinant DNA (site-specific mutagenesis) are being used in conjunction with this work. Furthermore, I am very grateful to have come upon this system because, in turn, it has stimulated my interest in the visual receptors and the chemistry and biochemistry of the visual processes. Thus, in the years ahead, bacteriorhodopsin, rhodopsin and related transmembrane proteins, and the biochemistry of the visual processes will continue to be my major interests.

Some Acknowledgements

The University of Wisconsin provided generous support and encouragement throughout the above research. In 1964 I was appointed to the newly created Conrad A. Elvehjem Professorship at the University.* It was most remarkable

*I remember seeing Van Potter soon after my appointment to the Elvehjem Professorship. All he said was, "You don't have to feel that you owe or have to do anything more for the University."

that the University never made the slightest demand on my time during those years.

Experimental work in my laboratory demanded teamwork. By the early '60's, the nucleic acid field had already begun to attract chemists and biochemists. Colleagues came to my laboratory from all over the world. They were stimulating and made contributions ensuring the continuity of our work. I conclude this account by including two annual group photographs. The photograph of 1965 (Figure 17.27) has in it many of the colleagues that made the major contributions to work on the genetic code. The second photograph (1970) (Figure 17.28) was taken at the successful conclusion of the alanine tRNA gene synthesis. The members of this group were mostly responsible for development and conclusion of the formidable project of gene synthesis.

References

Anderegg, R., Hayley, H., Biemann, K., Chang, S. H., Dunn, R. J., Gerber, G. E., Gray, C. P., Herlihy, W. C., Hojeberg, B., Huang, K.-S., Joshi, V. C., Kimura, K., Liao, M. J., Lind, C., London, E., Majumdar, A., Makabe, O., McCoy, J. M., Radhakrishnan, R., RajBhandary, U. L., Ross, A. H., Simsek, M., Takagaki, Y., Wildenauer, D., and Khorana, H. G. (1981) *Biomolecular Stereodynamics: Proceedings of a Symposium Held at the State University of New York at Albany* (Sarma, R. H., ed.), Vol. II, pp. 37–54. Adenine Press, Albany, New York.

Khorana, H. G. (1961) *Some Recent Developments in the Chemistry of Phosphate Esters of Biological Interest.* John Wiley and Sons, New York.

Khorana, H. G. (1968) Polynucleotide synthesis and the genetic code, in *The Harvey Lectures,* Series 62, p. 79. Academic Press, New York.

Khorana, H. G. (1978) Total synthesis of a gene. *Science* 203: 614.

Khorana, H. G. (1982) Synthesis in the study of nucleic acids, in *Some Historical and Modern Aspects of Amino Acids, Fermentations and Nucleic Acids, Proceedings of a Symposium* (Snell, E. E., ed.), pp. 93–141. Annual Reviews, Inc., Palo Alto, California.

Figure 17.27 Khorana laboratory, 1965. *Standing, left to right:* H. Kossel, E. Drissen, A. Stuart, U. L. RajBhandary, R. Lohrmann, D. Dammen, E. Ohtsuka, C. Agerjord, C. Fuchs, G. Stone, R. Faulkner, H. G. Khorana, T. M. Jacob, H. Hayatsu, H. Radloff, R. Promel, P. Witonsky, D. Soll, R. Wells. *Kneeling:* S. Nishimura, D. Jones, S. Narang, A. R. Morgan. *Not shown:* M. Moon.

Figure 17.28 Khorana laboratory, 1970. *Standing, left to right:* P. Besmer, R. Kleppe, R. Miller, H. Van De Sande, H. G. Khorana, G. Stone, K. Agarwal, R. Smecko, M. Caruthers, C. Agerjord, P. Cashion. *Seated, left to right:* K. Minamoto, K. Kleppe, L. Jones, T. Terao, A. Yamazaki, V. Sgaramella, A. Kumar.

Appendix A: Symposium Participants

Julius Adler
Kamaluddhin Ahmad
Laurens Anderson
A. Arnold
Alan Attie
L. Andrew Ball
C. E. Ballou
Carl A. Baumann
Helmut Beinert
Gregory R. Bellomy
N. J. Benevenga
Kevin Phillip Block
John Bloomenthal
Robert M. Bock
R. K. Boutwell
Paul Boyer
Robert W. Bray
Glen A. Broderick
Robert Bruckner
Robert H. Burris
Wilbur H. Campbell
Dena C. Cederquist
Carl O. Clagett
Mary Clagett
Wm. Wallace Cleland
Macie C. Coots
Robert H. Coots
Teresa Cornwell
Michael M. Cox
W. W. Cravens
Dawson C. Deese
Hector F. DeLuca
Carlton F. Dresden
Richard Eisenstein
Mrs. C. A. Elvehjem
Robert E. Feeney
Matthew A. Fisher
George Flentke
Perry A. Frey
Jack Garbutt
John C. Garver
Jack Gorski
Charles B. Grissom
John G. Halen
Arnold Hampel
Jeffrey C. Hansen
Donald R. Harkness
A. E. Harper
Colleen Hayes
LaVell M. Henderson
Martin Hochstrasser
Wm. G. Hoekstra

John Hofert
Ron Huss
Ross B. Inman
Mrs. Marvin J. Johnson
James D. Jones
Mary A. Jones
Paul Kaesberg
Roy Kanemoto
David J. Katz
H. Gobind Khorana
Judith Kimble
Janice A. Kolberg
Henry A. Lardy
John M. Konopka
Jean M. Lang
Sigrid Leirmo
Tom Link
Paul W. Ludden
Judith E. Lund
Patricia R. M. MacLeod
Robert A. MacLeod
Kingsley M. Mann
Gilbert J. Mannering
Sabeeha Merchant
Elizabeth C. Miller
Robert H. Miller
Scott Morrical
Portia M. Morris
Juan Navia
David L. Nelson
Sandra Neuendorf
Forrest H. Nielsen
Masayasu Nomura
Marion H. O'Leary
Stanley P. Owen
Griff Parks
Theresa E. Dowling Paul
Mrs. Paul Phillips
Leslie A. Piper
Mark R. Pope
Van R. Potter
Christian R. H. Raetz
Lakshman E. Rajagopalan
S. T. Rao
Bryan L. Ray
M. Thomas Record
Linda Jo Reinhard
Wm. S. Reznikoff
Ronald C. Roberts
Roland R. Rueckert
Arnold E. Schaefer
Howard A. Schneider
Marie Schneider
Heinrich K. Schnoes
B. S. Schweigert
Bruce R. Selman
Julie Senecoff
Mark N. Shahan
Fathieh M. Shahidsaless
James Shull
David P. Silva, Jr.
Thomas J. Skatrud
Esmond E. Snell
Mark A. Stahmann
Mrs. Harry Steenbock
Frank M. Strong
JoAnne Stubbe
Muttaiya Sundaralingam
M. L. Sunde
John W. Suttie
L. J. Teply
Duane A. Tewksbury
Jean Kring Tews
C. Ray Thompson
William Thomson
Jack C. Towne
Sudhakar S. Wagle
Joseph S. Wall
Robert Sam Wehbie
Paul M. Weiss
JoAnn Whitefleet-Smith
Marvin P. Wickens
Robert Wohlhueter
Lloyd Yu
Israel Zelitch

Appendix B: Faculty of the Agricultural Chemistry and Biochemistry Departments, 1888–1983*

Henry Prentiss Armsby
Yale University, 1879
1883–1887

Fritz Wilhelm Woll
University of Wisconsin, 1904
1885–1913

*Faculty are presented chronologically. The institution and the date the Ph.D. was conferred are indicated, as are the years of faculty service.

Stephen Moulton Babcock
University of Wisconsin, 1916
1888–1913

Edwin Bret Hart
1906–1945

Elmer Verner McCollum
Yale University, 1906
1907–1917

William Edward Tottingham
1910–1944

William H. Peterson
University of Wisconsin, 1913
1914–1951

Harry Steenbock
University of Wisconsin, 1916
1916–1956

Paul W. Boutwell
University of Wisconsin, 1916
1917–1920

Karl Paul Link
University of Wisconsin, 1925
1927–1971

Conrad Arnold Elvehjem
University of Wisconsin, 1927
1930–1962; Chairman 1944–1958; Dean 1946–1958; UW President 1958–1962

Paul H. Phillips
University of Wisconsin, 1933
1936–1964

Frank M. Strong
University of Wisconsin, 1932
1938–1972

Carl August Baumann
University of Wisconsin, 1933
1938–1974

Marvin J. Johnson
University of Wisconsin, 1932
1940–1972

Robert Harza Burris
University of Wisconsin, 1940
1944–1984

Esmond Emerson Snell
University of Wisconsin, 1938
1945–1953

Henry Arnold Lardy
University of Wisconsin, 1943
1945–present

Mark Arnold Stahmann
University of Wisconsin, 1941
1947–1979

Jesse N. Williams, Jr.
University of Illinois, 1948
1950–1959

Laurens Anderson
University of Wisconsin, 1950
1951–1986

Robert Manley Bock
University of Wisconsin, 1952
1952–present

Robert Lesh Baldwin
Oxford University, 1954
1955–1960

William George Hoekstra
University of Wisconsin, 1954
1955–present

Paul Joseph Kaesberg
University of Wisconsin, 1949
1955–present

Alfred Edwin Harper
University of Wisconsin, 1953
1956–present

John C. Garver
University of Wisconsin, 1955
1957–1981

William Wallace Cleland
University of Wisconsin, 1955
1959–present

Hector Floyd DeLuca
University of Wisconsin, 1955
1959–present; Chairman 1970–1986

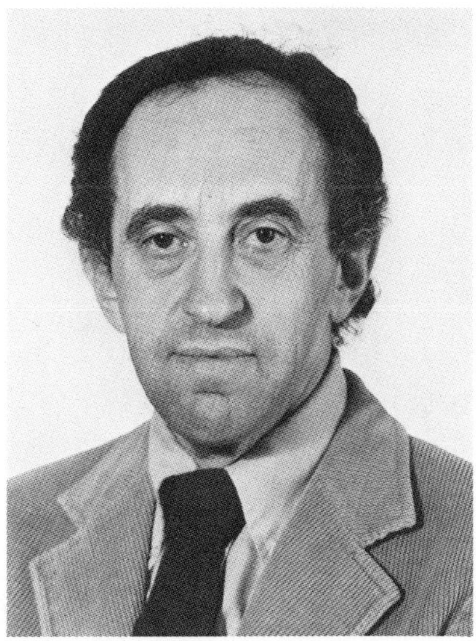

Julius Adler
University of Wisconsin, 1957
1960–present

George Harrison Echols
University of Wisconsin, 1959
1960–1969

John W. Suttie
University of Wisconsin, 1960
1961–present

Howard Rasmussen
Rockefeller Institute, 1959
1961–1965

Har Gobind Khorana
University of Liverpool, 1948
1962–1970

Roland R. Rueckert
University of Wisconsin, 1960
1965–present

Robert Harvey Rownd
Harvard University, 1963
1966–1981

Robert Dale Wells
University of Pittsburgh, 1964
1966–1981

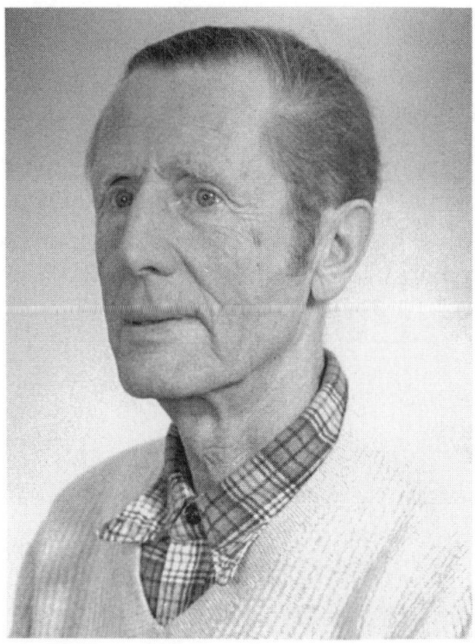

Helmut A. Beinert
University of Leipzig, 1943
1967–1984

Julian E. Davies
University of Nottingham, 1956
1967–1980

Ross Inman
University of Adelaide, 1960
1967–present

Heinrich Konstantin Schnoes
Massachusetts Institute of Technology, 1965
1967–present

Muttaiya Sundaralingam
University of Pittsburgh, 1961
1969–present

Masayasu Nomura
University of Tokyo, 1957
1970–1985

William H. Orme-Johnson
University of Texas-Austin, 1964
1970–1979

Sambhorao T. Rao
University of Madras, 1966
1970–1977

David Lee Nelson
Stanford University, 1969
1971–present

William Stanton Reznikoff
Johns Hopkins University, 1967
1971–present; Chairman, 1986–present

Jack Gorski
University of Wisconsin, 1958
1973–present

William A. Gibbons
University of Sheffield, 1964
1974–1982

Christian R. H. Raetz
Harvard University, 1973
1976–1987

Bruce R. Selman
University of Rochester, 1973
1976–present

Andrew L. Ball
Oxford University, 1969
1979–1987

Colleen E. Hayes
University of Michigan, 1973
1979–present

Marion H. O'Leary
Massachusetts Institute of Technology, 1966
1980–1988

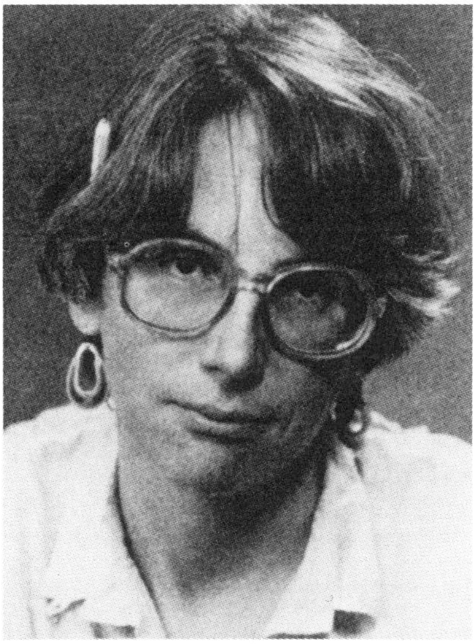

JoAnne Stubbe
University of California-Berkeley, 1971
1980–1987

Perry A. Frey
Brandeis University, 1967
1981–present

Paul Ludden
University of Wisconsin, 1977
1981–present

Alan D. Attie
University of California-San Diego, 1980
1982–present

Michael Matthew Cox
Brandeis University, 1979
1982–present

M. Thomas Record, Jr.
University of California-San Diego, 1967
1982–present

Judith E. Kimble
University of California-Berkeley, 1978
1983–present

Marvin P. Wickens
Stanford University, 1978
1983–present

Appendix C: Faces and Places

Early faculty of the Agricultural Chemistry Department. *Front:* Dr. and Mrs. Freudenberg (visiting), S. M. Babcock, E. B. Hart. *Rear:* W. H. Peterson, W. E. Tottingham, C. A. Elvehjem, K. P. Link, H. Steenbock, C. L. Christiansen (Dean, College of Agricultural and Life Sciences).

Nutrition seminar group, summer, 1945. *Front*: Walliker, J. E. Giese, T. M. Zepplin, E. Miller, L. F. Deemer, M. Ives Leonard. *Second row*: G. W. Newell, M. D. Cannon, W. A. Krehl, S. Banerjee, J. M. Cooperman, Glasinovic, K. B. McCall. *Third row*: C. K. Whitehair, P. R. Moore, R. K. Boutwell, B. S. Schweigert, R. R. Spitzer, A. W. Halverson, W. R. Ruegamer. *Rear*: F. M. Strong, C. A. Elvehjem, P. H. Phillips, E. B. Hart.

Nutrition seminar group, summer, 1946. *Front:* A. Ziegenhagen, T. M. Zepplin, E. P. Anderson, H. Nath, E. M. Sporn, S. H. Lipton, B. F. Steele, E. L. Pearce, A. N. Booth. *Second row:* W. H. Riesen, J. C. Rabinowitz, K. M. Mann, H. A. Lardy, J. M. Cooperman, W. A. Krehl, P. S. Sarma, R. P. Geyer. *Third row:* I. T. Greenhut, G. W. Newell, L. Anderson, L. M. Henderson, L. Lutwack, W. Jaffe, H. E. Sauberlich, P. H. Derse. *Fourth row:* D. R. Clandinin, A. R. Robblee, R. R. Grunert, J. J. Betheil, R. L. Potter, A. C. Griffin, R. M. Johnson, Jr., P. R. Moore, C. A. Baumann. *Fifth row:* A. Lepp, J. de la Huerga, J. K. Smith, W. C. Weir, O. G. Bentley, A. E. Schaefer, V. D. Wiebelhaus, C. C. Clayton. *Rear:* P. H. Phillips, E. B. Hart, C. A. Elvehjem, E. E. Snell.

Nutrition—metabolism seminar group, summer, 1953. 1. M. Winje, 2. J. Halliday, 3. J. Kring, 4. A. Haurowitz Sievert, 5. A. Bosch, 6. W. Rumsfeld, 7. R. Boldt, 8. A. Denton, 9. C. Beckner, 10. F. Taketa, 11. C. Alexander, 12. W. Leoschke, 13. W. Wilkinson, 14. R. Emerick, 15. D. Benton, 16. W. Wells, 17. J. Finlayson, 18. M. Schreiber, 19. W. Monson, 20. J. Bothwell, 21. H. Nino-Herrera, 22. J. H. Heinschke, 23. C. A. Elvehjem, 24. P. H. Phillips, 25. A. Harper, 26. J. Bixby, 27. J. Adler, 28. O. Abbott, 29. A. Kandutsch, 30. R. Miller, 31. C. A. Baumann, 32. J. Williams, 33. W. Hoekstra.

Three generations of departmental administrators. Carol Marth, Ann Terrio, Eloise Marsh, and Bonnie Jaedicke at the Centennial celebration.

Long-Time Departmental Associates

Seated front: Pat Bredlau, Sandy Anderson, Warren Dettinger, Al Jones, Fran Krings. *Rear:* Helen Frank, Bruce Selman, Zaia Mailes.

Willie Frey.

Art O'Keefe, Sandy Anderson, Fran Krings, Art Olson.

Agricultural Chemistry building (1917).

1957 wing of Biochemistry building.

1986 Biochemistry building addition.

Appendix D: Departmental Graduates*

Julius Adler

M.S.

1962 Sherman, John Roberts
1982 Kanemoto, Roy Haruo

Ph.D.

1962 Schade, Sylvia Marie
1967 Armstrong, John Brian
1968 DePamphilis, Melvin Lewis
1970 Mesibov, Robert Evan
1971 Tso, Wung-wai
1973 Larsen, Steven H.
1974 Toews, Myron Lee
1979 Kleene, Steven James
1980 Repaske, David Roy
1982 Black, Roy Alvin

Laurens Anderson

M.S.

1953 Docta, Lawrence, H.
1954 Leon, Arthur Sol
Swan, Emily Bishop
1956 Caldwell, Kathryn Anne
Coots, Robert Herman
Diedrich, Donald Frank
Kelley, John Joseph
1957 Post, Gerald George
1959 Weinhold, Paul Allen
1961 Howard, Charles F., Jr.
1964 Blocher, John Paul
1965 Mohlenkamp, (Marvin) Joseph, Jr.
1968 Parascandola, John Louis
Wu, Ming-Chi (Mickey)
1970 Conner, Anthony Hoyt
Espelie, Karl Edward
Meredith, Dennis Lee (*SW*)
1976 Mansaray, Yembeh Kellah

Ph.D.

1955 Drummond, George Ira
Halliday, June Wanda

*Joint degrees with other departments or programs are indicated as follows: *AH*, Animal Husbandry; *AS*, Animal Science; *B*, Bacteriology; *DFI*, Dairy and Food Industry; *DS*, Dairy Science; *E*, Entomology, *FP*, Forest Products; *FSI*, Food Science and Industry; *G*, Genetics; *HE*, Home Economics; *MAS*, Meat and Animal Sciences; *MS*, Medical Sciences; *NS*, Nutritional Sciences; *P*, Pathology; *PP*, Plant Pathology; *PS*, Poultry Science; *SW*, Science Writing; *VS*, Veterinary Science.

	Landel, Aurora Mamauag
1957	White, Frederick H. (Jt. with Williams)
1958	Coots, Robert Herman
1959	Diedrich, Donald Frank Kelley, John Joseph Aronson, John Noel Post, Gerald George
1960	Caldwell, Kathryn Anne
1961	Iyer, N. Thanukrishna Weinhold, Paul Allen
1962	Hsu, Robert Ying Hwang
1963	Howard, Charles Frank, Jr.
1964	Mao, Chieh Hsia
1968	Mohlenkamp, (Marvin) Joseph, Jr.
1970	Hixson, Susan Lynn Wu, Ming-Chi
1972	Espelie, Karl Edward
1973	Tang, Jiunn-Yann
1974	Holick (Teats), Sally Ann
1976	Wertz, Philip Wesley
1978	Slife, Charles Warren

Robert L. Baldwin

M.S.

1957	Nakatani, Misako
1959	Butzow, James J.

Carl A. Baumann

M.S.

1939	Jacobi, H. P. Lavik, Paul S.
1941	Field, John B. Miller, James A.
1942	Moore, Peter R.
1945	Giese, Janet E.
1946	Sauberlich, Howerde Edwin
1947	Johnson, Ralph M., Jr. Pearce, Ethel Lyon Clayton, Charles Curtis
1948	DeLeers, Vincent Joseph Tai, Tsong Kuang
1949	Swick, Robert Winfield
1950	Drysdale, George Robert Lih, Hwa
1951	Miller, William Louis, Jr. DeBey, Harold John Waibel, Paul E.
1952	Naber, Edward Carl (*PS*) Rumsfeld, Herbert William, Jr. Guzman, Jesus Garcia Kandutsch, Andrew August
1954	Miller, Edward Clarence (*PS*)
1955	Finlayson, John Sylvester Brooks, Samuel Corrall, Jr.
1956	Coleman, Douglas Leonard Gitler, Carlos (*PS*) Jones, James Donald
1958	Gaylor, James LeRoy Hardy, Ralph Wilbur
1959	Chindavanig, Supis
1961	Ganther, Howard E.
1963	Levander, Orville Arvid
1965	Paulson, Gaylord Dean (*MAS*)
1966	Byard, James Leonard

Ph.D.

1941	Jacobi, H.P. Lavik, Paul S.
1943	Miller, James A.
1945	Miller, Elizabeth C.
1947	Griffin, Amos Clark
1948	Johnson, Ralph M., Jr. Sauberlich, Howerde Edwin
1949	Noland, Jerre L. Clayton, Charles Curtis Steele, Betty Faye
1951	Linkswiler, Hellen Martha (*HE*) Johnson, Doris (*HE*) Swick, Robert Winfield
1952	Futrell, Mary Dean (*HE*) Lih, Hwa Miller, William Louis, Jr.
1953	Idler, David Richard Rumsfeld, Herbert William, Jr. Waibel, Paul Edward (*PS*)
1954	Naber, Edward C. (*PS*) Kandutsch, Andrew August
1955	Wells, William Wood

1956 Abbott, Okra (*PS*)
Jones, Evelyn M. (*HE*)
Steel, Dorothy L. (*HE*)
Abernathy (McTeer), Miriam L. (*HE*)
1957 Brooks, Samuel Carroll, Jr.
Finlayson, John Sylvester
1958 Coleman, Douglas Leonard
Gitler, Carlos
Braham, Jose Edgar (*PS*)
Jones, James Donald
Sullivan, Thomas W. (*PS*)
1960 Gaylor, James LeRoy
Hardy, Ralph Wilbur
1961 Rasmussen, Arlette I. (*HE*)
1962 Mahlum, (Daniel) Dennis
Hopkins, Leon Lorraine, Jr. (*MAS*)
1963 Neiderhiser, Dewey Harold
Page, Louise (*HE*)
Ganther, Howard Edward
Heth, Dwayne August (*PS*)
1964 Swan, Patricia Brintnall (*HE*)
Shortridge, Lucy J. (*HE*)
1965 Kellogg, Thomas Floyd
Levander, Orville Arvid
1966 Ewan, Richard Colin (*MAS*)
1967 Miller, Lorraine Theresa (*HE*)
Paulson, Gaylord Dean (*MAS*)
1968 Byard, James Leonard
1973 Orme-Johnson (Roberts), Nannette

Robert M. Bock

M.S.

1954 Ling, Nan-Sing
1956 Hoh, Fay
1959 Gilboe, Daniel Pierre
Stanley, Wendell M., Jr.
1960 Criddle, Richard S.
1961 Saponara, Arthur G.
Lichti, F. Ulrike
1962 Brieger, Ingetraud Rennenkampff
1963 Bruening, George Emil
1965 Hoffman, Jerald Lee
1966 Omilianowski, Daniel Roman
Lee, Se Yong
1967 Wang, Shih-Eh
1969 Chaffin, (Welda) LaJean
1978 Staggers, Kimberly Ann

Ph.D.

1957 Ling, Nan-Sing
Salegui (Nurnberg), Miriam
1959 Gilboe, David D.
Thakur, Vatsala Marayanrao
1960 Yin, Fay Hoh
1961 Hayman, Selma
1962 Criddle, Richard Sheffield
Adams, Emory Temple, Jr.
1963 Yang, Pauline Yun-Wo
Stanley, Wendell Meredith, Jr.
1964 Saponara, Arthur Gerald
Heintz, Roger Lewis
1965 Bruening, George Emil
1966 Lichti, (Frieda) Ulrike
1967 Hoffman, Jerald Lee
1968 Klagsbrun, Michael
1969 Lee, Se Yong
Leppla, Stephen Howard
Hampel, Arnold Everett
Kirkegaard, Leslie Harvey
1971 Omilianowski, Daniel Roman
Young, James Douglas
1972 Chaffin, (Welda) LaJean
1973 Schmidt, Francis John
1974 Johnson, Timothy John Albert
1975 Swaminathan, Santhanam
Laten, Howard Mark
Johnson (Bjoraker), Barbara Jeanne
1979 Toce, Joseph Anthony
1981 Sullivan, Mark Alan
1983 Cannon, John Francis

Paul W. Boutwell

M.S.

1921 Brunkow, O.R.

Robert H. Burris

M.S.

1946 Little, Henry Nelson
Stutz, Robert Eugene

1947	Kreko, Mary Jane
	Davis, Mary Shane
	Molnar, Dorothy May
1948	Tolbert, Nathan Edward
	Wagenknecht, Austin Clayton
	Fager, Earl E. C.
	Miller, Isaac Henry
	Storey, Robert Henry
1950	Eberts, Floyd Samuel, Jr.
	Firschein, Hilliard E.
1951	Williams, Arnold Eugene
	Newburgh, Robert Warren
	Zbinovsky, Vladimir
1952	Brummond, Dewey O.
	Dow, William Alexander, Jr.
	Krall, Albert Raymond
	Towne, Jack C.
1953	Harris, James Burdick
	Magee, Wayne Edward
1955	Bach, Michael Klaus
	Mann, Jay D.
1956	Hoch, George Edward
	Walker, Margaret Skorvaga
1957	Wasserman, Aaron Reuben
1958	Delhumeau-Arrecillas, Graciela
	Wang, Li Chuan
	Winter, Nancy S.
1960	Garcia-Rivera, Jean
1961	Plengvidhya, Prachote
	Burns, Richard C.
	Soulen, Thomas Kay
1962	Olson, Lee Charles
1963	Bohne, Mary Eleanor
	Lockshin, Arnold
1964	Klucas, Robert Vernon
	Winter, Harry Clark
1965	Brown, Ross Duncan, Jr.
	vonStedingk, Lars Victor Eugene
	Munson, Thomas Oren
1966	Vandecasteele, Jean-Paul
1967	Dravnieks, Dzintar Eric
	Siu, Florence Moon Ying
	Swank, Richard Tilghman
1970	Mague, Timothy Hall
	Sundquist, June Estelle
1973	Rivera-Ortiz, Jose Manuel
1975	Fiehweg, Robert Eugene
1981	Walsh, Kathleen O'Connell

Ph.D.

1946	MacVicar, Robert William
1947	Clagett, Carl Owen
1948	Olson, Oscar Edward
	Ghosh, Bhanu Prokash
	Little, Henry Nelson
	Mitchell, John E.
1950	Stutz, Robert Eugene
	Tolbert, Nathan Edward
	Wagenknecht, Austin Clayton
1951	Miller, Isaac H.
	Zelitch, Israel
1952	Hyndman, Lee Allen
	Eberts, Floyd Samuel, Jr.
	Aprison, Morris Herman
	Noll, Clifford Raymond, Jr.
	Sapp, Ralph W.
	Wall, Joseph Sennen
1953	King, Kendall Willard
	Wilson, David George
	Newburgh, Robert Warren
	Krall, Albert Raymond
1954	Brummond, Dewey O.
	Zbinovsky, Vladimir
1955	Allison, Russell Morris
	Towne, Jack C.
	Magee, Wayne Edward
	Haidri, S. Zafar Ali
	Mozen, Milton Michael
1957	Bach, Michael Klaus
	Bergmann, Fred Heinz
	Mann, Jay Donald
1958	Hoch, George Edward
	Mudd, J. Brian
	Neumann, Norbert P.
1959	Kurtzman, Ralph H., Jr. (*PP*)
1960	Wasserman, Aaron Reuben
1962	Garcia-Rivera, Jean
	Schneider, Kenneth Christian
1963	Hamilton, Ian Robert (*B*)
	Soulen, Thomas Kay
1964	Burns, Richard Charles
	Dua, Ramji Dass
	Plengvidhya, Prachote
1965	Olson, Lee Charles

1966 Lockshin, Arnold
1967 Winter, Harry Clark
Klucas, Robert Vernon
Ouellette, Claude Alain (*B*)
1968 Brown, Ross Duncan, Jr.
Munson, Thomas Oren
Hwang, Jiann-Chaur
Vandecasteele, Jean-Paul
1969 Swank, Richard Tilghman
Kratochvil, John Frank (*FP*)
1971 Dravnieks, Dzintar Eric
Mague, Timothy Hall
1972 Campbell, Wilbur Harold
Ljones, Torbjørn
1973 Tso, Man-Yin
Stombaugh, Ned Alan
1975 Erbes, David Lee
1976 Peterson, Richard Burnett
1977 Emerich, David William
Ludden, Paul W.
1979 Hageman, Robert Vern
1980 Stumpf, David Karl
Houchins, Jeffrey Peter
Arp, Daniel James
1983 Guth, Joseph Harry
Privalle, Laura Stein

William W. Cleland

M.S.

1962 Wratten, Craig Charles
1963 Henson, Carl Payne
1965 Kuwahara, Steven Sadao
1966 Garces, Edmundo
Haavik, Arne Goodwin
Zahler, Warren Leigh
Barden, Roland Eugene
Gulbinsky, Joseph Stephen
1980 Touton, Cynthia Ann

Ph.D.

1962 Sanchez-Quintanar, Estela
1965 Abou-Issa, Hussein Mohamed
Wratten, Craig Charles
1966 Henson, Carl Payne
1967 Kuwahara, Steven Sadao
Garces, Edmundo
1968 Zahler, Warren Leigh
1969 Barden, Roland Eugene
1970 Ainslie, G(eorge) Robert, Jr.
1973 Janson, Cheryl Ann
1975 Schimerlik, Michael Ira
1976 Raushel, Frank Michael
1978 Rife, James Ed
1979 Grimshaw, Charles Edward
Blanchard, John Sears
1981 Anderson, Vernon Emmett
1982 Dougherty, Thomas Michael
1983 Hermes, Jeffrey Donald

Julian E. Davies

M.S.

1970 Benveniste, Raoul Eric
1973 Biddlecome, Sandra Marie
1975 Ursic (Becker), Doris Brigitta

Ph.D.

1972 Helser, Terry Lee
1976 Grant, Paul Glen
Schindler, Daniel Gordon
1978 Smith, David I.
Haas, Michael Jeffrey
1979 Andresson, Olafur Sigmar
1982 Gray, Gary Stephen

Hector F. De Luca

M.S.

1960 Garbus, Joel
1961 Norman, Anthony W.
Williams, Virginia R.
1962 Madsen, Norman Peter
1963 Zull, James Elwood
1964 Lund, Judith Karman
Olson, Earl Burdette, Jr.
1965 Puno, Ma Cecilia E.
Martin, David Lee
1967 Weller, (Herman) Morris
1968 King, Margaret Sha-Chew
1970 Holick, Michael Francis
1971 Stacey, Dennis William
1974 Ahrens, Helga Mathilde
Harnden, Deborah Sharon
1978 Reeve, Lorraine Ellen
1980 Bednar, George John

1981 Silva, David Paul, Jr.
Zurcher, Jackie Ann
1982 Zicker, Steven Curtis

Ph.D.

1960 Reiser, Sheldon
Engstrom, George Wesley
1962 Garbus, Joel
1963 Smith, Janet Alice
Norman, Anthony Westcott
Thompson, Virginia R. Williams
1964 Reynolds, Marjorie K. Lavers (*HE*)
1966 Zull, James Elwood
Lund, Judith Karman
1967 Madiedo-Mendez, Gonzalo (*PS*)
Stohs, Sidney John
1968 Moore, Jerry Lamar (*FSI*)
Martin, David Lee
Roberts, Anita Bauer
1969 Olson, Earl Burdette, Jr.
1970 Bhattacharyya (Horsting), Maryka
1971 Drescher, Dennis George
Holick, Michael Frances
Hallick, Richard Bruce
1972 Frolik, Charles Alan
Chen, Tai Cheng
1973 Knutson, Joyce
1975 Brown, Donna Frances
Seymour, Jon Lisle
1976 Baxter, Lee Ann
1977 Ribovich, Martin Lee
1979 Esvelt, Robert Peter
Shepard, Richard Malcolm
Yoon, Poksyn Song
1980 LeVan, Leon William
Wichmann, Joseph Kurt
1981 Skare, Kevin Lynn
1982 Kabakoff, Bruce David
Sietsema, William Kendall

George H. Echols

M.S.

1965 Hill, Charles Whitacre
1967 Gingery, Roy Evans
1968 Butler, (William) Barkley

Ph.D.

1967 Adhya, Sankar Lal
Hill, Charles Whitacre
Joyner, Julia Anne
1968 Gingery, Roy Evans
1969 Butler, (William) Barkley
1970 McMacken, Roger L., Jr.
1971 Willard, Mark Benjamin
1972 Cheng, Pi-Yu

Conrad A. Elvehjem

M.S.

1933 Stare, Frederick J.
1934 Cohen, Eugene
Koehn, Carl J.
Schultze, Max
Sherman, William C.
1936 Arnold, Aaron
Potter, Van R.
Wasserman, L.S.
1937 Borgen, Daniel
Lipton, M.A.
Mickelson, Olaf
1938 Feinstein, R. N.
Frost, D. V.
1939 Runkel, LaVern E.
1940 Nielsen, Edward
Wagner, Joseph R.
1941 Black, Simon
Henderson, L. M.
Katzenelbogen, Eva
1942 McIntire, Junius M.
Potter, Richard L.
1943 Barki, Victor Hiam
Cooperman, Jack M.
Michaud, Laurent
Pilgrim, Francis J.
Schweigert, Bernard S.
Teply, Lester J.
1944 Luckey, T. D.
Herbst, Edward J.
Newell, Gordon W.
Zepplin, T. Marie
1946 Geyer, Beverly Ransone
Ziegenhagen, Anita
1947 Deodhar, Tara Dinkar
Lepp, Albert

Maass, Alfred Roland
Riesen, Willis H.
Sporn, Eugene Milton
Torbet, Nancy
Derse, Philip Henry
Halverson, Andrew Wayne
Mozingo, Alfred Kempton

1948 Maddy, Kenneth Hilton
Tove, Samuel B.
Brickson, William Lewis
Gershoff, Stanley Norton
Bhatt, Indira Navtam

1949 Ingram, George Reagan
Sunde, Milton Lester
Thompson, Hiram T.
Schurr, Paul Edward
Wykes, Arthur Albert

1950 Collins, Richard Andrew
Dietrich, LaRoy Seibert, Jr.
Lewis, Urban James
Sirny, Robert Jacob
Lewis, Eugene Walter
Denton, Arnold Eugene
Klungsøyr, Mathias

1951 Litwack, Gerald
Lyman, Richard Lee
Monson, William Joye
Tappan, Donald Vester
Lalor, Robert James
Meyer, Marian Louise

1952 Heinicke, Herbert Raymond
Schreiber, Manuel
Violante, Alice Marie
Brandt, Arthur Melvin
Leoschke, William LeRoy
Nino-Herrera, Hipolito
Bosch, Arthur James
Hawk, Edgar Allen

1953 Bixby, John Nelson
Boldt, Roger Earl
Palmer, Gertrude

1954 Beckner, Clyde William
Benton, Duane Allen
Svenneby, Gyda
Winje, Marie Ellen

1955 Methfessel, Alfred Henry
Aronson, John Noel
Quiros-Perez, Felipe

1956 Pokornijs, Ruta
Spivey, Herman Everette, Jr.

1957 Collins, Macie

Ph.D.

1934 Stare, Frederick J.

1936 Koehn, Carl J.
Schultze, Max
Sherman, William C.

1937 Arnold, Aaron

1938 Potter, Van R.
Stirn, Frank

1939 Axelrod, A. E.
Lipton, M. A.
Mickelson, Olaf
Randle, Stacy, B.
Waisman, H. A.

1940 Frost, D. V.
Krewson, Charles

1941 Anderson, H. D.
Conger, Theodore, W.
Moxon, A. L.

1942 Black, Simon
McKibbin, Jack M.
Sober, H. A.
Swingle, Karl F.

1943 Nielsen, Edward

1944 McIntire, Junius M.
Spector, Harry
Wagner, Joseph R.

1945 Cooperman, Jack M.
Teply, Lester J.

1946 Schweigert, Bernard S.
Luckey, Thomas Donnell
Sarma, P. S.
Geyer, Robert P.

1947 Moore, Peter Rogers
McCall, Keith Bradley
McLaren, Barbara Alberta
Henderson, LaVell M.
Schaefer, Arnold Edward

1948 Madden, Robert James
Newell, Gordon Wilfred
Nath, Haripada
Robblee, Alexander Robison
Ruegamer, William Raymond
Clandinin, Donald R.
Booth, Albert Nelson
Couch, James Russell
Greenhut, Irene Tatman

1949 Potter, Richard Louis
Barki, Victor Hiam
Halverson, Andrew Wayne
Nichol, Charles Adam

Carlisle, Edith M.
Hankes, Lawrence Valentine
Riesen, Willis Hugh

1950 Fung, Fung-Haan (*HE*)
Maass, Alfred Roland
Odland, Lura Mae
Register, Ulma Doyle
Tove, Samuel B.
Ingram, George Reagan
Sunde, Milton Lester

1951 Sirny, Robert Jacob
Gershoff, Stanley Norton
Smith, Samuel Corry
Feigelson, Philip

1952 Collins, Richard Andrew
Lewis, Urban James
Dietrich, LaRoy Seibert, Jr.

1953 Lewis, Eugene Walter
Litwack, Gerald
Tappan, Donald Vester
Denton, Arnold Eugene
Harper, Alfred Edwin
Monson, William Joye

1954 Leoschke, William LeRoy
Alexander, James Craig
Bixby, John Nelson (*DFI*, Jt. with Elvehjem and Swanson)

1955 Marquez, Lourdes R. (*HE*)
Schreiber, Manuel
Heinicke, Herbert Raymond

1956 Benton, Duane Allen
Jutton, Dorothy Reed (*HE*)

1957 Miller, Edward Clarence (*PS*)
Deshpande, Pandurang D.
Gupta, Jiban Dhan
Reddy, Saranya Kumari (*HE*)

1958 Nino-Herrera, Hipolito
Boldt, Roger Earl
Leong, Kam Choy (*PS*)
Bosch, Arthur J.

John C. Garver

M.S.

1961 Button, Don K.
Smith, James Henry

1962 Zust, James Richard
Rao, Paluri Sundara

1965 Peña, Carlos Enrique

1966 Stuiber, David Anthony

1967 Eifried, Gary
Jaeger, Douglas Albert

1969 Stitt, Paul Arthur

1970 Liu, Shu-len Huang

1974 Van Miller, John Paul

1980 Pitt, Munro Wesley

Ph.D.

1964 Button, Don Kuhner

1968 Sze, Thaddeus Tsun

1969 McNabola, William Thomas
Weinshank, Donald

1971 Peña, Carlos Enrique

1973 Pankratz, Thomas John (Jt. with Johnson)
Ordman, Alfred Bram
Lamb, Steve Charles

1979 Skatrud, Thomas Joseph

William A. Gibbons

M.S.

1978 Alper, Joseph Bernard

1979 Lee, Kar Sang Hermann

Ph.D.

1979 Kuo, Mei-Chang

1980 Ford, Joseph John

1982 Sikakana, Cynthia Nonhlanhla

Jack Gorski

M.S.

1975 Mazuski, John Edward

1977 Zmolek, Wesley Ward

Ph.D.

1976 Harris, Jack Nevil
Stone, Roger Tas

1978 Shupnik, Margaret Ann

1980 Wiklund, Julie Ann

1983 Kassis, Judith Ann
Stack, Gary Edward

Alfred E. Harper

M.S.

1958 Dakroury, Ahmed Mohammed

1959 Peraino, Carl

Seidel, John Charles
Stucki, William Paul
Spolter, Pari Naraghi
1960 Rogers, Quinton Ray
Cho, Yoon Sang
Wergedal, Jon E.
1961 Chen, Mou-liang
Becker, Radleigh Val
1968 Weiser, Karl Denny
Sando, Gloria Nathalie
1969 Hunter, James Edward
Schirmer, Marie Annette
1970 Shinnick, Frederick Lloyd, Jr.
1971 Aguilar-Fajardo, Tulio Sigfrido (*MAS*)
1973 Brand, Larry Milton
1974 Lutz, Josephine
1977 Good, Steven Spencer
1978 Cree, Theodore Clarence
1979 Williams, Bill Bradley

Ph.D.

1958 Freedland, Richard Allan
1959 Nath, Narindar
Vivian, Virginia May (*HE*)
1960 Morrison, Mary Alice (*HE*)
1961 Kumta, Umesh Shivrao
Adkins, James S. (*PS*)
Peraino, Carl
Seidel, John Charles
Spolter, Pari
Wiener, Ruta
1962 Featherston, (William) Roy (*PS*)
Waldorf, (Margaret) Ann (*HE*)
Methfessel, Alfred Henry
Stucki, William Paul
1963 Rogers, Quinton Ray
Savage, Jane Ramsdell (*HE*)
Wergedal, Jon Engval
1968 Perez, Lucy Cabahug (*HE*)
1969 Anderson, Helen Lester (*NS*)
1970 Wohlhueter, Robert Martin
1971 Peng, Yeh-shan
1972 Morris, Margot Lynn
Broderick, Glen Allen (*DS*)
1973 Ip, Clement
1974 Hunter, James Edward
1975 Brand, Larry Milton
Shinnick, Frederick Lloyd, Jr.
1977 Case, Gary Lee (*NS*)
1979 Haley, Carolyn Jane
1981 Zaplowski, Carol
1983 Peters, John Christopher

Edwin B. Hart

M.S.

1910 Carr, R. H.
1914 Suzuki, S.
1915 Bentley, W. H.
Bohn, R. M.
Brannon, William A.
Flint, E. M.
1916 Nelson, V. E.
1917 Morrison, F. B.
Schaal, A.
Wessels, Philip
1918 Clifford, C. W.
Sommer, H. H.
Willaman, J. J.
Woo, K.
1919 Lamb, A. R.
Whitmore, L. M.
1920 Ellis, N. R.
1923 Lepkovsky, S.
Stevens, Henry
1924 Elvehjem, Conrad A.
Griem, W. B.
1927 Scott, Henry T.
1930 Kemmerer, A. R.
Schafer, Herbert
Wilson, Logan T.
1931 Deobold, H. J.
Kline, O. L.
Schaaf, Dan
Todd, Wilbert
1932 Keenan, J. A.
Van Donk, E.
1933 Kline, Bernard
Streb, Bartholomew
Wormeli, Ben
1935 Haman, Kenneth
1936 Kohler, George O.
1937 Hove, Edwin
Wiese, A. C.

1938 Coombes, A. I.
1941 Booth, Albert N.
Wachtel, Louis W.
1942 Boutwell, Roswell K.
Mills, Russell C.
1944 Geyer, Robert P.
Lardinois, Clifford C.
Ruegamer, W. R.

Ph.D.

1910 Currie, J. N.
1913 Carr, R. H.
Peterson, W. H.
1916 Steenbock, Harry
Boutwell, Paul W.
1917 Roxas, R. A.
White, R. W.
1920 Strowd, W. H.
Sure, Barnett
1922 Sommer, H. H.
1925 Lepkovsky, S.
1926 Waddell, James
1927 Elvehjem, Conrad A.
Titus, R. W.
1928 Herrin, R. C.
1929 Scott, Henry T.
1930 Lamb, A. R.
1931 Wilson, Logan T.
1932 Kemmerer, A. R.
Kick, C. H.
Rupel, I. W.
1933 Kozelka, Frank
Phillips, Paul H.
Todd, Wilbert
1934 Keenan, J. A.
Kline, O. L.
Knowles, H. R.
1935 Deobold, H. J.
1936 Walker, A. H.
1937 Pearson, Paul B.
1938 Bird, H. R.
Kohler, George O.
1939 Christiansen, J. B.
Hove, Edwin
Schantz, Edward J.
1940 Cravens, William W.
Hegsted, D. M.
Wiese, A. C.
1941 Oleson, J. J.
Wegner, Marcus I.
1943 Teresi, Joseph D.
Wachtel, Louis W.
1944 Briggs, George M., Jr.
Mannering, Gilbert J.
Mills, Russell C.
Boutwell, Roswell K.

William G. Hoekstra

M.S.

1957 Welch, James G. (*AH*)
1959 Kirk, David L.
Ogilvie, Marvin L.
1960 Carpenter, Zerle Leon (*AH*)
Stevermer, Emmett Jerome (*AH*)
Kovacs, Martin F., Jr.
Priebe, Eugene H. (*AH*)
1961 Cassens, Robert Gene (*AH*)
1962 Theuer, Richard Charles
Reese, Nathan A. (*MAS*)
Rossmiller, John David
1963 Schefsky, Harvey W. (*MAS*)
1965 Lin, Chi-wei (*MAS*)
Becker, Wayne Marvin
Anand, Amarjit Singh (*MAS*)
1966 Nielsen, Forrest Harold
Pyke, Robert Erwin
1967 Greaser, Marion Lewis (*MAS*)
Dahmer, E. Joe (*MAS*)
1969 Rotruck, John Truman (*MAS*)
1981 Resvick, Robert Michael

Ph.D.

1958 Lewis, Paul K., Jr. (*AH*, Jt. with Phillips)
Briskey, Ernest J. (*AH*, Jt. with Phillips)
1959 Smith, Irvin D. (*AH*)
Welch, James G. (*AH*)
1960 Turk, Donald E. (*PS*)
Vetter, Richard Lloyd
1961 Kirk, David Livingstone
Kauffman, Robert G. (*AH*)
1962 Brungardt, Valerian H. (*MAS*)
Kienholz, Eldon Wesley (*PS*)
Sayre, Robert Newton (*MAS*)
Ogilvie, Marvin Lee (*MAS*)

Stevermer, Emmett Jerome (*MAS*)
1963 Cassens, Robert Gene (*MAS*)
Goll, Darrel Eugene (*MAS*)
1964 Radloff, Harold David (*DS*)
1965 Reese, Nathan A. (*MAS*)
Theuer, Richard Charles
Rossmiller, John David
1966 Menahan, Lawrence Albert (*DS*)
Beecher, Gary Richard (*MAS*)
Kastenschmidt, Lewis Leslie (*MAS*)
1967 Becker, Wayne Marvin
Nielsen, Forrest Harold
1968 Reimann, Erwin Max
Pyke, Robert Erwin
Westmoreland, Nelson Priddy
1969 Campbell, Lloyd Darwin (*MAS*)
Greaser, Marion Lewis (*MAS*)
Lin, Chi-Wei
1970 Coleman, Bruce Wayne (*MAS*)
Dahmer, E(van) Joe (*MAS*)
1971 Rotruck, John Truman (*MAS*)
1975 Jacobs, Richard Michael
Kasarskis, Edward Joseph, Jr. (*MS*)
Lawrence, Richard Aubrey
Swanson, Anne Barrett
Wagner, Patricia Ann (*NS*)
Hafeman, Dean Gary
1978 Dougherty, John James
1980 Sunde, Roger Allen
1982 Held, Douglas Dwight
1983 Ekstrom, Richard Charles
LeBoeuf, Robert Alexander

Ross B. Inman

Ph.D.

1977 Engler, Jeffrey Allen
1982 Funnell, Barbara Evelyn

Marvin J. Johnson

M.S.

1941 Koepsell, Harold J.
1942 Gailey, Franklin Bryan
1943 Colovos, George C.
1944 Park, James Theodore
1945 Olson, Birger, H.
1946 Jarvis, Francis George
1947 Mortimer, Donald Charles
Shah, Nagindas Raichand
Thorn, John Anthony
1948 Bernstein, Norman N.
1950 Daniel, John Warwick, Jr.
1951 Anderson, Ralph Frederick
Hosler, Peter
Maxon, William Densmore
Stein, Sylvia S.
Buelow, Grant H.
1952 Brewer, Glenn Aaron, Jr.
Soltero-Harrington, Fred F.
1953 Cleland, William Wallace
Davey, Vivian Frederick
Schepartz, Saul Alexander
Vavra, James Joseph
Törnqvist, Erik Gustaf Marcus
Chien, Hwei-Chiu
1954 Bhuyan, Bijoy Kumar
Mattoon, James Richard
Wright, Robert Errol
1955 Visser, Jeronimo
1956 Andres, William Wolcott
Hendershot, William Fred
Wideburg, Norman Earl
1957 Mohberg, Joyce
Sant, Rameschandra K.
Davis, Raymond Vincent
Jackson, William Thomas
Jensen, Alden Leo
1958 Phillips, Donald Howard
Nara, Takashi
1959 Walgren, Kenneth Lee
1960 Segel, Irwin Harvey
1961 Last, Jerold Alan
1962 Bloom, Sandra Josephine
Roberts, Joseph
1963 Pruess, David Louis
Thomas, James Arthur
1964 Miller, Thomas Lee
1966 Hernandez, Eovaldo
1967 Liu, Chao-Min
Vary, Patricia Susan

1969 Dobry, Duane Douglas
1971 Tanaka, Nobumasa

Ph.D.

1938 Berger, Julius
1943 Koepsell, Harold J.
1945 Perlman, David
1946 Gailey, Franklin Bryan
1948 Singh, Kesar
Olson, Birger H.
Jarvis, Francis George
Maass, Eleanor Anderson
Park, James Theodore
Peterson, Merlin Henry
Thorn, John Anthony
1950 Mortimer, Donald Charles
1951 Bomstein, Ralph A.
1952 Rao, Koppaka V.
1953 Anderson, Ralph Frederick
Chiao, Jui Shen (Jt. with Peterson)
Hosler, Peter
Maxon, William Densmore
Daniel, John Warwick, Jr.
1954 Brewer, Glenn Aaron
Soltero-Harrington, Fred V.
Holmlund, Chester Eric
Smith, Charles Giles
Whitmore, Lester McClellan, Jr. (Jt. with Peterson)
1955 Törnqvist, Erik Gustaf Marcus (Jt. with Peterson)
Cleland, Wm. Wallace
Owen, Stanley Paul
Schepartz, Saul Alexander
Vavra, James Joseph
1956 Bhuyan, Bijoy Kumar
Tardrew, Philip Leslie
1957 Mattoon, James Richard
Guenther, Karl Russell
1958 Schultz, Jerome S.
1959 Hendershot, William Fred
Woronick, Charles Louis
1960 Jackson, William Thomas
1961 Kelleher, William Joseph
Phillips, Donald Howard
1962 Mohberg, Joyce
Wylie, Edward Bernard
Segel, Irwin Harvey
1963 Clesceri, Lenore Stanke
1966 Miller, Thomas Lee
1967 Thomas, James Arthur
Hernandez, Eovaldo
Pruess, David Louis
1968 Borkowski, John David
Allin, Drusilla Ann
Wodzinski, Richard Stephen
1969 Wolff, Donald John
1970 Higgins, Thomas Engel
Liu, Chao-Min
Sheehan, Brian Talbot
1973 Pankratz, Thomas John (Jt. with Garver)

Paul J. Kaesberg

M.S.

1959 Fujimura, Robert Kanji
1960 Incardona, Nino Leonard
1961 Bockstahler, Larry Earl
1962 Stubbs, John Dorton
1963 Stubbs, Eleanor Ann
1965 Dumas, Lawrence Bernard
1966 Vermette, Raymond Edward
1967 Harwood, Steven John
Samuelson, Gay Marie
1970 Gupta, Phalguni

Ph.D.

1959 Mazzone, Horace Michael
1960 Yamazaki, Hiroshi
1961 Filmer, David Lee
Fujimura, Robert Kanji
1962 Incardona, Nino Leonard
1964 Mitra, Sankar
Bockstahler, Larry Earl
Enger, Merlin Duane
1966 Stubbs, John Dorton
1967 Knoblauch, Dorothy A.
Thirion, Jean-Paul Joseph Marie
1968 Dumas, Lawrence Bernard
1969 Samuelson, Gay Marie
1971 Lane, Leslie Carl
1972 Gupta, Phalguni
1973 Rohrschneider (Munro), Jean Marie

Lintner, Harold Joseph
1975 McClements, William Lytle
Palmenberg, Ann Carol
1978 Moosic, Joseph Paul
1980 Salerno, Theresa Ann

Henry A. Lardy

M.S.

1947 Betheil, Joseph Jay
Anderson, Laurens
1948 Robertson, Patricia Rose Marie
Johnson, Robert Bernal
1949 Armstrong, Katharine Leigh
Plaut, Gerhard Wolfgang Eugen
1950 Lampson, George P.
Pressman, Berton Charles
Lueptow, Arthur James, Jr.
Feldott, Gladys Ruth
Kuby, Stephen Allen
Kussi, Peter
1952 Wellman, Lenola Harlene
1953 Maley, Frank
1954 Adler, Julius
1955 Garcia-Hernandez, Mario
1956 Zile, Maija Helene
1957 Doskotch, Raymond Walter
1958 Lange, Jeanne Alice
Riiber, Anne Catherine
1959 Engber, Jerrold I.
1962 Kagawa, Tsuneo
1963 Morton, Bruce Eldine
1964 Lawrence, Paul Joseph
1964 Paetkau, Verner Henry
1966 Chiu, Chien-Ho
1969 Ogunbiyi, David Qlakunle Akanbi
1979 Rea, Douglas William
1981 Konopka, John Michael
1983 Scountzou, Ioanna

Ph.D.

1948 Wiebelhaus, Virgil Daniel
1949 Ghosh, Dharmabrata
Totton, Ezra Lester
Betheil, Joseph Jay
MacLeod, Patricia Rose Marie
Mann, Kingsley Meadows
1950 Anderson, Laurens
Brown, Kenneth Douglas
Johnson, Robert Bernal
1951 Plaut, Gerhard Wilhelm Eugen
1952 Drysdale, George Robert
1953 Berger, Magdalene
Kuby, Stephen Allen
Pressman, Berton Charles
Feldott, Gladys Ruth (*HE*)
Byrne, William Lawrence
1954 Parks, Robert Emmett, Jr.
1956 Maley, Frank
1957 Peanasky, Robert Joseph
Broberg, Patricia Louise
Adler, Julius
1959 Garcia-Hernandez, Mario
Zile, Maija Helene
Doskotch, Raymond Walter
1960 Wang, Shu-Fang
1962 Cha, Chung-Ja Mo
1965 Morton, Bruce Eldine
1967 Foster, Charles David Owen
Silhacek, Don LeRoy (*E*)
Lawrence, Paul Joseph
Paetkau, Verner Henry
1969 Veneziale, Carlo Marcello
Lin, Chien-Ho Chiu
1970 Hanson, Ronald Lee
McClure, William Robert
1971 Anderson, David Bennett
Johnston, James Bennett
Ferguson, Shelagh Mary Frances
Lambeth, David Odus
1972 Garbers, David Lorn (*MAS*)
Reed, Juta (Kuttis)
Lee, Yu-Bang (*MAS*)
1974 Ebel, Richard Eugene
1975 Bentle, Larry Andrew
Betz, Albert Lorris (*P*)
1976 Bosch, Anne Louise
Hutson, Susan Mary
Milkowski, Andrew Leszek
1977 Van Dop, Cornelis (*MS*)
Huang, Ming-Ta

1978 Colombo, Maria Giovanna Mognoni
Kauffman, Raymond Francis
1979 Chen, Jen-Ling James
Reinhart, Gregory Duncan
1980 Singh, Jai Pal
1981 Merryfield, Margaret Laue
1983 Hammond (Wernette), Mary Ellen
Riquelme, Patricio Transito

Karl P. Link

M.S.

1931 Burkhart, B. A.
Morell, Sam
1934 Kerst, Herman
1936 Harrower, R. J.
1938 Dimler, Robert J.
Joos, Lothar A.
1940 Burkert, G. M.
1941 Huebner, Charles F.
Lohmar, Rolland
Overman, R. S.
Smith, Frederick G.
1943 Graf, Lloyd Herbert
Olson, William J.
Spero, Leonard
1944 Roseman, Saul
1945 Bubl, Edward C.
Ikawa, Miyoshi
Pankratz, Russell A.
Bjornson, August G.
1947 Gangstad, Edward Otis
Volpert, Eugene Michael
1948 Scheel, Lester Dean
Wycoff, Harland DeWitt
Ballou, Clinton E.
Popenoe, Edwin Alonzo
Senn, Vincent John
1949 Clemmons, Jackson Joshua
Wu, Dorothy Lun
1951 Robertson, Dale Norman
Snyder, Jack Austin
1952 Eble, John Nelson
Schroeder, Collin Harold
1953 Hayashi, James Akira
1955 Titus, Edward Depue
1957 Wiener, Charles
1958 Walker, John Wiblur
1960 Zinkel, Duane F.
1961 Bednekoff, Alexander G.
Berg, David H.
West, Bruce D.
1962 Underwood, Chester Edward
1965 Carlson, Clarence Arnold
1968 Deckert, Fred Wolfgang

Ph.D.

1929 Dickson, Allan D.
1933 Schoeffel, E. W.
1934 Burkhart, B. A.
Morell, Sam
Niemann, C. G.
1935 Baur, L. S.
1936 Militzer, Walter E.
Worthington, O. J.
1937 Roberts, W. L.
Setter, F.
1938 Moore, Stanford
Sell, H. M.
1939 Campbell, Harold A.
1940 Dimler, Robert J.
1941 Stahmann, Mark A.
1942 Overman, R. S.
Sullivan, William R.
1943 Huebner, Charles F.
Lohmar, Rolland
Smith, Frederick G.
1944 Field, John B.
1948 Ikawa, Miyoshi
Olson, William J.
Roseman, Saul
Spero, Leonard
Doherty, David George
Graf, Lloyd Herbert
1949 Fletcher, T. Lloyd
Patterson, Ernest L.
Scheel, Lester Dean
Wilcox, Philip Edwin
1950 Ballou, Clinton E.
Popenoe, Edwin Alonzo
Seidman, Martin
1951 Senn, Vincent John
1952 Wycoff, Harland DeWitt
1953 Jeffay, Henry
Synder, Jack Austin

Stone, William Harold (*G*)
Robertson, Alice Ota
Robertson, Dale Norman
Eble, John Nelson

1954 Lin, Tien-Hui
Schroeder, Collin Harold
Hayashi, James Akira

1956 Kloss, Robert Albert
Clemmons, Jackson Joshua (*P*)

1960 Wiener, Charles

1961 Zinkel, Duane F.

1962 Bednekoff, Alexander George
West, Bruce David

1964 Berg, David Herbert

1965 Barker, Walter Marlin

1968 Hermodson, Mark Allen

1970 Deckert, Fred Wolfgang

Elmer V. McCollum

M.S.

1913 Hoagland, D. R.

1916 Drescher, H. A.
Kennedy, Cornelia

1917 Parsons, Helen

David Nelson

M.S.

1977 Riddle (McAmis), Lavon Lucille
Andrews, Douglas Raymond

1981 Glustrom (Watson), Leslie Jean

Ph.D.

1976 Browning, Jeffrey Lee
Walker, Frederick John

1977 Miller, Bonnie Ellen

1980 Brusilow, William Samuel

1982 Rauh, James Joseph

Masayasu Nomura

Ph.D.

1974 Sidikaro, Joseph

1976 Gette, William Robert

1979 Post, Leonard Edwin

1981 Yates, John Lee

William H. Orme-Johnson

M.S.

1973 Dunnette, Joel Harold

1981 Jacobson, Kathy Ann

Ph.D.

1974 Bale, Judith Robyn

1976 Jacob, Gary Steven

1978 Light, David Richard

1979 Rademacher, Thomas William

1980 Henzl, MichaelThomas
Rendina, Alan Ralph

W. H. Peterson

M.S.

1918 Verhulst, J.

1920 Churchill, Helen

1922 Anderson, J. A.
Dow, Odessa

1923 Schmidt, E. G.

1924 Parmele, H. B.

1925 Pederson, Carl
Stiles, Hugh R.

1926 Marten, E. A.

1927 Lindow, C. W.

1928 Allgeier, R. J.
Skinner, J. T.

1932 Wenck, Peter

1935 McNaughton, J.
Quackenbush, Forrest W.

1936 Snell, Esmond E.
Woolley, D. W.

1939 Fontaine, Francis E.

1941 Lampen, J. Oliver

1942 Corey, J. P.
McGregor, Margaret
Shull, Gilbert Malcolm

1943 Berl, Soll
Miller, Daniel Robert
McQuarrie, Ernest Barclay

1944 Stefaniak, Jerome J.
Thorne, Curtis Blaine

1945 Bowden, John P.
Higuchi, Kiyoshi
Snell, Junius F.
Krueger, Keatha Katherine

1948 Chang, Wei-Shen
Cheng, Yu Ching
Pine, Leo
Bomstein, Ralph A.

1949 Thoma, Richard William
Sternberg, Helen Esther

1950 Kita, Donald Albert
Larson, Lester Mikkel

1951 Doctor, Vasant Manilal

1957 Sant, Rameschandra (Jt. with Johnson)
Davis, Raymond Vincent (Jt. with Johnson)

Ph.D.

1923 Brunkow, O. R.

1924 Schmidt, E. G.

1925 Anderson, J. A.

1927 Parmele, H. B.

1928 Marten, E. A.

1929 Lindow, C. W.
Stiles, Hugh R.

1930 Allgeier, R. J.
Pruess, L. M.

1932 Johnson, Marvin J.
Skinner, J. T.
Wilson, Perry

1934 Georgi, Carl E.
Gorcica, H. J.
Tatum, E. L.
Wenck, Peter

1935 Conner, H. A.
Orcutt, Fred S.

1936 Langlykke, A. F.

1937 Pavcek, Paul

1938 Snell, Esmond E.
Woolley, D. W.

1939 Pan, S. C.

1940 Johnson, B. Connor
McIntire, F. C.

1941 Fontaine, Francis E.

1942 Hutchings, Brian L.

1943 Bohonos, Nester
Gardner, Josephine
Lampen, J. Oliver

1946 Hodgson, Roland
Rivett, Robert W.

1947 Leonard, Reid H.
Shu, Ping
Stefaniak, Jerome J.

1948 Krueger, Keatha Kathrine
Agarwal, Prem Narain
Bowden, John P.
Higuchi, Kiyoshi
Shull, Gilbert Malcolm
Thorne, Curtis Blaine

1949 Agarwala, Satish Chandra
Brown, William Everett
Snell, Junius F.

1950 Chang, Wei-Shen

1951 Sharpe, Robert W.
Kita, Donald Albert
Larson, Lester Mikkel
Thoma, Richard William

1953 Chiao, Jui Shen (Jt. with Johnson)

1954 Whitmore, Lester McClellan, Jr. (Jt. with Johnson)

1955 Törnqvist, Erik Gustaf Marcus (Jt. with Johnson)

Paul H. Phillips

M.S.

1941 Boyer, Paul D.
Lardy, Henry A.
Shaw, James H.

1944 Kirkwood, Sam

1945 Orr, Kathleen (*HE*)
Spitzer, Robert Ralph

1946 Hansen, Roger Gaurth

1947 Anderson, Elizabeth Potts
Badr, Mahmoud Foad
Bentley, Orville G.
Wakeman, Elizabeth Jane

1948 Maruyama, George Masao
Smith, James Knox
Verma, Indrapal Singh

1948 Campbell, Jeptha Edward, Jr.
Meyer, James Henry
Constant, Marguerite Anita

1950 Hentges, James F., Jr.
Shefchik, Ben Edward
Burns, Charles Henry
Nell, Ruth Elizabeth
Richard, Robert Max

1951 Perdue, Henry Stafford
O'Rourke, Willard Frank

1952 Gessert, Carl Frederick
Wilkinson, Walter Shepard (*AH*)
Haurowitz, Alice Roberta
Hoekstra, William George
Robajdek, Eugene Sylvester
Sievert, Herman William

1953 Madsen, Kenneth Olaf
Kelly, Robert Frank
Merkel, Robert Anthony
Miller, Russell Franklin

1954 Taketa, Fumito
Emerick, Royce Jasper (*AH*)

1955 Ontko, Joseph Andrew
Lewis, Paul Kermith, Jr. (*AH*)
Li, Yu-chen

1958 Faltin, Eugene Chester (*AH*)
Suttie, John Weston
Wuthier, Roy Edward

1959 Vogel, James John

1960 Wright, Paul Lee (*AH*)

1961 Gesteland, Raymond Fredrick
Thompson, David Jerome

1963 Heintz, Judith A. Fisher

1964 Chiemchaisri, Yongyut

Ph.D.

1939 Engel, R. W.
Evans, R. J.

1940 King, Willis A.

1942 Ross, O. Burr

1943 Boyer, Paul D.
Lardy, Henry A.
Shaw, James H.

1944 Cunha, Tony J.
Lundquist, Norman S.

1946 Pope, Arthur Lloyd (*AH*)
Grummer, Robert Henry

1947 Kirkwood, Sam
Spitzer, Robert Ralph
Weir, William C.
Whitehair, Charles Kenneth

1948 Zepplin, Theressa Marie
Hansen, Roger Gaurth
Niedermeir, Robert Paul (*DH*)
Michaud, Laurent

1949 Badr, Mahmoud Foad
Bray, Robert Woodbury
Blosser, Timothy H.
Bowland, John Patterson

1950 Grunert, R. Richard
Hale, William Harris
Bentley, Orville G.
Kastelic, Joe
Quicke, George Venn

1951 Meyer, James Henry
Campbell, Jeptha Edward, Jr.
Constant, Marguerite Anita
Shefchik, Ben Edward

1952 Burnside, Joseph E.
Hentges, James F., Jr.

1953 Moinuddin, Mohammed (*AH*)
Richard, Robert Max (*AH*)
Burns, Charles Henry
O'Rourke, Willard Frank (*PS*)
Wilson, George Donald (*AH*)

1954 Gaballah, Saeed (*DFI*)
Perdue, Henry Stafford
Wilkinson, Walter Shepard (*AH*)
Hoekstra, William George
Shaw, Ridley Kustaa

1955 Gessert, Carl Frederick
Kelly, Robert Frank (*AH*)
Miller, Russell Franklin

1957 Kropf, Donald Harris (*AH*)
Emerick, Royce Jasper (*AH*)
Merkel, Robert Anthony (*AH*)
Ontko, Joseph Andrew
Taketa, Fumito

1958 Lewis, Paul Kermith, Jr. (*AH*) (Jt. with Hoekstra)
Briskey, Ernest Joseph (*AH*) (Jt. with Hoekstra)
Madsen, Kenneth Olaf
Sievert, Alice Haurowitz
Sievert, H. William
Jenkins, Kenneth James W.

1960 Asplund, (John) Malcolm (*AH*)
DePape, Raymond George (*AH*)
Laerdal, Osmund Arnbjorn (*PS*)
Suttie, John Weston
Wuthier, Roy Edward

1961 Bunce, George Edwin
Vogel, James John
Wright, Paul Lee (*AH*)

1963 Thompson, David Jerome

1965 Chiemchaisri, Yongyut

Gerhard W. E. Plaut*

M.S.

1954 Broberg, Patricia Louise

Christian R. H. Raetz

Ph.D.

1980 Esko, Jeffrey David

1982 Ganong, Barry Reade

Sambhorao T. Rao

M.S.

1976 Ozanne, Steven Kenneth

Howard Rasmussen

M.S.

1963 Fang, Marie Ru-Yu

1965 Sze, (Paul) Yi Ling
Glass, John David, Jr.
Hawker, Charles Davis
Kimmick, George Arthur

Ph.D.

1966 Fang, Marie Ru-Yu

William S. Reznikoff

M.S.

1972 Corredor, Jorge Enrique

Ph.D.

1974 Barnes, Wayne Morris

1975 Aline, Robert Francis, Jr.
Jones, Barbara Ronald

1978 Jorgensen, Richard Alan
Postle, Kathleen

1979 Maquat, Lynne Elizabeth

1980 Rothstein, Steven Jay

1983 Munson, Lianna Marie
Stibitz, Earle Scott
Wray, Lewis Vastine, Jr.

Robert H. Rownd

M.S.

1970 Barnes, Marjorie Haxton

1972 Stickgold, Robert Allen

Ph.D.

1972 Mickel, Susan

1973 Stickgold, Robert Allen
Morris, Charles Frederick

1976 Appelbaum, Edward Robert
Womble, David Dale

1978 Farrelly, Frances Williams

1981 Barton, Carolyn Rebecca
Easton, Alan Michael
Huffman, Gary Allen

Roland R. Rueckert

M.S.

1969 Mak, Tak Wah (Edward)

Ph.D.

1971 Hall (McIntyre), Linda
Stoltzfus, Conrad Martin

1972 Butterworth, Byron Edwin

1975 Montelaro, Ronald Charles

1976 McLean, Charles Ernest, III
Yoshimura, Harold Hajime

1979 Duke, Robert Edward, Jr.
Shealy, David John

1982 Pallansch, Mark Alan

1983 Friesen, Paul Dean

Heinrich K. Schnoes

M.S.

1970 Christenson, Robert Allan

Ph.D.

1974 Semmler, Erich Joseph
Lam, Hing-Yat Peter

1976 Carter, Guy Thomas

1977 Ghazarossian, Vartan

1978 Onisko, Bruce Levander

1980 Boyer, Gregory Lynn

1981 Wichmann, Carol Jane Fix

1983 Koehn, Frank Erich
Lee, Seok Ho

---

*Enzyme Institute.

Bruce Selman

M.S.

1979 Dunham, Kristine Rene

1983 Preston, Glenn Geoffrey

Ph.D.

1980 Binder, Ross Gordon

1981 Dunham, Kristine Rene

1983 Merchant, Sabeeha

Esmond E. Snell

M.S.

1947 Rabinowitz, Jesse C.
Williams, William Lawrence

1948 Holden, Joseph Thaddeus

1950 Kitay, Estelle
Brown, Gene Monte
Metzler, David Everett

1951 Bruins, Hubert Williams
Kihara, Hayato
Peters, Vincent John
Shen, Madeline Siao-Fang

Ph.D.

1949 Broquist, Harry Pearson
Herbst, Edward John
McCollough, Willard George
McNutt, Walter Scott
MacLeod, Robert Angus
Rabinowitz, Jesse C.
Williams, William Lawrence

1951 Holden, James Thaddeus

1952 Kihara, Hayato
Metzler, David Everett
Prescott, John Mack

1953 Brown, Gene Monte
Peters, Vincent John

Mark A. Stahmann

M.S.

1950 Becker, Robert Richard
Burger, Warren Clark
Rubini, Joseph Raymond

1951 Colburn, Robert Warren

1952 Green, Maurice
Wetlaufer, Donald B.
Casida, John Edward
Rice, Robert Vernon

1953 Weinke, Karl Frederick

1955 Tauyuki, Emee
VanKley, Harold James

1958 Livesay, Richard Gilbert

1959 Kornguth, Steven Edward

1962 Stewart, Marcella Catherine
Smith, Marvin Artell

1963 Gerloff, Eldean Dale

1966 Ohno, Yasuhisa

1967 Sell, John Edward
Lavine, Lawrence Merion Mandel

1970 Abramson, David
Demorest, David Melville

1971 Abd-el Halim, Mostafa N.

1973 Woldegiorgis, Gebretateos

1981 Theobald, H(ans) Michael

Ph.D.

1952 Becker, Robert Richard
Burger, Warren Clark

1954 Green, Maurice
Wetlaufer, Donald Burton
Gothoskar, Sharad Shripad

1955 Rice, Robert Vernon
Tsuyuki, Hiroshi

1956 Weinke, Karl Frederick

1957 Kmetec (Dellert), Eva

1958 Miwa, Thomas Kanji
Van Kley, Harold James

1960 Wilding, (Morris) Dean
Richardson, Thomas

1961 Deese, Dawson
Kornguth, Steven E.
Slagel, Donald E.

1962 Knaak, James Bruce (*DH*)

1964 Stewart, John Wiley
Smith, Marvin Artell
Tewksbury, Duane Allan

1965 Wagle, Sudhakar Shantaram
LaBerge, Donald Emmanuel

1966 Akeson, Walter Roy
Hartman, Grant Henry, Jr. (*DFI*)

1967 Honold, Guy Robert
Guss, Paul LeRoy (*FSI*)
Marshall, Ronald Dean (*FSI*)
Ohno, Yasuhisa

1970 Oelshlegel, Fred James, Jr.

	Woodbury, William (*PP*)
1973	Spencer, Allen King
1979	Leatham, Gary Frank

Harry Steenbock

M.S.

1917	Gross, E. G.
1918	Koehler, A. E. Nelson, E. M. Pitz, Walter
1921	Bethke, R. M. Jones, J. H.
1922	Sell, Mariana Stewart, Lois
1923	Black, Archie Hoppert, C. A.
1924	Elmslie, W. P.
1926	Gunderson, Frank Kletzien, S. W. F.
1930	Hanning, Flora
1931	Baumann, Carl A. Haman, R. W.
1932	Lowe, J. T. Riising, B. M.
1935	Nordfeldt, S. B.
1936	Schneider, Howard A. Orlove, Harold
1938	Krieger, Carl
1939	Bunkfeldt, R. Fuhr, I.
1941	Cox, R. P. Kummerow, F. A. Thompson, C. R.
1942	Anthony, D. S.
1943	Snyder, Robert H.
1946	Gottlieb, Howard
1949	Shaw, Ina D.
1950	Eisner, Harold Johannes Bellin, Stuart Arnold
1952	Cramer, John Wesley Herting, David C.
1953	Guenther, Karl Russell Pileggi, Vincent Joseph DeLuca, Hector Floyd Ichioka, Patricia Shizuko
1955	Porrata-Doria, Elba Iris
1956	Benton, Pauline Noll
1957	Reiser, Sheldon
1959	Manatt, Margaret R. (Jt. with DeLuca)

Ph.D.

1919	Buell, Mary
1921	Gross, E. G. Koehler, A. E.
1922	Miller, H. G.
1923	Bethke, R. M. Nelson, E. M.
1924	Jones, J. H.
1925	Black, Archie Hoppert, C. A.
1928	Elmslie, W. P.
1929	Gunderson, Frank Kletzien, S. W. F. Thomas, B. H.
1930	Schrader, G. A.
1931	Kellerman, John
1932	Geiger, Beatrice J. Hauck, Hazel M.
1933	Baumann, Carl A. Templin, Vera
1934	Haman, R. W.
1935	Beeson, M. Hayward, J. W. Lowe, J. T.
1936	Reynolds, May S.
1937	Quackenbush, F. W.
1938	Dasler, W. Johnson, L. Margaret Nordfeldt, S. B. Schneider, Howard A.
1940	Glindeman, Portia M. Krieger, Carl H.
1943	Bunkfeldt, Rudolf Everson, Gladys Fuhr, I. Thompson, C. Ray
1944	Anthony, David S. Kummerow, Fred A.
1945	Cederquist, Dena
1949	Snyder, Robert H.
1950	Gottlieb, Howard
1952	Bellin, Stuart Edward
1954	Herting, David C. Meintzer, Roger B.

Pileggi, Vincent Joseph
1955 Cramer, John Wesley
DeLuca, Hector Floyd
1959 Guroff, Gordon (Jt. with DeLuca)
1960 Reiser, Sheldon (Jt. with DeLuca)

Frank M. Strong

M.S.

1940 Feeney, Robert E.
1941 Earle, Anne
1943 Leonard (Ives), Margaret
Krehl, Willard A.
1944 Price, James M.
1946 Deemer, Lena F.
Lutwack, Leo
1947 Lipton, Samuel Harry
Taylor, Walter Rowland
1949 El Gindi, Ibrahim M.
1950 Higgins, Harvey M., Jr.
1951 Clayton, Robert Allen
Handschumacher, Robert E.
Tener, Gordon M.
Buyske, Donald A.
Prier, Robert Frank
1952 Smeby, Robert Rudolph
1953 Lapidus, Milton
1954 Schilling Eugene Dudleigh
Sung, Rodney L.
1955 Siehr, Donald Joseph
1956 Garbutt, John Thomas
VonSaltza, Malcolm Hugo
Dombro, Roy S.
1957 Dresden, Carlton Francis
1958 Thomson, William A. B.
1960 Ehrhart, Leo Allen
1962 Brewer, Eugene Newton
Foster, George Andrew, Jr.
Morgan, David
1965 Rainey, Donald Paul
Ritter, Preston Otto
1967 Hinz, Robert Sidney
Prouty, Walter Francis
1969 Taylor, Paul Alan
1970 Ikediobi, Christopher Okwuolisa

Ph.D.

1942 Feeney, Robert E.
Porter, J. W.
1943 Carpenter, Lawrence E.
Neal, Arthur L.
1945 Leonard (Ives), Margaret
Krehl, Willard A.
1948 Lipton, Samuel Harry
1949 Ahmad, Kamaluddin
Bumpus, Merlin
Neilands, John Brian
Dunshee, Bryant Ray
Taylor, Walter Rowland
1952 Higgins, Harvey M., Jr.
Schneider, Horst Gerhard
1953 Clayton, Robert Allen
Handschumacher, Robert Edmund
Tener, Gordon Malcolm
Bruins, Hubert Williams (*PS*)
1954 Buyske, Donald Albert
Smeby, Robert Rudolph
1956 Lapidus, Milton
Schilling, Eugene Dudleigh
1957 Barnett, Bobby Dale (*PS*)
Siehr, Donald Joseph
1958 Liu, Richard Wen-chih
Dombro, Roy S.
Garbutt, John Thomas
Von Saltza, Malcolm Hugo
1959 Dresden, Carlton Francis
Ellison, Theodore (*VS*)
1960 Roy, Dwijendra Nath (*PS*)
Dickie, John Peter
Sehgal, Jag Mohan
1961 Thomson, William Alexander B.
Yueh, Mao Hsun
1962 Loomans, Maurice Edward
1963 Steinke, Frederick Herman (*PS*)
Frea, James Irving (*B*)
McDonald, Bruce Eugene (*PS*)
1964 Bingham, Robert J. (*DFI*)
Sprecher, Howard William (*DFI*)
Ehrhart, Leo Allen
Morgan, David

1965 Brewer, Eugene Newton
Farley, Thomas Michael
Gilgan, Michael Wilson

1966 Foster, George Andrew, Jr.

1967 Rainey, Donald Paul
Ritter, Preston Otto

1969 Bamburg, James Robert

1970 Watson, Theodore Raymond
Prouty, Walter Francis
Koch, Gilbert Henry

1971 Thompson, Richard Michael

1972 Hsu, Ih-Chang
Hinz, Robert Sidney

Muttaiya Sundaralingam

M.S.

1979 Emerson, Julie

Ph.D.

1974 Prusiner, Paul Edward

1976 Stout, Charles David

1977 Sprang, Stephen Robert

1978 Hogle, James Martin
Satyshur, Kenneth Andrew
McAlister, John Paul
Smith, Janet Louise

1982 Haromy, Tuli Patrick

1983 Bergstrom, Robert Walter

John W. Suttie

M.S.

1964 Carlson, James Roy

1967 Shearer, Thomas Robert
Coven, Charles Joseph
Partridge, Marian Jean

1969 LeSaouter, Joel Joseph

1972 Young, Kathryn Lynn

Ph.D.

1965 Satter, Larry Dean (*DS*)
Simkins, Karl LeRoy, Jr. (*DS*)
Zebrowski, Edward Joseph

1966 Carlson, James Roy

1969 Shearer, Thomas Robert

1970 Palmer (Thierry), Myrtle Jean

1971 Lorusso, David James

1972 Quissell, David Olin

1973 Haugen, David Allen

1974 Drescher, Marian Jean

1975 Grant, Gregory Alan

1976 Sadowski, James Albert

1978 Carlisle, Thomas Lyman

1979 Whitlon, Donna Sue

1980 Larson, Ann Elaine

1981 Repaske, Mary Gaile

1983 Hildebrandt, Ellen Frances
Swanson, Jhan Craig

William E. Tottingham

M.S.

1923 Gerhardt, O. H.
Link, Karl Paul
Rankin, E. J.

1924 Sullivan, W.

1925 Otterson, H.

1937 Chen, Shui-Che

1938 Dutton, H.

1939 Carson, R. B.

1941 Clagett, C. O.

Ph.D.

1925 Link, Karl Paul

1929 Otterson, H.

1930 Lowsma, Henry

1931 Stephens, H. L.

1935 Lease, Elmer J.

1936 Nagy, Rudolph

1937 Ross, A. F.

Robert D. Wells

M.S.

1969 Grant, Robert Charles

1970 Schendel, Paul Floren
Burd, John Frederick

1974 Brewer, Lorraine Carol

Ph.D.

1970 Harwood, Steven John

1971 Grant, Robert Charles

1972 Miller, Lois Kathryn
Sweet, Raymond Whitney

1974 Tamblyn, Toby Miriam
Schendel, Paul Floren

1975 Burd, John Frederick

Ryan, Michael Joseph
1976 Chan, Hardy Wai-Hong
Dodgson, Jerry Bruce
1979 Hardies, Stephen Charles
1980 Horn, Glenn Thomas
1982 Goodman, Thomas Carl
Stirdivant, Steven Milton
1983 Neuendorf, Sandra Kay

Jesse N. Williams, Jr.

M.S.

1951 Taborda, Armando Rodrigues
1952 Bothwell, James Walter
Tews (Kring), Jean P.
Schwartz, Morton Allen
Ebisuzaki, Kaney
Chen, Lillian Sangone
Kennedy, William Robert
1954 Salegui, Miriam
1955 Gilboe, David Dougherty
1956 Hayman, Selma

Ph.D.

1954 Bothwell, James Walter
Tews (Kring), Jean P.
Cheslock, Katherine E. (*HE*)
Schwartz, Morton Allen
Ebisuzaki, Kaney
1956 Arata, Dorothy Ann
Chaloupka, Marilyn M. (*HE*)
Iyengar, Melkote Raja
Monder, Carl
Prigmore, John Raulston
1957 White, Frederick H. (Jt. with Anderson)

Index of Names